Market Orientation

'The only way we can get out of this mess is for us to become customer driven or market oriented. I'm not even sure what that means, but I'm damn sure that we want to be there.'
Benson P. Shapiro, *What the Hell is 'Market Oriented'?*

Market Orientation

Transforming Food and Agribusiness
around the Customer

Edited by
ADAM LINDGREEN,
MARTIN K. HINGLEY,
DAVID HARNESS
and
PAUL CUSTANCE

GOWER

Published by
Gower Publishing Limited
Wey Court East
Union Road
Farnham
Surrey
GU9 7PT
England

Gower Publishing Company
Suite 420
101 Cherry Street
Burlington
VT 05401-4405
USA

www.gowerpublishing.com

British Library Cataloguing in Publication Data
Market orientation : transforming food and agribusiness
  around the customer. -- (Food and agricultural marketing)
  1. Farm produce--Marketing--Management. 2. Food industry
  and trade.
  I. Series II. Lindgreen, Adam.
  381.4'1-dc22

  ISBN: 978-0-566-09208-4 (hbk)
         978-0-566-09236-7 (ebk)

Library of Congress Cataloging-in-Publication Data
Market orientation : transforming food and agribusiness around the customer / by Adam Lindgreen ... [et al.].
     p. cm. -- (Food and agricultural marketing.)
  Includes index.
  ISBN 978-0-566-09208-4 (hbk.) -- ISBN 978-0-566-09236-7
(ebook) 1. Agricultural industries--Management. 2. Food industry and trade--Management.
3. Farm produce--Marketing. 4. Marketing research. I. Lindgreen, Adam.
  HD9000.5.M353 2009
  630.68'8--dc22

                                                                    2009029715

Mixed Sources
Product group from well-managed
forests and other controlled sources
www.fsc.org Cert no. SA-COC-1565
© 1996 Forest Stewardship Council

Printed and bound in Great Britain by
MPG Books Group, UK

Contents

List of Figures

List of Tables

About the Editors

Adam Lindgreen

After graduating with degrees in engineering, as well as equivalent degrees in chemistry and physics, Dr Adam Lindgreen completed an MSc in food science and technology at the Technical University of Denmark. He also finished an MBA at the University of Leicester, as well as a One-Year Postgraduate Program at the Hebrew University of Jerusalem. Professor Lindgreen received his Ph.D. in marketing from Cranfield University. Since May 2007, he has served as a Professor of Strategic Marketing at Hull University Business School.

Professor Lindgreen has been a Visiting Professor with various institutions, including Georgia State University, Groupe HEC in France, and Melbourne University; in 2006, he was made an honorary Visiting Professor at Harper Adams University College. His publications include more than 70 scientific journal articles, 8 books, more than 30 book chapters, and more than 80 conference papers. His recent publications have appeared in *Business Horizons, Industrial Marketing Management*, the *Journal of Advertising*, the *Journal of Business Ethics*, the *Journal of the Academy of Marketing Science*, the *Journal of Product Innovation Management, Psychology and Marketing*, and *Supply Chain Management*; his most recent books are *Managing Market Relationships* (Gower Publishing, 2008), *Memorable Customer Experiences* (Gower Publishing, 2009), *The New Cultures of Food* (Gower Publishing, 2009), and *The Crisis of Food Brands* (Gower Publishing, 2009). The recipient of the 'Outstanding Article 2005' award from *Industrial Marketing Management* and the Christer Karlsson Award at the 2007 International Product Development Management conference, Professor Lindgreen also serves on the board of several scientific journals; he is the editor of the *Journal of Business Ethics* for the section on corporate responsibility and sustainability. His research interests include business and industrial marketing management, experiential marketing, and corporate social responsibility.

Adam Lindgreen has discovered and excavated settlements from the Stone Age in Denmark, including the only major kitchen midden – Sparregård – in the south-east of Denmark; because of its importance, the kitchen midden was later excavated by the National Museum and then protected as a historical monument for future generations. He is also an avid genealogist, having traced his family back to 1390 and published widely in scientific journals related to methodological issues in genealogy, accounts of population development, and particular family lineages.

Martin K. Hingley

Dr Martin K. Hingley has degrees from three UK universities: He first graduated (Agricultural and Food Marketing, BSc Honours) from the University of Newcastle upon Tyne; he has an MPhil in marketing from Cranfield University; and he received his Ph.D. in marketing from the Open University. Dr Hingley is a reader in marketing and supply chain management at Harper Adams University College, the leading university in the United Kingdom specialising in agri-food business. He is also a Visiting Fellow to the University of Hull Business School and has previously held a Fellowship endowed by Tesco Plc. Dr Hingley has wide-ranging business experience in the international food industry and has spent some time in the provision of market and business analysis with the Institute of Grocery Distribution, a leading UK research and training organization.

Dr Hingley's research interests are in applied food industry marketing and supply chain relationship management. He has presented and published widely in these areas, including in the *British Food Journal, Entrepreneurship and Regional Development, Industrial Marketing Management*, and the *Journal of Marketing Management*. Also, he has published *The New Cultures of Food* (Gower Publishing, 2009) and *The Crisis of Food Brands* (Gower Publishing, 2009). He serves on the board of several scientific journals and also regularly guest edits such journals.

David Harness

Dr David Harness holds an undergraduate degree in management from Aston University, an MPhil from Birmingham City University, and a Ph.D. from Huddersfield University. His Ph.D. thesis focused on managing service product elimination. Dr Harness is currently a senior lecturer in strategic and international marketing at Hull University Business School, having previously worked for Leeds University Business School. His commercial experience was gained in retail banking, and he has conducted consultancies in a range of industries in the areas of service product management, value marketing, customer care, and relationship marketing. His research interests are in service product management, specifically product elimination, service quality, and, more recently, corporate social responsibility. Dr Harness has published in the *International Journal of Bank Marketing*, the *Journal of Business Ethics*, the *Journal of Product and Brand Management*, and the *Services Industries Journal*; also, he has written chapters in edited books covering sports marketing and service product management.

Paul Custance

Dr Paul Custance graduated with a BA (Honours) in agricultural economics from the University of Nottingham. This degree was followed by a Ph.D. in Economics at the same university. For the past 20 years, he has been a principal lecturer in marketing at Harper Adams University College based in Shropshire (United Kingdom). Dr Custance is the former director of ruralconsultancy.com, which undertakes industry-orientated research and consultancy and which produced more than 250 reports for a wide range of clients during his time as director. These clients included government bodies, regional development agencies, multinational companies, and local and regional small and medium-sized enterprises. In addition to teaching undergraduate and master's courses, Dr Custance has interests in the development of work-based learning. His research interests are in food and agricultural marketing (particularly branding and supply chain management) and social marketing applied to the rural sector. He has presented papers at conferences for more than 30 years and refereed papers for several academic journals. He has recently published papers in *Business Strategy and the Environment*, the *Journal of Business and Industrial Marketing, Journal of Marketing Management*, and the *Journal of Rural Enterprise and Management*.

About the Contributors

Atanu Adhikari

Atanu Adhikari is Assistant Professor at the ICFAI Business School, India, and was previously visiting researcher at the Whitman School of Management, Syracuse University. He has published chapters in edited books published by John Wiley & Sons and Cambridge Scholar Press and articles in journals such as *Journal of Creative Communication*, among others. His research interest centres on pricing of experience products, agricultural marketing, the application of Bayesian econometrics in marketing research, and applications of conjoint analysis.

Luís Kluwe Aguiar

Luís Kluwe Aguiar is a Senior Lecturer in Marketing and International Business at the Royal Agricultural College and an associate lecturer at the Open University Business School. Mr. Aguiar received his MSc in agricultural economics from Wye College at the University of London. Prior to becoming a full-time academic, he worked for many years in both the private and public sectors and lectured on a part-time basis. Mr. Aguiar has extensive international research and consultancy work. His main research interests focus on marketing and consumer studies, especially relating to ethical consumerism.

Lucia Bailetti

Lucia Bailetti, a food engineer, is the director of the Italian Centre for Sensory Analysis. She received her MBA at the University of Buenos Aires. Previously, she worked in research and development for various multinational companies. She taught business management at the University of Buenos Aires; in 1996, she received the Mario Hirsch Award for educational development at that university. Lucia Bailetti is a professional member of the Institute of Food Technologists, USA; the Italian Society of Sensory Science; and the Argentine Association of Food Engineers.

Márcia Dutra Barcellos

Dr Marcia Dutra Barcellos is a Lecturer in Business Management at the Catholic University in Brazil. She obtained her Ph.D. from the Federal University of Rio Grande do Sul and currently holds a post-doctoral position at the Institute for Marketing and Statistics at Aarhus School of Business, University of Aarhus, Denmark. Dr Barcellos has published in several international journals. Her research interests include innovation in the food chain, chain management, and consumer studies.

Michael B. Beverland

Dr Michael Beverland is Professor of Marketing at RMIT University. Michael received his Ph.D. from University of South Australia. He has published in *Business Horizons, European Journal of Marketing, Industrial Marketing Management, Journal of Advertising, Journal of Business & Industrial Marketing, Journal of Business Research, Journal of Consumer Research, Journal of Management Studies, Journal of Product Innovation Management,* and *Journal of the Academy of Marketing Science,* among others. His research interests include the design–brand marketing interface, consumer interactions with design, the marketing of authenticity, consumer responses to design aesthetics, creative processes associated with new product design, and brand management. He serves on the boards of *Industrial Marketing Management* and *Journal of Business and Industrial Marketing.*

Jos Bijman

Dr Jos Bijman is Assistant Professor of Management and Organisation in the Department of Business Administration at Wageningen University. Dr Bijman received his Ph.D. from Rotterdam School of Management at the Erasmus University. He has published in several journals, including the *American Journal of Agricultural Economics,* the *European Review of Agricultural Economics,* the *Journal on Chain and Network Science,* the *International Journal of Cooperative Management,* and *Science and Public Policy.* At Wageningen University, Dr Bijman teaches economic organisation theory, institutional economics, project management, and management and organisation. His research interests include supply chain management, the internationalisation of business, entrepreneurship, corporate governance, and the economic organisation of international agri-food chains. Currently, most of his research focuses on the interaction between horizontal collaboration and vertical coordination in agri-food supply chains.

Freddy Brofman

Freddy Brofman obtained his MBA from the Royal Agricultural College and is presently a Ph.D. student at the University of Kent. He has previously researched, as well as published in specialised trade journals on, issues regarding radio frequency identification. Mr. Brofman's current research is about economic evaluations of technology innovation in food chain traceability systems.

Stefanie Bröring

Dr Stefanie Bröring is Head of Marketing and Innovation of Bröring Group, a family business in the animal feed sector in Germany. Also, she is Professor in Food Change

Management at the University of Applied Sciences Osnabrück. Prior to this position, she was a senior consultant with RoelfsPartner Management Consultants, where she headed the Competence Center Agribusiness. She has a wide array of consulting and industry experience in the fields of food, food ingredients, and agribusiness. She obtained her Ph.D. from the University of Münster in Germany and was appointed as visiting researcher at the University of Quebec at Montreal, where she carried out research on innovation in converging industries with a focus on the emerging nutraceuticals and functional foods industry. Her research interests involve innovation management, market entry strategies, and new business development with a focus on agri-business and life sciences. She has published several articles in the *British Food Journal*, the *Biotechnology Journal*, the *Creativity and Innovation Management Journal*, the *European Journal of Innovation Management*, the *International Food and Agribusiness Management Review*, the *International Journal of Product Development*, and the *R&D Management Journal*. Recently, she was appointed as part-time lecturer in innovation management at the Wageningen University and Research Centre in the Netherlands.

Alessio Cavicchi

Dr Alessio Cavicchi is a researcher in agricultural economics. He received his Ph.D. in the economics of food and environmental resources from the University of Naples 'Parthenope' and a MSc in food economics and marketing from the University of Reading in the United Kingdom. His research has been published in several journals, including *Agribusiness*, the *British Food Journal*, *Food Quality and Preference*, the *Journal of Agricultural Economics*, and the *International Journal of Wine Business Research and Food Economics*. Previously, he worked in finance and had an internship at the Economic and Monetary Affairs Commission of the European Parliament in Brussels. His main fields of research interest are consumer food choice, economics of food quality and safety, and innovation in the agri-food sector.

Armando Maria Corsi

Armando Maria Corsi is a Ph.D. student in wine economics and rural development. He graduated in tropical and subtropical agricultural sciences and then studied for a master's degree specialising in rural development and sustainable techniques at the University of Florence. He completed a traineeship at the F.A.O. (Italy) and was a visiting student at the University of Reading in the United Kingdom. He also was a visiting academic at the University of South Australia. His main fields of interest are consumer choice behaviour and the interaction/correlation effects of product attributes in determining customers' loyalty, with particular attention on wine.

Marie-Noëlle Duquenne

Dr Marie-Noëlle Duquenne is Assistant Professor in Statistical and Econometric Methods for Spatial Analysis at the University of Thessaly's Department of Planning and Regional

Development. She received her Ph.D. from the University of Paris-X-Nanterre. Dr Duquenne has published in *Economie Rurale*, *Espace-Population-Sociétés*, *Revue géographique des pays méditerranéens*, *Sociologia Ruralis*, and *Topos*, among others. She is the co-author of four books. Dr Duquenne's research interests include statistical analysis and econometric models of spatial phenomena, macro-economic models, forecasting methods, and regional and local development.

Christos Fotopoulos

Christos Fotopoulos is a full time Professor of Marketing Management and the head of the Department of Business Administration of Food and Agricultural Enterprises at the University of Ioannina. His scientific contributions include the completion of 27 research projects. His scientific work includes more than 35 papers, many of which were published in international peer-reviewed journals; 13 monographs; and more than 30 peer-reviewed conference papers. His current research interests include, but are not limited to, management, agri-business, food marketing (especially of organic products), total quality management, supply chain management, and the promotion and marketing of PDO products and aspects of contemporary marketing of agricultural businesses.

Andrea Insch

Dr Andrea Insch is a lecturer and researcher at the University of Otago, New Zealand. She has a Ph.D. in international business and Asian studies from Griffith University in Brisbane, Australia. Andrea has published in the *British Food Journal*, the *Journal of Marketing Communications*, the *Journal of Place Management and Development*, and *Place Branding*, among others. Her research interest include the use and acceptance of country-of-origin indications on food products, communication about environmental citizenship, and dynamics of value creation and delivery in food marketing systems, with a particular emphasis on historical research methods.

Menno Keizer

Menno Keizer is the Sustainable Agriculture Chain Development Specialist in East Africa for a Belgian development agency, Vredeseilanden (VECO). For almost 10 years, he has worked abroad in Africa and South East Asia training local people in various aspects of agricultural production and marketing. He has presented his research findings in forums in Indonesia, the Philippines, Thailand, and Vietnam. His research interests are in developing pro-poor agricultural market chains, marketing strategy development, and multi-stakeholder dialogues.

Marja-Riitta Kottila

Marja-Riitta Kottila is finalising her Ph.D. studies within the Department of Applied Biology, Faculty of Agriculture and Forestry, at the University of Helsinki. She has a multifaceted experience in organic food chains from her work as executive director for Finfood Luomu, in which context she organised and promoted organic food consumption and organic labels in Finland, as well as thereafter as a researcher focusing on organic foods. She is currently conducting research into supply chain management, collaboration and knowledge sharing, as well as the sustainability of food chains. She has published widely, including in the *British Food Journal* and at international conferences.

Beata Kupiec-Teahan

Dr Beata Kupiec-Teahan is an agri-food economist in the Food Marketing Research Team at the Scottish Agricultural College (SAC) and is based in Edinburgh. She has a Ph.D. in food science from Krakow's Agricultural University. Her experience and interests are in the areas of marketing and consumer studies.

Philip Leat

Philip Leat is a senior economist in the Food Marketing Research Team at the Scottish Agricultural College and is based in Aberdeen. He has a MSc in agricultural economics from the University of Aberdeen and more than 30 years of experience, working and publishing in the area of agri-food marketing at the Scottish, UK, and EU levels.

Adam Lindgreen

Dr Adam Lindgreen is Professor of Strategic Marketing at Hull University Business School. Dr Lindgreen received his Ph.D. from Cranfield University. He has published in *Business Horizons, Industrial Marketing Management, Journal of Advertising, Journal of Business Ethics, Journal of Business and Industrial Marketing, Journal of Marketing Management, Journal of Product and Innovation Management, Journal of the Academy of Marketing Science,* and *Psychology & Marketing,* among others. His most recent book is *Managing Market Relationships* (Gower Publishing, 2008). His research interests include business and industrial marketing, consumer behaviour, experiential marketing, relationship and value management, and corporate social responsibility. He serves on the boards of many journals.

George Maglaras

George Maglaras is a Ph.D. student and research assistant in the Department of Business Administration of Food and Agricultural Enterprises at the University of Ioannina. He has previously worked in marketing. His current research interests include consumer behaviour and research, food marketing, and (food) supply chain management.

Smitha Nair

Dr Smitha Nair is Academic Associate in Marketing at the Sadanam Institute of Commerce and Management Studies, India. Dr Smitha received her Ph.D. from Cochin University of Science and Technology, India. She also has worked as a senior research fellow in the Indian Council of Agricultural Research–funded research project on the development of a total quality management system for sustainable growth of the seafood trade from India. Her research interests include market orientation, total quality management, market research, and consumer behaviour.

Hernán Palau

Hernán Palau, an agricultural engineer with a master's degree in food and agri-business, is a research area coordinator in the Food and Agribusiness Program, School of Agronomy, University of Buenos Aires. He is also Postgraduate Professor of Economy and Strategy in Food and Agribusiness and Graduate Professor in Food and Agribusiness Marketing Strategies at the School of Agronomy, University of Buenos Aires. He has presented at several congresses (IAMA, PENSA, and Chain Conference) and published in books edited by the School of Agronomy and the Uruguayan Government. His research interests include food and agri-business marketing, beef agri-business, consumer behaviour, and networks. He serves on the board of the *International Food and Agribusiness Management Association Review*.

Marco Platania

Dr Marco Platania is an agricultural economics researcher at the Mediterranea University, Reggio Calabria (Italy), where he teaches models and theories of food consumption, agricultural economics, and agro-industrial economics. He received his Ph.D. from the University of Catania. A member of the Society Italian of Agricultural Economists (SIDEA) and the European Association of Agricultural Economists (EAAE), Dr Platania has authored several papers in Italian and international publications. His research interests include information and communication technologies applied to the agricultural sector, quality food product markets, agricultural marketing, consumer behaviour, and local and rural development.

Donatella Privitera

Dr Donatella Privitera, an agricultural economy researcher, teaches marketing of agricultural products, the economy of landscape, and the economy of food product markets in the Faculty of Agriculture at the University Mediterranea, Reggio Calabria. A member of the Society Italian of Agricultural Economists (SIDEA) and the European Association of Agricultural Economists (EAAE), she has authored more than 45 scientific publications, mainly associated with her research in the following areas: information and communication technologies in agriculture; agri-business in Italy; quality food product markets and strategies; consumer behaviour and trends in agri-food marketing; multifunctional agriculture and its implications for agri-tourism; and landscape marketing. At present, she is participating in cooperative research projects developed between the Calabria Regional Authority and University of Reggio Calabria.

S.P. Raj

S.P. Raj is the Distinguished Professor of Marketing at the Whitman School of Management, Syracuse University, where he served as senior associate dean. He was also a tenured Professor of Marketing at Cornell University and visiting faculty at Northwestern University. Professor Raj's research on marketing strategies, their influences on customer behaviour, and managing new product development has been cited extensively and recognised with awards such as the John D.C. Little Award by the Institute for Management Sciences and the Donald R. Lehmann Award by the American Marketing Association. His research has also been supported by the NSF/Corporate Center for Innovation Management Studies. He has authored several teaching cases and won the EFMD 2006 award for best case on managerial issues in transitory economies. He has consistently received high commendations for his teaching in full-time and executive programmes internationally. Professor Raj has published in journals such as *European Journal of Operational Research*, *IEEE Transactions on Engineering Management*, *Journal of Consumer Research*, *Journal of Marketing*, *Journal of Marketing Research*, *Journal of Product Innovation Management*, and *Marketing Science*. Professor Raj earned a bachelor's degree with distinction in electronics engineering from the Indian Institute of Technology, Madras, and master's and Ph.D. degrees in industrial administration from Carnegie-Mellon University.

Cesar Revoredo-Giha

Dr Cesar Revoredo-Giha is a senior food marketing economist and the team leader of Food Marketing Research at the Scottish Agricultural College (SAC), based in Edinburgh. He received his Ph.D. in agricultural and resource economics from the University of California, Davis. His areas of specialisation and interest are the industrial organisation of food market industries and international trade.

Päivi Rönni

Päivi Rönni is a Ph.D. student in the Department of Economics and Management, Faculty of Agriculture and Forestry, University of Helsinki. She has multi-faceted experience in food chains, based on her work as a marketing manager and executive director for Finfood Luomu, which organises and promotes organic food consumption and organic labels in Finland, and thereafter as a project manager for the Union of Agricultural Producers and Forest Owners in the province of Häme. Her research interests include supply chain management, customer relationship management, collaboration, and knowledge sharing. She has published widely, including in the *British Food Journal* and in international conferences.

Cristina Santini

Dr Cristina Santini received her Ph.D. in economics and management of enterprises and local systems from the University of Florence; previously, she received a master's degree in marketing and management in wine business from that same university. She has lectured on (international) management at the University of Florence. Previously, she worked as marketing manager in a Tuscan winery and has had work experience in a multinational company. Dr Santini currently works in research and is a consultant with the Italian Centre for Sensory Analysis in Italy. Her research interests are in entrepreneurship, innovation and small business.

Maria Rosaria Simeone

Dr Maria Rosaria Simeone is a researcher in agricultural economics. She received her Ph.D. in economics of food and environmental resources from the University of Naples 'Parthenope' and a MSc in food economics and marketing from the School of Agriculture Policy and Development, University of Reading in the United Kingdom. She has taken several courses in sensory analysis and consumer behaviour, as well as experimental methods in agri-food matters. She also has taught several such courses. Recently, she was appointed to the Ph.D. board in economics of food and environmental resources at the University of Naples 'Parthenope.' Her research interests include consumer food choice and innovation in the agri-food sector.

Sebastián Senesi

Sebastián Senesi, an agricultural engineer with a master's degree in food and agribusiness, is the sub-director in the Food and Agribusiness Program, School of Agronomy, University of Buenos Aires. He is Postgraduate Professor of Economy and Strategy in Food and Agribusiness and Graduate Professor in Food and Agribusiness Marketing Strategies at the School of Agronomy, University of Buenos Aires. He has presented at several congresses (IAMA, PENSA, Chain Conference) and published in books edited by the School of Agronomy. His research interests include food and agribusiness management, social capital, and bio-fuels.

Nguyen Thi Le Thuy

Nguyen Thi Le Thuy was previously the director of the Coconut Research Center in Ben Tre province, Vietnam, for almost 17 years, where she was engaged in different coconut-based projects. Most of the projects related to development of poor coconut farmers in rural areas. Currently, she is the Deputy Director of the Ben Tre Department of Science and Technology, where her main focus is on the sustainable development of coconut and other commodities.

Erik M. van Raaij

Dr Erik M. van Raaij is Assistant Professor of Purchasing & Supply Management at Rotterdam School of Management, Erasmus University (Rotterdam). He holds an engineering degree in business and a Ph.D. in marketing from the University of Twente, the Netherlands. His work has been published in practitioner outlets such as the *Holland Management Review*, as well as international academic journals such as *Computers & Education*, the *European Journal of Marketing*, the *Journal of Business Research*, the *Journal of Business Ethics*, the *Journal of Purchasing and Supply Management*, *Industrial Marketing Management*, *Marketing Letters*, *Supply Chain Management*, and *Marketing Intelligence & Planning*. Erik M. van Raaij is the associate editor of the *Journal of Purchasing and Supply Management*. He is a member of the Erasmus Research Institute of Management, a visiting lecturer at Cass Business School (London), and a member of the International Purchasing and Supply Education and Research Association (IPSERA) and the Dutch Purchasing Association (NEVI). Current teaching includes the course Strategic Sourcing (MSc program) at Rotterdam School of Management and the course Marketing (Executive MBA program) at Cass Business School. Dr van Raaij has received several awards and commendations for teaching excellence. His current research interests include new technologies in purchasing and marketing, value, cost and profitability in business-to-business relationships, and market orientation.

Fernando Vilella

Fernando Vilella, an agricultural engineer, is the director of the Food and Agribusiness Program and the dean of the School of Agronomy at the University of Buenos Aires. Professor Vilella is a graduate professor in the School of Agronomy. He serves on the IAMA board. He has presented at congresses (IAMA, PENSA, and Chain Conference) and published in journals and in books edited by the School of Agronomy. His research interests include food and agribusiness management and bio-fuels.

Ilias P. Vlachos

Dr Ilias P. Vlachos is a lecturer at the Agricultural University of Athens, Greece. He holds a Ph.D. from Cranfield University, United Kingdom. His research interests include e-business, food supply chain management, and business growth.

Georges Vlontzos

Dr George Vlontzos is a lecturer in agricultural economics at the Thessaly University's Agricultural Sciences School, where he also received his Ph.D. He has published in several journals, including *Current Agriculture Food and Resource Issues*, the *Journal of International Law and Trade Policy*, and the *Mediterranean Journal of Economics, Agriculture, and Environment*, among others. He is the co-author of three books. Dr Vlontzos's research interests include international trade of agricultural products, international and European agricultural policy, and rural development issues.

Vincenzo Zampi

Vincenzo Zampi is Professor of Management at the University of Florence. He is a member of the Accademia dei Georgofili and the Accademia della Vite e del Vino. Vincenzo Zampi serves on the editorial board of the *International Wine Business Research Journal*. He also has had experience in wine business consulting.

Reviews for Market Orientation

'*Adam Lindgreen and his colleagues have written a seminal book on the role of market orientation in driving business performance and customer satisfaction in the agribusiness industry. It is a must read for agribusiness executives and academics who are looking to shift their perspective from a supply-side, production model to a demand-side, customer-lead model.*'

Dr Bernie Jaworski (Monitor Executive Development, Los Angeles, USA.)

'*This book provides a thoughtful treatment of market orientation research, and its applications in the food and agriculture industries. It is a 'must read' for executives in these industries.*'

Professor Ajay Kohli, Ph.D. (Editor, *Journal of Marketing*, and Professor of Marketing at Goizueta Business School, Emory University, USA.)

'*Food. Everyone needs it and everyone buys it. But who directs it? Lindgreen, Hingley, Harness and Custance have put together the first comprehensive volume to address the issue of the role of the customer in the food supply chain. From the beginnings of the human agricultural industry one thing has been certain: if customers won't buy it, farmers won't grow it. This is a basic consequence of being customer- or market-driven. As the authors point out, firms that stick to the basic premise of listening to their customers lead in sales and profit growth. Market Orientation consists of twenty articles by leading scholars in the marketing of agricultural products. The articles in the three major sections of the book: Marketing Agricultural Products, Market Orientation in the Downstream Food Chain, and Market Orientation for Specialty Products all provide exceptional insight to understanding how market orientation can benefit food suppliers and how it is really essential for long-term success. Anyone in the food industry will be better managers after reading the superbly written articles contained in Market Orientation.*'

Prof. Peter LaPlaca, Ph.D. (Editor, *Industrial Marketing Management*, and Professor of Management and Marketing, Barney School of Business, University of Hartford, USA.)

'*This is a unique and valuable set of papers that specifically address applications of market orientation in the context of food and agricultural markets. We have come a long way from the days when all agricultural products were seen as simply commodities with few or no opportunities for creating additional customer value, thereby leading all participants in the channel of distribution to be merely price takers rather than value creators. Clearly, when market-oriented practices are fully and carefully implemented, multiple opportunities for creating customer value are discovered and appropriate routes to exploit them are developed.*

The results, as the analyses in this book suggest, are win-win opportunities for producers, distributors, retailers, and consumers. This collection of empirical analyses will be valuable to managers as well as to researchers as they both seek to understand the implications of market orientation and the steps to implement it.'

Professor emeritus John Narver (Michael G. Foster School of Business, University of Washington, USA) and Professor John Slater (College of Business, Colorado State University, USA.)

Foreword and Acknowledgments

Agriculture has often been considered production-, rather than market-, oriented, whether in a developed or a developing country context, but any in-depth analysis of agricultural businesses must show that the entrepreneurial spirit is alive and well in rural communities. Further down the supply chain, there is an increased focus on consumer needs as a means to identify potential market opportunities and provide customer satisfaction. Within the field of food and agriculture, there are tremendous challenges and opportunities that arise from developing, monitoring, and measuring market orientation programmes. The overall objective of this book is to provide a comprehensive collection of cutting-edge research on the challenges and opportunities presented to those seeking to adopt a market orientation, in particular with regard to the consequences for food and agricultural businesses and the development of appropriate strategic marketing plans for those businesses. The book will consider a number of different issues pertaining to markets, products and services, which result in the development of a market-oriented perspective. The context involves international food and agricultural marketing. Contributors to the book present research from Europe, Asia, Australasia and South America.

Implementing Market Orientation

In the book's first chapter, Van Raaij considers the marketing concept's value in the past 50 years as a useful indicator of performance differentials among firms. Although advances on the theory side include many papers with definitions, the issues of measurement, models, implementation and market orientation have been less well researched. This area is the focus of this first chapter, which reviews research on the subject from the past 20 years. The author draws the conclusion that though various approaches for implementing market orientation exist, there is a real need for comparative studies of implementation success.

In Chapter 2, a similar theme gets advanced by Beverland and Lindgreen through two case studies (from New Zealand) of the implementation of market orientation in agri-industrial firms. They develop a useful model of the stages in implementing a market orientation and offer advice to organisations about to how to implement marketing strategies, with an emphasis on the need to encourage employee risk-taking and make a bottom-up commitment to market-oriented change.

Marketing Agricultural Products

In the first contribution in this section (Chapter 3), Bröring looks at market orientation from the viewpoint of the farm-input industry, with illustrations from the German feed industry. This industry, at the very front of the food chain, has suffered not only raw material shortages but also increasing expectations from both retailers and consumers regarding food safety. Once considered a commodity (production-oriented) sector, the feed industry has become market oriented, as the author notes by considering its raw material purchases, as well as the interests of both its direct customers (farmers) and the players further downstream.

A more detailed exploration of one facet of market orientation – that of brand equity – is undertaken by Beverland in Chapter 4. He identifies six components of brand orientation that can be combined with existing brand and market orientation scales to examine the relationship between brand orientation and long-term business performance. A longitudinal case study of a New Zealand global marketer of raw wool illustrates the analysis.

Chapter 5 (by Leat, Revoredo-Giha and Kupiec-Teahan) examines whether reform of the European Union's Common Agricultural Policy, and particularly the introduction of the Single Payment Scheme, has made Scottish farmers more market-oriented and competitive. The case study of *Qboxanalysis*, a performance-related communication system, shows that in addition to improving farmers' market orientation, the meat supply chain has benefited overall by reduced wastes and cost savings.

The meat sector is again the subject of Chapter 6 by Palau, Senesi and Vilella, which focuses on Argentina. A case study of Prinex, founded in 1989 by a group of traditional producers, demonstrates the effectiveness of adding value to beef for export markets. However, restrictions have continued to challenge the organisation in its efforts to develop export markets.

In Chapter 7, Bijman investigates the restructuring of agricultural cooperatives during the past 15–20 years, mainly as a result of their need to become more market-oriented. The increased focus on consumer and retailer needs has led to both strategic reorientations and organisational restructuring in traditional cooperatives, as well as the establishment of new types of producer-owned firms. Examples come from north-western Europe, citing organisations the author believes are suitable and viable business forms for marketing agricultural products.

The cooperative theme gets further developed by a second contribution from Beverland, in this case in reference to brand building by New Zealand cooperatives. Chapter 8 identifies why agricultural cooperatives must build brands and explores the viability of traditional and new generation cooperatives in supporting a market-oriented strategy, with special reference to brands and long-term relationships with channel buyers. Although both types of organisations can develop innovative marketing programmes, the new generation cooperatives achieve more sustained long-term success because of their more flexible ownership structures.

In the final chapter in this section (9), Nair investigates the adoption of market-oriented principles by Indian seafood firms. This adoption is necessary, Nair believes, as a result of the essentially commoditised trade of Indian seafood products and the very competitive nature of competing emerging market suppliers. The adoption of market-oriented principles by Indian producers should enhance their business performance.

Market Orientation in the Downstream Food Chain

The opening contribution to this section (Chapter 10) is by Kottila and Rönni, who offer a case study of two organic food chains in Finland. The organic food market is hindered in its development by several factors. A study of the various actors in the food chain, from farmers to consumers, employs efficient consumer response (ECR) as a platform.

Chapter 11 (Cavicchi, Simeone, Cristina and Bailetti) considers the contribution of market research and sensory analysis to market orientation in the food industry. A literature review of the use of sensory analysis for designing, testing, launching and rethinking food products leads into an investigation of the use of two tools to assist innovation: the 'House of Quality' and the 'Buyer Utility Map.'

Another further contribution that refers to India appears in Chapter 12, where Raj and Adhikari focus on efforts to move agri-business from a commodity (product) orientation toward a market orientation. The chapter is illustrated by a case study of the efforts of a firm in the cold storage business to break out of the commodity trap; it also contains a discussion of public policy initiatives to facilitate an economy-wide transition.

Australia is Insch's focus in Chapter 13, including an investigation of the characteristics and consequences of becoming market-oriented in meat retailing. To illustrate the challenges facing meat retailers in Australia, especially complacency and indifference to changing consumer lifestyles, a case study of two meat retailers investigates specific market-oriented behaviours, such as innovative value-added solutions, consumer understanding, leading (or driving) the market, flexible response, and whole-chain coordination.

The role of technology in assisting market orientation is the topic in Chapter 14, by Aguiar, Brofman and Barcellos. Radio frequency identification (RFID) might enhance not only the movement of products through the food chain but also customer relationships, thus providing true market orientation. Certain negative perceptions persist among consumers, and ethical issues still influence RFID, which may limit its widespread adoption.

In Chapter 15, Vlontzos and Duquenne study the interrelationship between ethnicity and international trade with reference to the Greek virgin olive oil trade. The authors present a set of proposals and suggestions for increasing the competitiveness of the sector, especially in terms of quality and safety. The efforts by Greek producers to achieve PDO (protected designation of origin) or PGI (protected geographical indication) status indicate evidence of moves away from anonymous commodity production and toward market orientation. Findings from the application of a gravity model and a SWOT (strengths, weaknesses, opportunities and threats) analysis produce recommendations regarding a revised marketing mix for olive oil producers.

Finally, Platania and Privitera focus on organic wine in Chapter 16. The sector is expanding but still occupies a limited share of the wine market. These authors undertake a survey of Italian consumers and provide data that show that consumers are influenced less by the organic nature of the wine than by factors such as its origin and brand name.

Market Orientation for Specialist Products

The first contribution to this section (Chapter 17) provides a further analysis of market orientation in Italy. Cavicchi and Corsi analyse consumer values and the choice of specialty foods, with special references to olive oil. In a case study of *Oliva Ascolana del Piceno*, development of PDO-PGI products in Italy seems to promote the recovery of historic values and traditions in local communities while also encouraging greater appreciation of consumers' needs and thus a drive to increase market orientation among specialist olive oil producers.

Another niche industry receives attention in Chapter 18. Fotopoulos, Vlachos and Maglaras present a case study of moving from a product to a market orientation in Greece. The Chios Mastiha Growers' Association offers a unique product, and through a combination of good management support, leadership, strong social cohesiveness among the producers, effective supply chain management, and a strong brand image, the cooperative has created a new vision for its product and a market-oriented strategy.

Chapter 19 (Keizer and Le Thuy) investigates market orientation using the example of a high-value coconut product from Vietnam. Small-farmers have often been impeded by a lack of market information and access to markets, which reinforces poverty in many coconut-growing communities. The authors study the marketing chain for coconut leaf-based gift baskets and identify the various marketing strategies employed by actors in the supply chain to show how the incomes of these specialist rural producers have been enhanced.

The final chapter (20) is by Cristina, Cavicchi and Zampi and pertains to the level of market orientation of old world wineries. By investigating the competitive strategies of Italian wine producers, the authors undertake a study of the changing market for wine and the rise of new world producers. The key weaknesses in the level of market orientation of Italian wine producers emerge from focus group investigations. The authors show that the product orientation exhibited by both Italian wine entrepreneurs and producers has insulated them from rapid changes in the global wine industry.

Closing Remarks

The double-blind process for selecting entries required the services of a number of reviewers who dedicated their time and effort to providing helpful feedback to the authors. We greatly appreciate their work, which has helped improve the chapters included in this book. We extend a special thanks to Gower Publishing and particularly its staff, who have been extremely helpful throughout the entire process. Equally, we warmly thank all of the authors who submitted their manuscripts for consideration for this book. They have exhibited the desire to share their knowledge and experience with the book's readers – and a willingness to put forward their ideas and evidence for possible challenge by their peers. Finally, we thank Harper Adams University College and Hull University Business School for their support in this venture. Special thanks go to Elisabeth Nevins Caswell for her editorial assistance.

We are hopeful that the chapters in this book fill knowledge gaps for readers but also that they stimulate further thoughts and actions regarding issues of market orientation in the ever-changing agricultural and food marketing environment. To help extend the boundaries of scholarship in this area, the chapter authors have provided an eclectic mix of both theory and practice. Using a wide range of methodologies, they offer contemporary examples of market orientation – both successful and less so. The division of the book into four sections reflects the diverse ways in which global food and agricultural markets have achieved the adoption of successful market orientation. To assist the reader, each chapter contains an abstract and a detailed list of references. We hope that these contributions add to and initiate further dialogue that advances research on this important topic.

Professor Dr Adam Lindgreen, Hull
Reader Dr Martin K. Hingley, Newport
Senior lecturer Dr David Harness, Hull
Principal lecturer Dr Paul R. Custance, Newport

Implementing Market Orientation

CHAPTER 1

Making the Transformation Toward a Market-orientated Organisation: A Review of the Literature

BY ERIK M. VAN RAAIJ*

Keywords

market orientation, implementation theory, organisational learning, literature review, evolution of theory

Abstract

Since its inception 50 years ago, the marketing concept has served as marketing's implicit theory of the firm by relating performance differentials among firms to their degree of market orientation. Despite significant advances in the development of market orientation theory, a large void remains in the literature when it comes to studies that tackle the implementation of market orientation. This chapter assesses the current status of market orientation literature and reviews the most prominent implementation approaches published since the 'Rediscovery of the Marketing Concept' in 1988.

Introduction

Many organisations aspire to become more market oriented. Various developments fuel this drive to instil a more outward orientation in organisations. First of all, increased competition, shorter product life cycles, and more demanding customers drive firms toward

* Dr Erik M. van Raaij, Department of Management of Technology and Innovation, Rotterdam School of Management, Erasmus University, Burg. Oudlaan 50, 3062 PA Rotterdam, the Netherlands. E-mail: eraaij@rsm.nl. Telephone: + 31 – 10 408 1948.

establishing a focus on customers, competitors, and the market in general. Examples of such environmental pressures to become market-oriented appear in this book, including Bijman's review of agricultural cooperatives, Bröring's assessment of the livestock feed industry, and Santini et al.'s insights into the wine industry. Second, performance that falls short of targets may drive top management to take corrective action, and this unsatisfactory performance may be ascribed to a lack of external focus, as Fotopoulos et al. describe in their chapter in this book. Third, macro-economic developments can be a driving force behind the search for a customer or market focus. Privatisation, deregulation, and cutbacks on government subsidies push organisations toward a more conscious consideration of customer demands and competitive advantages. Examples include the privatisations of railways, postal services, and airports;[1] deregulation in the utilities sector, health care, and the food industry;[2] and subsidy cutbacks for cultural organisations and charities.[3]

At the same time, while academics as well as practitioners can easily connect to the market orientation theme, they are often talking about concepts other than what has been defined as a 'market orientation' in scholarly literature. As Lear has noted,[4] if we can use the term market orientation so loosely, we can *say* a lot about market orientation while in fact *doing* quite different things.

The market orientation concept has its origin in a management philosophy known as 'the marketing concept.' The marketing concept has been a cornerstone of the marketing discipline since Drucker first argued that '[t]here is only one valid definition of business purpose: *to create a customer*' and described marketing as 'the whole business seen from the point of view of its final result, that is, from the customer's point of view.'[5] Over the years, it has served as marketing's implicit theory of the firm by relating performance differentials between firms to their degree of market orientation.[6] The marketing concept has appealed to generations of managers and been one of marketing's most influential ideas. And yet, formal research into the concept has been lacking until the academic 'rediscovery' of the concept,[7] which led to a stream of recent research papers. This contemporary market orientation literature deals with four issues:

1. The definition issue, focusing on the conceptualisation of the construct. This literature addresses the question: What is a market orientation?[8]
2. The measurement issue, focusing on the development of scales. This literature is concerned with how the market orientation construct can be operationalised and assessed.[9]
3. The model issue, focusing on the antecedents and consequences of a market orientation. This literature deals with the causes and effects of a market orientation.[10]
4. The implementation issue, focusing on approaches for managerial action to implement a market orientation. This literature addresses the question: How can firms become more market oriented?[11]

The emphasis of this chapter is on the fourth issue: implementing a market orientation. I start with a discussion of the historical roots of market orientation. Then, I summarily discuss the definition issue, the measurement issue and the model issue, all with a particular focus on the implementation of a market orientation. Subsequently, I concentrate on the implementation issue itself, presenting the most prominent implementation approaches published to date. The chapter concludes with an evaluation of the current status of the market orientation concept and the outlook of the field.

Historical Roots of the Market Orientation Concept

The literature provides wide support for the idea that the market orientation concept, as it is discussed in today's marketing journals, originates from the 'marketing concept.'[12] Drucker defines the central tenet of the marketing concept as 'marketing is the whole business seen from the customer's point of view' and illustrates the marketing concept with the example of General Electric, where a reorganisation in the early 1950s provided marketing with the authority to direct engineering, design and manufacturing on the basis of their market knowledge.[13] Sachs and Benson trace the central theorems of the marketing concept back to two philosophical strains of the eighteenth century:[14] rationality of man and utilitarianism. Rational consumers seek to maximise utility, and based on this notion, companies should be created in which marketing identifies the consumers' utility functions and 'will establish for the engineer, the designer, and the manufacturing man what the consumer wants in a given product, what price he is willing to pay, and where and when it will be wanted.'[15]

The advent of the marketing concept has been attributed to the maturing of the United States economy in the 1950s. The transformation from a production economy to a consumption economy was characterised by an abundance of suppliers and brands and an increasingly affluent consumer.[16] Keith, at that time executive vice-president of The Pillsbury Company, describes how a marketing revolution took place in his organisation as a result of the emancipation of consumers and the consequent necessity to align the company with the needs and wants of current and potential customers.[17] According to Webster,[18] the marketing concept was pushed into the background by the advent of financial planning and strategic planning in the 1960s and 1970s, but it was 'rediscovered' in the late 1980s. Houston adds that the misapprehension and the misuse of the marketing concept (e.g., the misguided advice that companies should depend on current, expressed needs and wants of customers) also contributed to its temporary decline in popularity.[19]

Two influential publications in the *Journal of Marketing* that emanated from Marketing Science Institute–sponsored research heralded a new era.[20] Subsequent works use these publications as a reference point, and in the academic debate, the term 'marketing concept' has been replaced by the term 'market orientation.' Kohli and Jaworski prompted this replacement by using the term market orientation to mean the implementation of the marketing concept: organisations that implement the marketing concept (a management philosophy) possess a specific trait, which is called market orientation.[21] The Kohli and Jaworski and Narver and Slater papers made an impressive impact on the marketing literature. They were the first attempts to define market orientation, develop scales for measuring the construct, and formulate propositions that linked market orientation to business performance. A formidable body of literature has been built on the foundations laid by Kohli and Jaworski and Narver and Slater. I first turn this chapter to the way in which 'market orientation' is defined in existing literature.

The Definition Issue

In this section, I will focus on definitions of *market orientation* and thus leave out definitions and descriptions of the *marketing concept* before 1988. The most influential definitions are undisputedly those of Kohli and Jaworski and Narver and Slater.[22] But to put these definitions in perspective, a wider set of definitions is provided here:

- A company is market oriented if 'information on all important buying influences permeates every corporate function,' 'strategic and tactical decisions are made interfunctionally and interdivisionally,' and 'divisions and functions make well-coordinated decisions and execute them with a sense of commitment.'[23]
- 'Market orientation is the organisationwide *generation* of market intelligence pertaining to current and future customer needs, *dissemination* of the intelligence across departments, and organisationwide *responsiveness* to it.'[24]
- Market orientation is defined as 'the business culture that most effectively and efficiently creates the necessary behaviors for the creation of superior value for customers.' Market orientation 'consists of three behavioral components – customer orientation, competitor orientation, and interfunctional co-ordination – and two decision criteria – long-term focus and profitability.'[25]†
- 'The level of market orientation in a business unit [is] the degree to which the business unit (1) obtains and uses information from customers; (2) develops a strategy which will meet customer needs; and (3) implements that strategy by being responsive to customer needs and wants.'[27]
- Customer orientation‡ is 'the set of beliefs that puts the customer's interest first, while not excluding those of all other stakeholders such as owners, managers, and employees, in order to develop a long-term profitable enterprise.'[28]
- 'Market orientation represents superior skills in understanding and satisfying customers.'[29]

All definitions entail an external focus with the customer as the primary focal point. All except for the definition of Deshpandé and colleagues have a clear action component, that is, being responsive to customers. And all definitions except for Ruekert's suggest that market orientation involves more than just a focus on customers. The major differences lie in the organisational element emphasised in each definition: Shapiro emphasises the decision-making processes, Kohli and Jaworski the information-processing activities, Narver and Slater the business culture as a set of behavioural components, Ruekert the organisational strategy process, Deshpandé and colleagues the business culture as a set of beliefs, and Day emphasises skills and organisational capabilities. When it comes to the implementation of a market orientation, different definitions of market orientation suggest different types of interventions for enhancing the degree of market orientation.

Over the years, the majority of authors in market orientation literature have been using either Kohli and Jaworski's or Narver and Slater's definition, such that Narver and Slater's approach appears to be more popular, though Narver and Slater have used slightly different definitions over the years. Two main perspectives on market orientation have emerged as a result: a behavioural perspective based on Kohli and Jaworski and a cultural

---

† In a 1995 publication, these same authors provide a different definition of market orientation. A market orientation is defined as 'the culture that (1) places the highest priority on the profitable creation and maintenance of superior customer value while considering the interests of other key stakeholders; and (2) provides norms for behaviour regarding the organisational development of and responsiveness to market information.'[26]

‡ These authors do not use the term market orientation but use customer orientation instead. They refer to the same concept however, because they 'see customer and market orientations as being synonymous.'

perspective based on Narver and Slater.[30] Homburg and Pflesser have recently shown, however, that these two perspectives need not be treated as opposing perspectives.[31] Using a multi-layer model of organisational culture, these authors show that market-oriented behaviours can be modelled as one of four layers of a market-oriented culture. For the purpose of implementing a market orientation, the definitions of market orientation point to the following change levers: certain organisational *behaviours*, such as information processing, decision making, and strategy formation; specific *skills* to enable those behaviours, such as market sensing and customer linking; and elements of *culture* to drive the desired behaviours, such as beliefs, values, and norms.

The Measurement Issue

With the variety of definitions of market orientation comes a multitude of scales for measuring the construct. Only a few scales have reached the status of being reused by other researchers. Wrenn mentions a number of studies that have used Kotler's marketing effectiveness scale,[32] but these studies have not been published in the major marketing journals. Since the early 1990s, the measurement batteries of Narver and Slater (the MKTOR scale[33]) and Kohli and Jaworski (the MARKOR scale[34]) have been reused often, either as they are or as the basis for adapted scales. Both the MKTOR and MARKOR are Likert-type scales. Underlying MKTOR's 15 items are three components of market orientation: customer orientation, competitor orientation and interfunctional co-ordination. A business's market orientation score is the simple average of the scores of the three components.[35] Underlying MARKOR's 20 items are three components: intelligence generation, intelligence dissemination and responsiveness. The third component is composed of two sets of activities: response design and response implementation.[36] Similar to MKTOR, the market orientation score is an unweighted sum of the three components. The received view of measuring market orientation is one of self-administered questionnaires directed at senior executives, which measure self-evaluations of business unit activities using Likert-type scales.

Both scales have been criticised from a methodological scale-development perspective,[37] for their single informant strategy[38] and for their common reliance on the focal organisation only.[39] More important for this review in light of the implementation issue is the critique of the scales' usefulness as a diagnostic tool for managers.[40] Both MKTOR and MARKOR were designed to assess levels of market orientation across companies. Bisp, Harmsen and Grunert argue that these instruments are less suitable in cases in which a single company wants to assess its current level of market orientation and its potential for increasing market orientation.[41] They employ intervention theory that hypothesises that successful intervention for increased market orientation rests on identifying and reducing three gaps: (1) gaps between attitudes toward market orientation and actual market-oriented behaviours; (2) gaps between the degree of market orientation as perceived by different groups in the organisation (top management, sales/marketing, production); and (3) differences between orientations toward three external groups (end users, customers and competitors). Therefore, they develop a measurement instrument that measures attitudes and behaviours related to an orientation towards these three external groups, administered to three different groups of organisational members. Finally, they propose pre- and post-change measurements of market orientation to assess the effectiveness of the change effort.

The MKTOR and MARKOR scales represent the current state of the art in measuring market orientation. These scales are not without critiques, however, and for the purpose of implementing a market orientation they specifically lack actionability. Possible directions for developing more actionable scales are presented in the final section of this chapter.

The Model Issue

The third issue in market orientation research is the model issue. The model issue focuses on the antecedents and consequences of a market orientation, as well as variables that might moderate or mediate the relationships between market orientation and its consequences. Although the marketing concept has been readily accepted as an article of faith for better performance since its inception in the 1950s, a vast amount of studies have researched the relationship between the degree of market orientation and business performance. This relationship has been studied for large firms and small firms, for manufacturers as well as service suppliers, industrial firms and consumer goods companies, for profit and not-for-profit organisations, and in industrialised economies and in transition economies.[42]

Contrary to what is often reported,[43] there is no consistent finding that being market-oriented improves business performance.[44] Despite these 'anomalies,' the dominant view is that an organisation's degree of market orientation has a positive effect on business performance, especially profitability.[45] Market orientation is also believed to have positive consequences for employees.[46] Ruekert finds a positive effect on job satisfaction, trust in leadership and organisational commitment, Jaworski and Kohli find a positive effect on employees' esprit de corps and organisational commitment, and Siguaw, Brown, and Widing find a negative effect on role stress but positive effects on job satisfaction and the organisational commitment of salespeople.[47] The chapter by Nair in this book finds positive effects of a market orientation on business performance, employee consequences and customer consequences in the Indian seafood processing industry. It is important to note that these studies are based on self-reported measures of performance. A recent study using objective performance measures found only very weak associations between market orientation and performance.[48]

Several variables have been introduced as potential moderators in the market orientation-performance relationship, such as market turbulence, technological turbulence, competitive intensity, market growth, and buyer power.[49] The results from these moderator studies are equivocal. Wrenn has reviewed the extant literature and concludes that moderators have little effect on the positive impact of market orientation on firm performance.[50] Harris, in contrast, comes to the conclusion that the picture is mixed but that moderators play a significant role.[51] Table 1.1 summarises the empirical evidence to date.

Other studies have focused on variables that mediate the market orientation–performance relationship. Although some researchers have conceptualised organisational learning as either a mediator or a moderator,[53] Day argues that being market oriented represents a particular type of learning, namely, learning about markets.[54] A strong case is being built for another mediating variable: innovation. The underlying rationale is that market-oriented organisations have a knowledge advantage over their competitors

Table 1.1   Empirical evidence of moderators' effects on the market orientation (MO)–performance relationship[52]

Study	Moderator	Influence
Slater and Narver	Market growth	MO – sales growth (-)
	Technological turbulence	MO – new product success (-)
	Market turbulence	MO – return on assets (-)
Greenley	Technological change	MO – new product success (-)
	Market turbulence	MO – return in investment (-)
	Customer power	MO – sales growth (+)
Appiah-Adu	Market dynamism	MO – return in investments (-)
	Competitive intensity	MO – sales growth (+)
Harris	Market turbulence	MO – profitability (-)
	Competitive hostility	MO – sales growth (+)

and that this knowledge helps them become more proficient in their new product development activities.[55] Empirical studies have substantiated this mediating role of innovation: Atuahene-Gima finds positive effects of a market orientation on various measures of innovation characteristics and innovation performance.[56] Han and colleagues find that a market orientation affects performance only through innovativeness, not directly.[57] Studies by Langerak and colleagues and Vazquez, Santos and Alvarez mirror these results with their findings of no direct relationship between market orientation and performance,[58] though they uncover an effect mediated by new product development proficiency and innovativeness. Conceptual work suggests that apart from new product development, other mediating factors may include proficiency in customer relationship management and supply chain management.[59]

In addition to the consequences, mediators and moderators, the antecedents of a market orientation have been studied. A distinction can be made between external antecedents (i.e., environmental factors that stimulate a firm's adoption of a market orientation) and internal antecedents (i.e., organisational factors that enable the adoption of the market orientation concept). External antecedents that have been proposed in the literature are market dynamism and competitive intensity.[60] Kohli and Jaworski argue that in a stable environment, few adjustments to the marketing mix are needed, requiring a low level of market orientation. Furthermore, the lower competitive intensity, the more a firm can 'get away with' a low level of market orientation.[61]

Various internal antecedents have been proposed and empirically tested by Ruekert and Jaworski and Kohli.[62] Ruekert identifies three organisational processes that foster a market orientation: recruiting and selecting customer-focused individuals; market-oriented training; and market-oriented reward and compensation systems. All three factors correlated positively with market orientation. Jaworski and Kohli advance eight antecedents: top management emphasis on market orientation; top management risk aversion; interdepartmental conflict; interdepartmental connectedness; degree of formalisation; degree of centralisation; degree of departmentalisation; and reliance on market-based factors for evaluations and rewards. Not all hypothesised relationships were empirically supported. Top management emphasis, interdepartmental conflict and connectedness, and reward systems appear to be the most important antecedents.[63] Within the set of internal antecedents, strategies (mentioned as an antecedent by Pelham and Wilson[64]) play a specific role in that they do not *enable* the adoption of market-oriented behaviours, but specific strategies (e.g., differentiation strategy) *necessitate* the adoption of market-oriented behaviours.[65] Strategies are, in that sense, more like the external antecedents, increasing the need to adopt more market-oriented behaviours.

Summarising the model issue as it is put forward in the literature, a market orientation leads to improvements in financial, customer and employee outcomes. Studies suggest that employee outcomes influence customer outcomes,[66] and customer outcomes drive financial outcomes.[67] Several variables have been proposed as mediators (e.g., proficiency in innovation) and moderators (e.g., market turbulence) to the relationships between market orientation and performance. The adoption of a market orientation is stimulated by certain environmental factors and specific organisational strategies. The ability to become market oriented depends on specific organisational factors, which can act as either enablers or barriers to market orientation. Empirical support for the relationship between market orientation and financial outcomes is rather strong, whereas empirical support for the other relationships is more sketchy.

The Implementation Issue

The fourth issue in market orientation research is the implementation issue. This literature addresses the question of how firms can become more market oriented. In recent years there has been an increase of papers dealing specifically with implementing a market orientation. The year 1988, when Webster's 'The Rediscovery of the Marketing Concept' was published, serves as the start for this overview of published implementation approaches.

Implementation approaches can be classified as systems-oriented, culture-oriented, or hybrid. The first is focused on changing the organisational system within which people operate. The idea is that by changing the environment in which people operate, their behaviours will adapt and change, and if executed well, these behaviours should become more market oriented. The culture-oriented approach is focused on changing the norms, beliefs, and values of organisational members. The idea behind these approaches is that market orientation is ultimately an organisational culture. These approaches focus on training and competence development. The hybrid approaches try to combine the best of both worlds. A classification of implementation approaches appears in Table 1.2, presented in chronological order of first publication in an international academic journal. All three types of approaches have been advocated throughout the 20 years of this review.

THE NEW MARKETING CONCEPT APPROACH BY WEBSTER

Webster's 'The Rediscovery of the Marketing Concept' concludes with a paragraph dedicated to the five basic requirements for developing a market-driven, customer-focused business:[68] (1) customer-oriented values and beliefs supported by top management; (2)

integration of market and customer focus into the strategic planning process; (3) the development of strong marketing managers and programmes; (4) the creation of market-based measures of performance; and (5) the development of customer commitment throughout the organisation. Webster's approach starts with the CEO signalling market-oriented values and beliefs. Webster further outlines marketing's role in increasing its involvement in strategic planning and in making the entire business aware of service and customer satisfaction. Furthermore, it is the role of human resource management to develop marketing competence and design market-based reward systems. The author's later work echoes these views with a list of 15 guidelines for implementing the new marketing concept.[69] Yet most of the guidelines are mere should-dos, without specification of the interventions a manager could carry out to make that happen.

THE MANAGEMENT DEVELOPMENT APPROACH BY PAYNE

The timing of Payne's programme to increase market orientation coincided with Webster's 'The Rediscovery of the Marketing Concept.'[70] Payne's approach consists of three steps: (1) understanding the mix of orientations in the organisation; (2) identifying the present levels of marketing effectiveness; and (3) implementing a plan to improve market orientation.[§] The third step consists of five sub-steps: (3a) understand the organisational and cultural dimensions of the problem; (3b) identify a marketing champion; (3c) conduct a training needs analysis; (3d) design a marketing training and development programme; and (3e) organise key support activities. Steps 1, 2, 3a and 3c together comprise the diagnostic stage of the implementation approach. Payne also proposes using Kotler's marketing audit to diagnose current marketing effectiveness (step 2) and the well-known 7S framework to understand the organisational and cultural dimensions of the problem (step 3a).[71]

The main interventions that Payne suggests appear in steps 3d and 3e. Through a management development programme involving marketing staff and executives from other functions, the training programme will explain the knowledge, skills, and attitudes necessary for the development of a market-driven organisation. Key support activities that need to be considered include an organisation around markets, recruitment of marketing talent, promotion of market-oriented executives, development of a marketing information system, and installation of an effective marketing planning system. Although its main thrust is management development and training, this approach attempts to cover almost all aspects of organisation defined by McKinsey's 7S framework.

THE INFORMATION PROCESSING APPROACH BY KOHLI AND JAWORSKI

Kohli and Jaworski define market orientation as 'the organisationwide generation of market intelligence pertaining to current and future customer needs, dissemination of the intelligence across departments, and organisationwide responsiveness to it.'[72] In a subsequent publication, they define responsiveness as 'being composed of two sets of activities – response design (i.e., using market intelligence to develop plans) and response implementation (i.e., executing such plans).'[73]

---

§   Payne uses both 'market orientation' and 'marketing orientation' in his text. This chapter sticks to the term market orientation throughout.

Kohli and Jaworski explicitly provide suggestions for implementing a market orientation based on their exploratory study.[74] The authors have identified antecedents to a market orientation (the antecedents mentioned in the previous section, 'The model issue'), and because senior managers can control these factors, they argue that the deliberate engendering of a market orientation is possible. First, senior managers must communicate their commitment to a market orientation to junior employees. This requirement means that junior employees need to witness behaviours and resource allocations that reflect that commitment. Second, the interdepartmental dynamics need to be managed. Connectedness must be improved while reducing interdepartmental conflict. Interdepartmental activities and the exchange of employees are examples of specific interventions. Third, changing organisation structure (less formalisation and centralisation) and instituting more market-based reward systems should facilitate the implementation of a market orientation. In later work, these authors show that all antecedents, except for formalisation, have a significant correlation with market orientation.[75] The authors suggest that their measurement instrument MARKOR can be used for an initial diagnosis of the current degree of market orientation, as well as for post-intervention measurements of market orientation.[76]

THE NORM-BASED APPROACH BY LICHTENTHAL AND WILSON

Lichtenthal and Wilson provide a social structure perspective to implementing a market orientation.[77] Their main focus is on the inter-functional co-ordination aspect of market orientation. According to Lichtenthal and Wilson, norms prescribe individual behaviours and, in the context of organisations, the behaviour of individuals in relation to others within the organisation. For an organisation to act in accordance with the marketing concept, it must inculcate and transmit the appropriate values and create a set of norms to guide behaviour.

To change, an organisation must first define its current value system that creates the norms that drive behaviour – the diagnostic phase in Lichtenthal and Wilson's approach. Next, it can select which values need to be altered and instigate changes in these values. Norms can change with respect to their prevalence, rigidity, frequency of activity, directionality, specificity or diffuseness, and object of orientation. The firm's first task is to develop a list of desired behaviours. The norms that drive these behaviours can be derived and programmes can be developed per department to change these norms or create new ones. These changes are best made from the top down. The activities of change include revising job descriptions, educational programmes, communication programmes and incentive programmes to reward appropriate behaviour. These activities constitute the intervention phase. Lichtenthal and Wilson do not outline any post-intervention evaluation tools or activities. However, Homburg and Pflesser find, in an empirical study, that the establishment of market-oriented norms will not produce market-oriented behaviours unless those norms are supported by artefacts, such as stories, rituals and language.[78]

THE STRATEGY AND SUPPORT PROCESSES APPROACH BY RUEKERT

Ruekert defines the level of market orientation as the degree to which the business unit (1) obtains and uses information from customers; (2) develops a strategy that will meet customer needs; and (3) implements that strategy by being responsive to customers' needs and wants.[79] In the discussion of his empirical study, Ruekert provides recommendations for diagnosis, intervention and evaluation. To assess the current degree of market orientation, Ruekert recommends that corporate management use an approach as outlined in the study: a questionnaire with Likert-type items administered to all managers, as well as a sample of sales reps and sales managers of each business unit. The questionnaire developed by Ruekert includes sub-scales of market orientation practices and behaviours, organisational systems of recruiting and selection, training, and reward and compensation, individual outcomes, and business unit performance. This assessment should make it possible for corporate management to design initiatives to improve customer responsiveness at the business unit level.

According to Ruekert, interventions for increased market orientation should be found in the organisational support systems. The findings of the study suggest that changes in an organisation's behaviour in the marketplace need to be supported by organisational structures and processes that serve to guide the activities of the business unit. That is, 'While a temporary change in behaviours may be accomplished without the corresponding changes in organisational systems, the long term shift toward a market orientation probably requires a more permanent shift in organisational processes as well.'[80] These support systems, in Ruekert's view, are first and foremost human resource systems of recruiting and selection, training, and reward and compensation.

In terms of evaluating the change process, Ruekert notes that his study showed a positive relationship between the degree of market orientation and the attitudes (job satisfaction, commitment to the organisation, and trust in management) of managers. Moreover, the study revealed a positive relationship between market orientation and long run financial performance. Ruekert warns however that there is merely an association between the degree of market orientation and the outcomes, both individual and organisational. A repeated assessment of the degree of market orientation using the same instrument might also serve as an evaluation of the interventions.

THE CAPABILITIES APPROACH BY DAY

According to Day, market orientation represents superior skills in understanding and satisfying customers.[81] Therefore, market-driven organisations have superior market sensing, customer linking, and channel bonding capabilities.[82] Day offers a comprehensive change programme inspired by the total quality management (TQM) literature and aimed at enhancing these capabilities. This change programme includes (1) the diagnosis of current capabilities; (2) anticipation of future needs for capabilities; (3) bottom-up redesign of underlying processes; (4) top-down direction and commitment; (5) creative use of information technology; and (6) continuous monitoring of progress. The diagnostic stage of Day's approach involves an analysis of current capabilities and anticipated future capabilities. It is not clear from Day's work what these capabilities exactly are, because the description of this capability assessment actually is an assessment of the business processes.

The interventions Day proposes in his 1990 publication entail a four-step programme. The focus of these interventions is on aligning strategy, structure, people and programmes, as well as redesigning performance measures to encourage and reward market-driven behaviour. In his later work, the author focuses on business process redesign, either radically or gradually from the bottom up, combined with top-down signalling of commitment and stretching of improvement targets.[83] The use of information technology enables firms to do things they could not do before. Day suggests defining key performance indicators (KPIs) and using them to monitor progress and evaluate the results of the interventions. Biemans has extended Day's model,[84] resulting in a more detailed implementation framework and enriching it with the obstacles and facilitators to implementing a market orientation.

More recently, these recommendations have been incorporated in a larger suggested programme for creating a market-driven organisation.[85] Day now defines the market-driven organisation as one that has superior skills in understanding, attracting, and keeping valuable customers, and he proposes a change programme that aligns culture, capabilities, and configuration with the creation of superior customer value. The role of top management is emphasised as the initiator and driver of the change programme. Day makes two points that are specifically worth mentioning: (1) management should focus on the conditions that enable employees to produce good results, and (2) change happens by altering behaviour patterns, and eventually, these changes in behaviour will be absorbed into the underlying norms, beliefs, and mind sets.

THE COGNITIVE BELIEFS APPROACH BY ALLEN ET AL.

Allen, McQuarrie and Feldman Barr view market orientation not as a trait of an organisation or SBU, but as a set of subjective beliefs individual members of an organisation possess to varying degrees.[86] They prefer the term 'customer focus' over 'market orientation' and define customer focus as 'an individual's beliefs about the value of direct customer contact for achieving desired performance outcomes in his or her own job.'[87] The authors suggest that training lies at the heart of implementing customer focus beliefs in individuals. Individuals with strong customer focus beliefs are expected to engage more in customer contact, which is posited to result in higher responsiveness on the part of the employee. Through rewards stemming directly from this responsive customer interaction, as well as from formal institutional rewards, customer focus beliefs get reinforced. The authors also mention top-down value dissemination through senior executive's modelling of desired practices as a strong antecedent to customer focus beliefs. An instrument for measuring current beliefs is presented and tested in the article, but the suggested implementation approach is not backed up by empirical data.

THE CULTURE CHANGE APPROACH BY NARVER AND SLATER

Narver and Slater's ideas about implementing a market orientation are concentrated in one publication.[88] The argument can be summarised as follows: Creating a market orientation is about organisation members' learning how to create superior customer value continuously. This learning can be achieved in two ways: (1) through a priori education, which the authors call the programmatic approach, or (2) through experiential learning, called the market-back approach. The authors claim that most businesses fail to create a market orientation because they favour the more popular a priori learning over experiential learning, but experiential learning is required to realise a culture change, and the role of a priori education is only that of preparing for hands-on problem solving and experimentation in a results-driven continuous improvement process. An empirical study by Farrell lends some support to these propositions, because he finds that both planned and emergent approaches to organisational change are associated with higher degrees of market orientation.[89] Hennestad's description of an in-depth, longitudinal case study provides ample illustrations of how experiential learning can increase the level of market orientation in an organisation.[90]

THE SYSTEMS-BASED APPROACH BY BECKER AND HOMBURG

Becker and Homburg add a systems perspective to the debate and introduce the term 'market-oriented management.'[91] They have identified five managerial systems that influence the degree of market-oriented management: (1) the organisation system; (2) the information system; (3) the planning system; (4) the controlling system; and (5) the human resource management system.[¶] For each of these systems, the authors provide suggestions about how to redesign them for increased market orientation. These suggestions are numerous, and they vary considerably in their level of detail and value for practitioners. The suggestion to 'fill key management positions with employees having a marketing background' is clear and implementable, but the suggestion to 'empower customer contact employees' is vague, and the suggestion to 'collect and disseminate market information, and store it in accessible information systems' to increase market orientation is tautological. In terms of empirical backing for these suggestions, the authors show that more market-oriented configurations of the five managerial systems are associated with higher performance, but the empirical data do not necessarily indicate that the suggested interventions relate to successful organisational change. The measurement instrument developed in this paper can be used to diagnose the current degree of market-oriented management or evaluate the impact of interventions for increased market orientation.

THE MANAGEMENT BEHAVIOUR APPROACH BY HARRIS

After exhibiting an initial preference for Kohli and Jaworski's behavioural conceptualisation of market orientation,[93] Harris and colleagues have turned to cultural change for the implementation of a market orientation.[94] With a focus on UK retailing organisations, this work includes extensive empirical research into the barriers to implementing a market orientation.[95] More recently, it has been complemented with empirical studies into the process of becoming more market oriented.[96] In these studies, management behaviour is recognised as a crucial factor in the implementation process. First, conflict-laden, formalised, and political behaviours have a negative impact on market orientation, while vertical communication positively affects market orientation. Second, participative and supportive leadership styles help implement a market orientation, whereas an instrumental leadership style should be avoided. Both sets of recommendations are supported by empirical survey data. Of particular interest is a complementary study that explores why and how executives, managers, and employees intentionally sabotage market-oriented culture change.[97]

¶   A later publication identifies information systems, accounting systems, reward systems, and human resource management systems from 50 interviews with managers as determinants of customer-focused organisational structure.[92]

THE CULTURAL TRANSFORMATION APPROACH BY KENNEDY, GOOLSBY AND ARNOULD

Kennedy, Goolsby and Arnold report on a comparative study of two schools: one struggling to implement a customer orientation, and one successfully transforming itself to an award-winning, customer-centred school.[98] By contrasting high and low levels of success, the authors derive three pillars of successful implementation. First, for staff members to internalise a customer orientation, they must experience an unbroken chain of committed leadership from top management to local managers. The adoption of a customer orientation can be encouraged by reducing ambiguity about leadership commitment. Second, well-articulated customer requirements and fact-based performance feedback help unify individuals and departments into a coordinated effort to deliver customer value. Third, the collection, dissemination and use of market intelligence (both from internal and external customers) ensures that a customer orientation becomes self-reinforcing. Although Kennedy and colleagues do not explicitly address diagnosis, intervention and evaluation as three stages in organisational transformation, it is clear from their descriptions that fact-based customer data play a crucial role in the diagnosis and evaluation stages, and leadership and cross-functional improvement efforts prove key in the transformation effort.

THE CULTURAL TRANSFORMATION APPROACH BY BEVERLAND AND LINDGREEN

Beverland and Lindgreen develop a three-stage cultural transformation process based on Lewin's three-stage model of planned change: unfreezing, movement, and refreezing.[99] They use two case studies to illustrate this change process. The first stage involves unfreezing and challenging long-held cultural assumptions about products, markets, customers, value, and relationships. Market intelligence is needed to challenge the status quo and create a sense of urgency, as is a conscious building of coalitions and support for change. The second stage is about moving organization members toward market orientation through the reconfiguration of strategy, systems, and processes. The third stage requires refreezing through institutionalisation and continuous reinforcement of market-oriented assumptions (e.g., market-oriented reward systems).

THE CULTURAL TRANSFORMATION APPROACH BY GEBHARDT, CARPENTER, AND SHERRY

Building on ethnographic studies at seven firms, Gebhardt, Carpenter and Sherry develop a four-stage model of the process of creating a market orientation: (1) initiation, (2) reconstitution, (3) institutionalisation, and (4) maintenance.[100] The stages are path dependent, and each stage includes multiple steps or activities. The initiation stage is about powerful stakeholders recognising a threat and then creating coalitions to plan and implement change efforts. Reconstitution comprises five steps: (1) demarcation, (2) value and norm development, (3) reconnection with the market, (4) removal of dissenters and hiring of believers, and (5) collaborative strategy. The institutionalisation stage entails formalisation of organisational structures and processes, alignment of rewards, and cultural indoctrination through training. The maintenance stage finally involves reinforcing market-oriented culture through three processes: cultural screening of new members, culture maintenance rituals, and ongoing market connection activities to update market schemas and validate market-oriented process schemas. According to the authors, firms creating a market orientation embrace six cultural values: trust, openness, keeping promises, respect, collaboration, and viewing the market as the raison d'être. These values are the basis for market-oriented behaviours.

Table 1.2 summarises the implementation approaches discussed herein. Because no empirical studies have made comparisons of these various approaches, there is no basis for advocating the use of one approach over another.

Table 1.2   A classification of implementation approaches

Author(s)	Approach	Recommendations
Webster	Hybrid	Instil customer-oriented values and beliefs Integrate marketing in strategic planning Develop strong marketing managers and programmes Create market-based measures of performance Develop customer commitment throughout the organisation
Payne	Hybrid	Understand mix of current orientations Identify present level of marketing effectiveness Understand organisational and cultural dimensions Conduct training needs analysis Design management development programme (with a focus on knowledge, skills, and attitudes) Organise key support activities (structure, recruitment, promotions, information systems, planning systems)
Kohli and Jaworski	Systems-oriented	Instil senior management commitment Improve interdepartmental connectedness and reduce interdepartmental conflict Redesign organisation-wide systems (organisation structure, reward systems)
Lichtenthal and Wilson	Culture-oriented	Diagnose current organisational value system Develop list of desired behaviours Develop top-down programmes to change norms and/or create new norms
Ruekert	Systems-oriented	Diagnose current behaviours, systems, individual outcomes, and business performance Adapt systems for recruitment and selection Adapt systems for training Adapt systems for rewards and compensation
Day	Systems-oriented	Diagnose current market-sensing, customer-linking, and channel-bonding capabilities Anticipate future needs for capabilities Redesign business processes Signal management commitment Use information technology creatively Stretch improvement targets and monitor progress continuously

Table 1.2   *Continued*

Author(s)	Approach	Recommendations
Allen et al.	Culture-oriented	Diagnosis with a questionnaire that measures individuals' beliefs Training to affect customer focus beliefs and coach how to act on those beliefs Introduction of reward systems to reinforce customer focus beliefs Top-down value dissemination via senior executive's modelling of desired practices
Narver and Slater	Culture-oriented	Use a priori education to gain commitment to the continuous creation of superior customer value Use experiential learning to create an understanding of how to implement this norm
Becker and Homburg	Systems-oriented	Reduce number of hierarchical levels, appoint key account managers, and fill key management positions with employees having a marketing background Increase inter-functional integration Empower customer contact employees and involve customers in process redesign Collect and disseminate market information, and store it in accessible information systems Set market-based objectives, engage in environmental scanning, and involve customer contact personnel and customers in decision making Measure and analyse performance on the basis of market data Recruit people with a customer orientation, use training to disseminate market information, use customer satisfaction for performance assessment and rewards, and use marketing skills as the basis for career development
Harris and colleagues	Culture-oriented	Recognise and confront negative organisational behaviours (such as conflict-laden, formalised, and political behaviours) Identify and foster positive organisational behaviours (such as communication) Use a participative or a supportive leadership style to implement market orientation, and avoid an instrumental leadership style Use recruitment and training to establish the appropriate leadership styles
Kennedy, Goolsby, and Arnould	Culture-oriented	Ensure an unbroken circuit of passionate, sincere, unified, and committed leadership from top levels to local managers, 'walking the walk' of customer orientation Use customer requirements and performance feedback to instil a culture of interdepartmental connectedness Collect, disseminate, and use data from external and internal customers so that a customer orientation becomes self-reinforcing

Table 1.2   *Concluded*

Author(s)	Approach	Recommendations
Gebhardt, Carpenter, and Sherry	Hybrid	Once a threat to the organization is recognized, a group of empowered managers needs to create a coalition to plot the change process
		A complete transformation of the organisation must be planned, the larger organisation must be mobilized, and a cultural shift created through a process of value and norm development, reconnecting organisation members with customers, and removal of dissenters and hiring of believers
		Formal changes, such as alignment of rewards and indoctrination and training, should follow informal ones
		Cultural screening of new hires, culture maintenance rituals, and ongoing market connections should be used to sustain the new orientation of the organisation
Beverland and Lindgreen	Hybrid	Unfreeze and challenge long-held cultural assumptions about products, markets, customers, value, and relationships
		Move organisation members toward market orientation through reconfiguration of strategy, systems, and processes
		Refreeze through institutionalisation and continuous reinforcement of market-oriented assumptions (e.g., market-oriented reward systems)

Conclusions and implications

This chapter has reviewed literature with respect to how market orientation has been defined, how the degree of market orientation can be measured, how market orientation relates to its antecedents and consequences, and, above all, how a market orientation can be implemented in an organisation.

DEFINITION, MEASUREMENT, AND MODEL

At present, there is still no shared understanding of what constitutes a market orientation exactly. There are definitions of market orientation as a culture, as a set of activities, or as a set of skills. Nevertheless, the differences are not insurmountable, and most authors on the subject agree that it contains elements of market intelligence generation, dissemination, and use with the aim to create differential value for customers. Yet though marketing scholars who write about market orientation appear to have a common understanding of the concept, its central ideas are often misunderstood by those who are outsiders to the discussion. These misunderstandings lead to unjust criticisms as well as unjustified usage of the term. Attacks on the normative implications of the market orientation concept are often based on just such misinterpretations. One of the recurring problems appears to be that 'being market oriented' gets equated with 'being led by the demands of current customers.' When market oriented is equated with customer led, it is easy to attack the market orientation concept on the grounds that it leads to short-term, reactive management behaviour, that it stifles innovation, and that it represents a sure road to bankruptcy.[101] Many years of repeated explanations of what market orientation is and what it is not have apparently failed to communicate a relatively simple message: Market-oriented organisations are organisations that are well-informed about the market and that have the ability to use that information advantage to create differential customer value.[102] There is nothing in this message that suggests that organisations should follow customers and not lead them, that they should focus on current customers and not listen to potential customers, or that they should ignore technology push in favour of market pull.

With respect to the measurement issue, MARKOR by Kohli and colleagues and MKTOR by Narver and Slater are still the most frequently used scales,[103] either as they are or as bases for adapted scales, despite the methodological critiques and limited statistical validity of the scales.[104] The established scales are not very useful as a tool for managers to assess the degree of market orientation of their organisation and thereby start an improvement programme,[105] because MARKOR and MKTOR are too general in their wording and cover only a small part of total business operations. If a measurement scale is to be used as a diagnostic tool for a particular organisation, it might be worthwhile to take the organisation's core business processes as the starting point and develop items about the collection, dissemination, and use of market information for each business process. In this way, scores of items link directly to processes and thus to process improvement projects.

Of the three types of outcomes that have been associated with a market orientation – financial outcomes, customer outcomes and employee outcomes – financial outcomes still overshadow the others. An increased focus on the antecedents of a market orientation would help managers in their efforts to implement a market orientation. Recent studies have focused on barriers to a market orientation,[106] which is already a step toward more practitioner-oriented models of market orientation. New studies on relationships between human resource practices and a market orientation also have valuable implications for practitioners.[107] The biggest problem that remains, however, is that so many factors have been forwarded as antecedents that it becomes very difficult to develop an understanding of what the most important antecedents of a market orientation are.

One issue that has not yet been resolved is that of the decreasing marginal returns of market orientation, and, related to it, the question whether performance should be conceptualised as a monotonic increasing function of market orientation or if negative marginal returns exist. Narver and Slater suggest that 'the basic law of economics applies: for every business, at some point the incremental costs to increase its market orientation will exceed the incremental benefits.'[108] Steinman and colleagues add: '[T]here is an implicit acknowledgement that becoming market oriented involves real investment in a set of capital-intensive processes and activities. This raises the issue of how much market orientation is enough?'[109] The problem is that to calculate an optimal level of market orientation, marketers need to know the costs per increment in market orientation, as well as the performance advantage per increment. Currently, no empirical studies provide those insights. The existence of negative marginal returns would mean that an increase in market orientation beyond a certain point would make performance outcomes decrease. The costs that offset the gains of better market intelligence can be either the costs of improving the degree of market orientation (implementation costs) or those costs associated with too much market intelligence, such as information overload, paralysis through analysis, or data asphyxiation. The question of when to invest in market orientation – and perhaps more informatively, when *not* to invest in more market orientation – has only been raised, not yet answered.

THE IMPLEMENTATION ISSUE

In recent years, several thoughtful publications have addressed the question of how managers can make their organisations more market oriented. The implementation issue of market orientation has been enriched with both quantitative and qualitative empirical research. Several case studies describe change processes toward more market orientation. These qualitative empirical studies range from comparisons between overall change programmes[110] to rich descriptions of organisational reorientations[111] to the responses of individuals to cultural change programmes.[112] This book contains a number of new in-depth case studies of market-oriented transformations, such as Fotopoulos and colleagues', Insch's, and Leat and colleagues' chapters. Quantitative empirical studies show that both planned and emergent change strategies relate positively to market orientation,[113] that some management behaviours and leadership styles are more conducive to the implementation of a market orientation than others,[114] and that artefacts (e.g., stories, rituals, language) drive market-oriented behaviours.[115]

These important advances notwithstanding, empirically, we are only at the beginning of understanding the *what* and *how* of the change process. Because various approaches for implementation now exist, we are in dire need of comparative studies of implementation success. Furthermore, an additional challenge that has not been addressed extensively in existing literature is the implementation of a market orientation at the level of a supply chain instead of the single organization. Several chapters in this book, such as Bröring's and Kottila's, present particular issues pertaining to market-orientated transformations of supply chains.

References

1. Advani, A. and Borins, S. (2001), 'Managing Airports: A Test of the New Public Management', *International Public Management Journal*, vol. 4, no. 1, 91–107.
2. Fitzgerald, K. (2003), 'Changes in the New Zealand Apple Industry', *British Food Journal*, vol. 105, no. 1/2, 78–95.
3. Gainer, B. and Padanyi, P. (2002), 'Applying the Marketing Concept to Cultural Organisations: Am Empirical Study of the Relationship Between Market Orientation and Performance', *International Journal of Nonprofit and Voluntary Sector Marketing* 7(2), 182.
4. Lear, R.W. (1963), 'No Easy Road to Market Orientation', *Harvard Business Review* 41(5), 53–60.
5. Drucker, P.F. (1954) *The Practice of Management*. New York, NY: Harper and Row, pp. 37, 39.
6. Stoelhorst, J.W. and Van Raaij, E.M. (2004), 'On Explaining Performance Differentials: Marketing and the Managerial Theory of the Firm', *Journal of Business Research* 57(5), 462–77.
7. Webster, F.E. (1988), 'The Rediscovery of the Marketing Concept', *Business Horizons*, 29–39.
8. Day, G.S. (1994b), 'The Capabilities of Market-Driven Organizations', *Journal of Marketing* 58, 37–52; Jaworski, B.J. and Kohli, A.K. (1996), 'Market Orientation: Review, Refinement, and Roadmap', *Journal of Market-Focused Management* 1(2), 119–35; Kohli, A.K. and Jaworski, B.J. (1990), 'Market Orientation: The Construct, Research Propositions, and Managerial

Implications', *Journal of Marketing* 54, 1–18; Narver, J.C. and Slater, S.F. (1990), 'The Effect of a Market Orientation on Business Profitability', *Journal of Marketing* 54, 20–35.

9.   Deng, S. and Dart, J. (1994), 'Measuring Market Orientation: A Multi-Factor, Multi-Item Approach', *Journal of Marketing Management* 10, 725–42; Deshpandé, R. and Farley, J.U. (1998), 'Measuring Market Orientation: Generalization and Synthesis', *Journal of Market-Focused Management* 2(3), 213–32; Deshpandé, R., Farley, J.U. and Webster, F.E. (1993), 'Corporate Culture, Customer Orientation, and Innovativeness in Japanese Firms: A Quadrad Analysis', *Journal of Marketing* 57, 23–37; Kohli, A.K., Jaworski, B.J. and Kumar, A. (1993), 'MARKOR: A Measure of Market Orientation', *Journal of Marketing Research* 30, 467–77; Narver and Slater, op. cit.; Wrenn, B.W. (1997), 'The Market Orientation Construct: Measurement and Scaling Issues', *Journal of Marketing Theory and Practice* 5(3), 31–54.

10.   Jaworski, B.J. and Kohli, A.K. (1993), 'Market Orientation: Antecedents and Consequences', *Journal of Marketing* 57, 53–70; Kirca, A.H., Jayachandran, S. and Bearden W.O. (2005), Market orientation: A Meta-Analytic Review and Assessment of Its Antecedents and Impact on Performance, *Journal of Marketing*, 69(2), 24–41; Narver and Slater, op. cit.

11.   Beverland, M.B. and Lindgreen, A. (2007), 'Implementing Market Orientation in Industrial Firms: A Multiple Case Study', *Industrial Marketing Management* 36(4), 430–42; Day, G.S. (1999), 'Creating a Market-Driven Organization', *Sloan Management Review*, 11–22; Harris, L.C. and Ogbonna, E. (2001a), 'Leadership Style and Market Orientation: An Empirical Study', *European Journal of Marketing* 35(5/6), 744–64; Narver, J.C., Slater, S.F. and Tietje, B. (1998), 'Creating a Market Orientation', *Journal of Market-Focused Management* 2(3), 241–55.

12.   Kohli and Jaworski, op. cit.; Lafferty, B. and Hult, G.T.M. (2001), 'A Synthesis of Contemporary Market Orientation Perspectives', *European Journal of Marketing* 35(1/2), 92–109; Ruekert, R.W. (1992), 'Developing a Market Orientation: An Organizational Strategy Perspective', *International Journal of Research in Marketing* 9, 225–45.

13.   Drucker, op. cit. See also Lusch, R.F. and Laczniak, G.R. (1987), 'The Evolving Marketing Concept, Competitive Intensity and Organizational Performance', *Journal of the Academy of Marketing Science* 15(3), 1–11.

14.   Sachs, W.S. and Benson, G. (1978), 'Is It Time to Discard the Marketing Concept?', *Business Horizons*, 68–74.

15.   Ibid., p. 68, quoting General Electric's 1952 Annual Report, p. 21.

16.   Webster (1988), op. cit.

17.   Keith, R.J. (1960), 'The Marketing Revolution', *Journal of Marketing* 24, 35–38.

18.   Webster (1988), op. cit.

19.   Houston, F.S. (1986), 'The Marketing Concept: What It Is and What It Is Not', *Journal of Marketing* 50, 81–87.

20.   Kohli and Jaworski, op. cit.; Narver and Slater, op. cit.

21.   Kohli and Jaworski, op. cit.

22.   Kohli and Jaworski, op. cit.; Narver and Slater, op. cit.

23.   Shapiro, B.P. (1988), 'What the Hell Is "Market Oriented"?', *Harvard Business Review* 66, 119–125.

24.   Kohli and Jaworski, op. cit.

25.   Narver and Slater, op. cit.

26.   Slater, S.F. and Narver, J.C. (1995), 'Market Orientation and the Learning Organization', *Journal of Marketing* 59, 63–74. See p. 67.

27.   Ruekert, op. cit.

28.   Deshpande et al., op. cit., p. 27.

29. Day (1994b), op. cit.

30. Dreher, A. (1994), 'Marketing Orientation: How to Grasp the Phenomenon' in M.J. Baker (ed.), *Perspectives on Marketing Management*, New York, NY: John Wiley and Sons, 149–70; Griffiths, J.S. and Grover, R. (1998), 'A Framework for Understanding Market Orientation: The Behavior and the Culture', paper presented at the American Marketing Association Winter Educators Conference, Austin, February 21–24.

31. Homburg, C. and Pflesser, C. (2000), 'A Multiple Layer Model of Market-Oriented Organizational Culture: Measurement Issues and Performance Outcomes', *Journal of Marketing Research* 37, 449–62.

32. Kotler, P. (1977), 'From Sales Obsession to Marketing Effectiveness', *Harvard Business Review* 55, 67–75; Wrenn, op. cit.

33. Deshpandé, R. and Farley, J.U. (1996), *Understanding Market Orientation: A Prospectively Designed Meta-Analysis of Three Market Orientation Scales*, MSI Report 96–125. Cambridge, MA: Marketing Science Institute.

34. Kohli et al., op. cit.

35. Narver and Slater, op. cit.

36. Jaworski and Kohli (1993), op. cit.

37. Farrell, M.A. and Oczkowski, E. (1997), 'An Analysis of the MKTOR and MARKOR Measures of Market Orientation: An Australian Perspective', *Marketing Bulletin* 8, 30–40; Gabel, T.G. (1995), 'Market Orientation: Theoretical and Methodological Concerns', paper presented at the American Marketing Association Summer Educators Conference, American Marketing Association, Chicago, IL.

38. Wensley, R. (1995), 'A Critical Review of Research in Marketing', *British Journal of Management* 6, S63-S82.

39. Gabel, op. cit.; Steinman, C., Deshpandé, R. and Farley, J.U. (2000), 'Beyond Market Orientation: When Customers and Suppliers Disagree', *Journal of the Academy of Marketing Science* 28(1), 109–19.

40. Van Bruggen, G.H. and Smidts, A. (1995), 'The Assessment of Market Orientation: Evaluating the Measurement Instrument as a Tool for Management', paper presented at the Annual Conference of the European Marketing Academy, ESSEC, Cergy-Pontoise, May 16–19.

41. Bisp, S., Harmsen, H. and Grunert, K.G. (1996), 'Improving measurement of market orientation – An attitude/activity based approach', paper presented at the Annual Conference of the European Marketing Academy, University of Economic Sciences, Budapest, May 14–17.

42. For overviews, see González-Benito, Ó. and González-Benito, J. (2005), 'Cultural vs. Operational Market Orientation and Objective vs. Subjective Performance: Perspective of Production and Operations', *Industrial Marketing Management* 34, 797–829; Kirca et al., op. cit.

43. Jaworski and Kohli (1996), op. cit.

44. Appiah-Adu, K. (1998), 'Market Orientation and Performance: Empirical Tests in a Transition Economy', *Journal of Strategic Marketing* 6, 25–45; Dawes, J.G. (2000), 'Market Orientation and Company Profitability: Further Evidence Incorporating Longitudinal Data', *Australian Journal of Management* 25(2), 173–199; Greenley, G.E. (1995a), 'Forms of Market Orientation in UK Companies', *Journal of Management Studies* 32(1), 47–66; Langerak, F. (2003), 'An Appraisal of Research on the Predictive Power of Market Orientation', *European Management Journal* 21(4), 447–64.

45. Jaworski and Kohli (1993), op. cit.; Kirca et al., op. cit.; Pelham, A.M. and Wilson, D.T. (1996), 'A Longitudinal Study of the Impact of Market Structure, Firm Structure, Strategy, and Market Orientation Culture on Dimensions of Small-Firm Performance', *Journal of the Academy*

of Marketing Science 24(1), 27–43; Slater, S.F. and Narver, J.C. (1994), 'Does Competitive Environment Moderate the Market Orientation-Performance Relationship?', *Journal of Marketing* 58, 46–55; Wrenn, op. cit.

46. Jaworski and Kohli (1993), op. cit.; Kirca et al., op. cit.

47. Jaworski and Kohli (1993), op. cit.; Ruekert, op. cit.; Siguaw, J.A., Brown, G. and Widing, R.E., II (1994) 'The Influence of the Market Orientation of the Firm on Sales Force Behavior and Attitudes', *Journal of Marketing Research* 31, 106–116.

48. Haugland, S.A., Myrtviet, I. and Nygaard, A., (2007), 'Market Orientation and Performance in the Service Industry: A Data Envelopment Analysis', *Journal of Business Research* 60, 1191–1197.

49. Appiah-Adu, op. cit.; Greenley, G.E. (1995b), 'Market Orientation and Company Performance: Empirical Evidence From UK Companies', *British Journal of Management* 6(1), 1–13; Harris, L.C. (2001), 'Market Orientation and Performance: Objective and Subjective Empirical Evidence From UK Companies', *Journal of Management Studies* 38(1), 17–43; Jaworski and Kohli (1993), op. cit.; Slater and Narver (1994), op. cit.

50. Wrenn, op. cit.

51. Harris (2001), op. cit.

52. Appiah-Adu, op. cit.; Greenley (1995b), op. cit.; Harris (2001), op. cit.; Slater and Narver (1994), op. cit.

53. Baker, W.E. and Sinkula, J.M. (1999), 'The Synergistic Effect of Market Orientation and Learning Orientation on Organizational Performance', *Journal of the Academy of Marketing Science* 27(4), 411–27; Slater and Narver (1995), op. cit.

54. Day (1994b), op. cit.

55. Cooper, R.G. (1979), 'The Dimensions of Industrial New Product Success and Failure', *Journal of Marketing* 43, 93–103; Han, J.K., Kim, N. and Srivastava, R.K. (1998, 'Market Orientation and Organizational Performance: Is Innovation a Missing Link?', *Journal of Marketing* 62, 30–45; Langerak, F., Hultink, E.J. and Robben, H.S.J. (2004), 'The Impact of Market Orientation, Product Advantage, and Launch Proficiency on New Product Performance and Organizational Performance', *Journal of Product Innovation Management* 21(1), 79–94.

56. Atuahene-Gima, K. (1996), 'Market Orientation and Innovation', *Journal of Business Research* 35, 93–103.

57. Han et al., op. cit.

58. Langerak et al., op. cit.; Vazquez, R., Santos, M.L. and Alvarez, L.I. (2001, 'Market Orientation, Innovation and Competitive Strategies in Industrial Firms', *Journal of Strategic Marketing* 9(1), 69–90.

59. Day, G.S. (1994a) 'Continuous Learning About Markets', *California Management Review* 36, 9–31; Martin, J.H. and Grbac, B. (2003), 'Using Supply Chain Management to Leverage a Firm's Market Orientation', *Industrial Marketing Management* 32(1), 25–38; Srivastava, R.K., Shervani, T.A. and Fahey, L. (1999), 'Marketing, Business Processes, and Shareholder Value: An Organizationally Embedded View of Marketing Activities and the Discipline of Marketing', *Journal of Marketing* 63, 168–79.

60. Avlonitis, G.J. and Gounaris, S.P. (1999), 'Marketing Orientation and its Determinants: An Empirical Analysis', *European Journal of Marketing* 33(11/12), 1003–1037; Kohli and Jaworski, op. cit.; Pelham and Wilson, op. cit.

61. Pelham and Wilson, op. cit., p. 31.

62. Jaworski and Kohli (1993), op. cit.; Ruekert, op. cit.

63. Kirca et al., op. cit.

64. Pelham and Wilson, op. cit.
65. Cf. Song, X.M. and Parry, M.E. (1999), 'Innovation Strategy: The Impact of Market Orientation on Firm Performance', paper presented at the ECIS Opening Conference, Eindhoven Centre for Innovative Studies, Eindhoven, July.
66. Heskett, J.L., Jones, T.O., Loveman, G.W., Sasser, W.E. and Schlesinger, L.A. (1994), 'Putting the Service-Profit Chain to Work', *Harvard Business Review* 72, 164–74.
67. Becker, J. and Homburg, C. (1999), 'Market-Oriented Management: A Systems-Based Perspective', *Journal of Market-Focused Management* 4(1), 17–41; Homburg and Pflesser, op. cit.
68. Webster (1988), op. cit.
69. Webster, F.E. (1994), *Market-Driven Management: Using the New Marketing Concept to Create a Customer-Oriented Company*. New York, NY: John Wiley and Sons.
70. Payne, A. (1988), 'Developing a Marketing-Oriented Organization', *Business Horizons*, 46–53; Webster (1988), op. cit.
71. Kotler, op. cit.; Waterman, R.H., Peters, T.J. and Phillips, J.R. (1980), 'Structure Is Not Organization', *Business Horizons*, 14–26.
72. Kohli and Jaworski, op. cit., p. 6.
73. Jaworski and Kohli (1993), op. cit., p. 54.
74. Kohli and Jaworski, op. cit.
75. Jaworski and Kohli (1993), op. cit.
76. Kohli et al., op. cit.
77. Lichtenthal, J.D. and Wilson, D.T. (1992), 'Becoming Market Oriented', *Journal of Business Research* 24, 191–207.
78. Homburg and Pflesser, op. cit.
79. Ruekert, op. cit.
80. Ibid., p. 243.
81. Day, G.S. (1990), *Market Driven Strategy: Processes for Creating Value*. New York, NY: The Free Press.
82. Day (1994b), op. cit.
83. Ibid.
84. Biemans, W.G. (1995), 'Implementing Market-Oriented Product Development', *Technology Review* 83, 47–53.
85. Day (1999), op. cit.
86. Allen, C.T., McQuarrie, E.F. and Feldman Barr, T. (1998), 'Implementing the Marketing Concept One Employee at a Time: Pinpointing Beliefs About Customer Focus as a Lever for Organizational Renewal', *Journal of Market-Focused Management* 3(2), 151–70.
87. Ibid., p. 157.
88. Narver et al., op. cit.
89. Farrell, M.A. (2000), 'Developing a Market-Oriented Learning Organization', *Australian Journal of Management* 25(2), 201–22.
90. Hennestad, B.W. (1999), 'Infusing the Organisation With Customer Knowledge', *Scandinavian Journal of Management* 15(1), 17–41.
91. Becker and Homburg, op. cit.
92. Homburg, C., Workman, J.P. and Jensen, O. (2000), 'Fundamental Changes in Marketing Organization: The Movement Towards a Customer-Focused Organizational Structure', *Journal of the Academy of Marketing Science* 28(4), 459–78.
93. Harris, L.C. (1996), 'Benchmarking Against the Theory of Market Orientation', *Management Decision* 34(2), 25–29.

94. Harris, L.C. (1998b), 'Cultural Domination: The Key to Market-Oriented Culture?', *European Journal of Marketing* 32(3/4), 354–373; Harris, L.C. and Ogbonna, E. (1999), 'Developing a Market Oriented Culture: A Critical Evaluation', *Journal of Management Studies* 36(2), 177–196.

95. Harris, L.C. (1998a), 'Barriers to Market Orientation: The View From the Shopfloor', *Marketing Intelligence and Planning* 16(3), 221–8; Harris, L.C. (2000), 'The Organizational Barriers to Developing Market Orientation', *European Journal of Marketing* 34(5/6), 598–624; Harris, L.C. and Ogbonna, E. (2000), 'The Responses of Front-Line Employees to Market-Oriented Culture Change', *European Journal of Marketing* 34(3/4), 318–40.

96. Harris, L.C. (2002a), 'Developing Market Orientation: An Exploration of Differences in Management Approaches', *Journal of Marketing Management* 18, 603–32; Harris and Ogbonna (2001a), op. cit.; Harris, L.C. and Piercy, N.F. (1999), 'Management Behavior and the Barriers to Market Orientation', *Journal of Services Marketing* 13(2), 113–31.

97. Harris, L.C. (2002b), 'Sabotaging Market-Oriented Culture Change: An Exploration of Resistance Justifications and Approaches', *Journal of Marketing Theory and Practice* 10(3), 58–74.

98. Kennedy, K.N., Goolsby, J.R. and Arnould, E.J. (2003), Implementing a Customer Orientation: Extension of Theory and Application, *Journal of Marketing*, 67(4), 67–81.

99. Beverland and Lindgren, op. cit.; Lewin, K. (1951) *Field Theory in Social Science*, New York, Harper and Row.

100. Gebhardt, G.F., Carpenter, G.S. and Sherry J.F. (2006), 'Creating a Market Orientation: A Longitudinal, Multifirm, Grounded Analysis of Cultural Transformation', *Journal of Marketing* 70(1), 37–55.

101. Christensen, C.M. and Bower, J.L. (1996), 'Customer Power, Strategic Investment, and the Failure of Leading Firms', *Strategic Management Journal* 17(3), 197–218.

102. Day (1994b), op. cit.; Houston, op. cit.; Jaworski and Kohli (1996), op. cit.; Narver and Slater, op. cit.; Shapiro, op. cit.

103. Kohli and Jaworski, op. cit.; Narver and Slater, op. cit.

104. Caruana, A. (1999), 'An Assessment of the Dimensions and the Stability of Items in the MARKOR Scale', *Marketing Intelligence and Planning* 17(5), 248–53; Farrell and Oczkowski, op. cit.; Gabel, op cit.; Langerak, op. cit.

105. Van Bruggen and Smidts, op. cit.

106. Bisp, S. (1999), 'Barriers to Increased Market-Oriented Activity: What the Literature Suggests', *Journal of Market-Focused Management* 4(1), 77–92; Harris (1998a, 2000), op. cit.

107. Conduit, J. and Mavondo, F.T. (2001), 'How Critical Is Internal Customer Orientation to Market Orientation?', *Journal of Business Research* 51, 11–24; Harris, L.C. and Ogbonna, E. (2001b), 'Strategic Human Resource Management, Market Orientation, and Organizational Performance', *Journal of Business Research* 51, 157–66.

108. Narver and Slater, op. cit., p. 33.

109. Steinman et al., op. cit., p. 100.

110. Day (1999), op. cit.

111. Ballantyne, D. (1997), 'Internal Networks for Internal Marketing', *Journal of Marketing Management* 13(5), 343–66; Beverland and Lindgren, op. cit.; Hennestad, op. cit.; Kennedy et al., op. cit.

112. Harris (2002b), op. cit.; Harris and Ogbonna (2000), op. cit.

113. Farrell, op. cit.

114. Harris and Ogbonna (2001a), op. cit.; Harris and Piercy, op. cit.

115. Homburg and Pflesser, op. cit.

2 *Implementing Market Orientation in Industrial Firms: A Multiple Case Study[i]*

BY MICHAEL B. BEVERLAND* AND ADAM LINDGREEN†

Keywords

market orientation, organisational change, case studies

Abstract

Literature on market orientation is silent on the process of change involved in moving firms to a market orientation. Understanding this process is important for commodity sellers or industrial organisations with a traditional sales focus. We examine the change programmes of two New Zealand-based agricultural organisations. Drawing upon Lewin's three-stage change process model (unfreezing–movement–refreezing), we note that the creation of a market orientation involves uncovering long-held assumptions about the nature of commodity products, the nature of production and marketplace power, and the commodity cycle. Moving the firm toward a new set of values involves changes in the role of leadership, the use of market intelligence, and organisational learning styles. To refreeze these values, supportive policies are needed that form closer relationships between the organisation and the marketplace. The degree of refreezing affects the quality of market orientated outcomes, with less effective refreezing leading to sub-optimal market-oriented behaviours.

[i]   This study was first published in *Industrial Marketing Management* 2007, vol. 36, no. 4, pp. 430–42. Reproduced with permission from Elsevier.

*   Professor Michael B. Beverland, School of Economics, Finance and Marketing, RMIT University, GPO Box 2476V, Melbourne, Victoria 3001, Australia. E-mail: michael.beverland@rmit.edu.au. Telephone: + 61 416 102 492.

†   Professor Adam Lindgreen, Hull University Business School, Cottingham Road, Hull HU6 7RX, the UK. E-mail: a.lindgreen@hull.ac.uk. Telephone: + 44 1482 463 096.

Introduction

The development of a market orientation can lead to a number of positive performance outcomes.[1] Although research has shown that business-to-business firms are less likely to adopt a market orientation than business-to-consumer firms,[2] this same research also indicates that the relationship between market orientation and performance is stronger for industrial companies than for business-to-consumer firms. To date, no studies have examined the implementation of a market orientation in business-to-business firms. Because many business-to-business firms have adopted sales orientations,[3] the implementation of a market orientation is likely to be difficult and require top-down revolutionary changes to long-held practices and beliefs.[4]

Despite the identified importance of a market orientation to firm performance, the implementation of a market orientation is an issue that has remained largely unexplored in the literature.[5] Harris states, 'the topic of 'market orientation' will remain perplexing to theorists and continue to be illusive for practitioners'[6] unless studies start to examine the processes and dynamics of developing a market orientation. Narver and colleagues identify two paths for organisations to move toward a market orientation,[7] though no empirical research examines the process of change associated with adopting a market orientation. To date, only one study has examined the implementation of a market orientation: Kennedy and colleagues identify three strategies – leaders' support for change, inter-functional coordination, and the use of market intelligence – as assisting with the implementation of a market orientation.[8] However, they do not focus on the actual *process of change* involved in adopting a market orientation. Such a focus would advance our knowledge substantially, as it would identify practical implications for marketing managers and the importance of different support strategies at different stages in the change process.[9] Although change management literature is replete with advice on the process of change *per se*, none of it addresses the specific processes involved in moving toward a market-oriented culture, a focus that involves specific subtleties for marketing researchers.[10]

This chapter addresses the process of change involved in moving toward a market orientation in two New Zealand-based agricultural cooperatives. This effort responds to calls for in-depth studies of firms that, with or without success, have been involved in market orientation implementation efforts,[11] and it extends current research by examining the implementation of a market orientation in new contexts.[12] We address two research questions: (1) How do firms deliberately change to a market orientation? and (2) What strategies are most effective during different stages of the change process in relation to affecting a market-oriented culture? This chapter adopts the following structure: First, we review aspects of market orientation, including issues of implementation. Second, we review various change theories, placing emphasis on Lewin's three-stage model of planned change.[13] Third, we provide details on the two cases developed for this study. Fourth, we present the findings. Fifth, we identify theoretical and managerial contributions.

Literature Review: Implementing Market Orientation

Market orientation must be understood as a culture, rather than a set of behaviours and espoused values,[14] because culture mediates between strategy and implementation.[15]

Bisp states that the form and intensity of market orientation are manifestations of cultural commitment and strategic clarity.[16] Market orientation is defined as 'a culture in which all employees are committed to the continuous creation of superior value for customers.'[17] A culture is 'the pattern of shared values and beliefs that help individuals understand organisational functioning and thus provide them norms for behavior in the organization.'[18] Little research exists regarding the development of a market orientation *per se*, whether cultural or behavioural.

Traditionally, New Zealand agribusinesses saw their responsibility for the product end when their produce left the farm/orchard gate.[19] Marketing for agricultural products was controlled by 'single-desk' sellers that engaged in generic country-of-origin marketing programmes on behalf of the industry as a whole and resulted in the inability of many agribusinesses to develop diverse or innovative marketing strategies.[20] Agricultural producers have been under increasing pressure to develop new forms of competitive differentiation to break out of commodity price cycles.[20] The culture of commodity production represents the opposite of a market orientation.[21] Moving from commodity production to a market orientation involves significant changes in culture, strategic outlook, and marketing practices.[22] Therefore, programmes seeking to reposition commodities to create greater and sustainable market value represent a rich context for studying the planned implementation of market orientation.

IMPLEMENTING A MARKET-ORIENTED CULTURE

To our knowledge, no research has focused on implementing a market-oriented *culture*, though Homburg and Pflesser's work examining the cultural characteristics of market orientation suggests firms must adopt new (or make changes to existing) artefacts, values and deeply held cultural assumptions.[23] Kennedy and colleagues have conducted the only empirical examination of implementing a market orientation, though Narver and colleagues also propose two paths toward developing a market-orientation.[24] The findings and propositions of both studies and the implications for our research are identified in Table 2.1.

Several authors propose that senior management is critical to the successful implementation of a market-oriented culture.[25] However, questions remain. Are certain behaviours more effective at different stages of the change process than others? If it is important early on to challenge long-held cultural assumptions about products to drive an understanding 'that there is no such thing as a commodity'[26] for market-oriented change to occur, then what role do leaders play in this process? Also, what strategies do they use? Do they develop mission and value statements first, and use these statements to drive change, or do values and missions emerge throughout the process? Triggering change may involve outside help, top management directives, and formal education programmes.[27] Although necessary to trigger market-oriented change in commodity firms, the development of a shared vision is also vital to effecting market-orientation implementation so that employees, through market-back learning, come to adopt new assumptions as part of their day-to-day work behaviours, eventually operating on these assumptions sub-consciously.[28] This development suggests that the behaviours of senior management, and their influence on the effectiveness of market-orientation programmes, change during the duration of the change process.

When is cascading leadership necessary? It is likely that cascading needs occur because bottom-up buy-in is necessary for market-back learning and the development of widespread cultural acceptance of change.[29] Likewise, is emotional commitment more critical during early stages of change, given the likelihood of barriers to change and resistance to new approaches?[30] When is driving commitment to change more relevant in effecting market-oriented change? With regard to inter-functional coordination, to effect market-oriented change, established functions and their interrelationships may need to be reconfigured to support a market-oriented outlook.

Finally, does the content and role of market intelligence differ throughout the change process? For example, early market intelligence identifying poor performance, the effectiveness of alternative practices, and the continued decline of the firm's performance could be used to bolster the case for change and challenge under-the-surface assumptions. Later market intelligence identifying the positive effects of the adopted programme (e.g., short-term wins) may help reinforce market-oriented assumptions and market-back learning, essential to the adoption of a market-orientation.[32] We next examine research into the change process.

Table 2.1   Research results on implementing market orientation[31]

Focus	Results	Implications
Kennedy and colleagues compare two schools' attempts to implement market orientation and offer explanations as to why one succeeded and the other failed.	Senior leadership support consisting of: connectivity to ownership for change; high degree of commitment intensity and emotion; cascading leadership; driving commitment to change. Inter-functional coordination consisting of: complex interlocking customer orientation; internalised shared mission and vision. Market intelligence consisting of: extracting causality from robust stakeholder data; tying operational performance to customer requirements.	Are certain leadership behaviours more effective during each stage of the change process? What role does leadership have in culture change? Do leadership style, intensity, and commitment need to change throughout the change process? When do shared visions and missions need to be developed? Do these emerge through the change process, or do they drive it? Also, is this process top-down or bottom-up (or a combination)? Will the content and role of market intelligence be different throughout the change process?
Narver and colleagues propose that firms sit on a continuum from commodity focused to market oriented.	Firms further away from market orientation would require greater degrees of change, which is more likely to be driven by top management. Firms closer to market-oriented end likely to go through evolutionary change that is bottom up.	At what stage of change do managers need to move from a top-down approach to a bottom-up one? What strategies will achieve this?

APPROACHES TO ORGANISATIONAL CHANGE CONSISTENT WITH ADOPTING MARKET ORIENTATION

This chapter examines the processes underpinning planned[‡] change efforts to achieve a market-oriented culture. Such a focus is critical given calls for greater attention to the cultural elements of market orientation, as opposed to activities and behaviours associated with the construct.[34] The adoption of a market orientation is likely to begin with a planned process for commodity-style firms,[35] because the greater the distance between the firm and a market orientation, the more likely such change will need to be driven (at first) from the top down. As a result, we adopt as a starting point Lewin's three-stage model of change,[36] the most relevant change model for radical, planned change, including moving from a commodity focus to a market orientation.[37]

LEWIN'S THREE-STAGE CHANGE THEORY

This chapter examines the adoption of a market orientation from the perspective of Lewin's force field model of change.[§] This model characterises change as a 'state of imbalance between driving forces (pressures for change) and restraining forces (pressures against change).'[39] To effect change, managers must change the equilibrium between driving and restraining forces by creating pressure in favour of change.[40] This change requires managers to *unfreeze* past practices associated with the *status quo*. As a result, unlearning is critical to a learning orientation and the development of a market orientation.[41] Unlearning involves the ability to question past assumptions, which requires uncovering long-held, unchallenged, cultural assumptions about the 'right way to do things.'[42] Because these assumptions are often sub-conscious, they must first be resurfaced through a change intervention. This unfreezing process may involve heated debate and energise forces against change.[43] Barriers to the development of a market orientation include threats to stability, fear of change, a belief that market orientation is inappropriate for the firm, and a fear of marketing-driven myopia (i.e., focusing on serving the customer may result in the firm losing sight of its core values).[44] In summary, unfreezing involves surfacing and challenging past assumptions and practices.

Should long-held cultural assumptions be surfaced and unfrozen, managers need to *move* the firm to a new set of assumptions.[45] The identification of the need to adopt a market orientation is just the start of the change process.[46] Lafferty and Holt propose that incentives and training in the use of gathering and using market-based information are needed to operationalise these values.[47] Narver and colleagues predict that movement will involve the following practices:[48] deliberating role modelling; paying attention to, measuring, and controlling organisational phenomena; reacting to critical incidents and crises; and creating creative tension.

[‡]   In their review of change theories, Van de Ven and Poole identify four broad approaches to studying change: teleological, lifecycle, dialectical, and population ecology. We adopt as our starting point the teleological view of change, which focuses on planned change, whereby the desired end result set by managers in goals and plans drives the trajectory of the firm. Other theories deal with macro-level impacts on industry clusters (population ecology), deterministic cumulative and conjunctive stages in firm development (life cycle theory), or internal conflicts as drivers of new organisational forms (dialectical). Such models are less relevant for studies of planned change processes.[33]

[§]   In thorough reviews of theories of change, both Burnes and Wilson conclude that though many planned change models exist, they retain the essential characteristics of Lewin's original model.[38]

Managers must then *refreeze* cultural assumptions to affect a new state.[¶] Depending on the degree of change necessary, refreezing may involve wider changes to firm structure and systems,[50] which is also likely to involve 'market-back learning' (learning from doing), because it reinforces the values of market orientation[51] and is consistent with the requirements of cascading leadership to effect market-oriented change.[52] Bottom-up buy-in is necessary for the development of effective market orientation.[53] A learning orientation is necessary to ensure the refreezing of market-oriented cultural values.[54] For example, Baker and Sinkula propose that a learning orientation is a resource that influences the quality of market-oriented behaviours, such that 'Firms may have a market orientation, but the quality of their market-oriented behaviors may be weak relative to other firms.'[55] In this scenario, employees can learn how to learn (generative learning), which involves constantly reflecting on past strategies and approaches to business rather than just learning through adaptation (trial and error).[56] Baker and Sinkula also propose that the adoption of such a learning process will result in higher quality market-oriented outcomes.[57] Yakimova and Beverland support this claim by identifying how firms with the behavioural characteristics of a market orientation cannot effectively act upon market-driven information due to their poor learning styles.[58] At this stage, leaders are likely to play a lesser role in driving change,[59] because employees must reinforce market-oriented assumptions through practices.[60] In summary, refreezing involves institutionalising assumptions and practices consistent with a market orientation.

Methodology

The use of qualitative methods is appropriate when studying complex processes.[61] A multiple-case approach was chosen due to the complex nature of the phenomenon at hand and the need to take into account a large number of variables.[62] This study uses the multiple case study approach of Eisenhardt, who proposes that richer theory can be generated with multiple case studies as opposed to one single case.[63] Eisenhardt also contends that the use of secondary data and multiple interviews in each case would help develop rich insights and provide the basis for greater transferability of the findings to other contexts. These methods are adopted in this study.

Cases were selected using theoretical sampling.[64] Two cases were selected for study – Merino NZ (MNZ) and the New Zealand Game Industry Board (NZGIB) – because of their high-profile marketing successes, which assisted both industries' turnaround in their ailing fortunes. Also, the simplicity of each case's competitive scenarios and strategic responses, relative to larger, more complex producer boards, made these two cases more attractive. Finally, each case was selected because its programmes had largely been developed, and each was now undergoing evolutionary change, whereas at the time of the data collection, many other boards were subject to government-sponsored reviews or undergoing substantial changes processes, for which the outcomes were less certain.

Prior to each interview, publicly available secondary material and promotional information provided by each Board was reviewed to increase the first author's familiarity

---

[¶]   The refreezing process can result in institutionalising practices that may need to be unfrozen due to subsequent environmental shifts, yet Lewin saw the refreezing process as one that only approached equilibrium rather than being a stable state.[49]

with the case. The first author conducted interviews at each Board's head office in New Zealand. In total, nine in-depth interviews were conducted (three at MNZ, six at NZGIB). Each interview lasted for, on average, four hours (range: three to eight hours). Given the control that both Boards had over their respective industries, gaining insights from Board representatives and full-time employees was judged to yield the richest source of information of the process of change, because each participant had been involved in the change programme. At MNZ, three interviews were conducted (this Board is substantially smaller than NZGIB) with the CEO, marketing manager, and marketing representative. At NZGIB, six interviews were conducted with the brand developer, current chairperson, largest exporter, marketing manager, and three past chairs who had been involved in the change programme (all were deer farmers). Interviewing stopped when saturation occurred, that is, when extra interviews began to yield few new insights.[65]

Questions focused on gaining a descriptive history of the motivation for change, the pressures for and against change, major objections, supportive programmes, and ultimate successes or reasons for failure. Following the primary interviews, further information provided by the interviewees or other sources was examined. This step involved widespread searches of government documents, mandated reports, industry conference proceedings, books, conference papers, and consultant's reports on each case, as well as a search of a local NewsIndex database covering 10 years of history of each Board. This process enabled us to examine farmers' and customers' views of each Board's respective programmes (due to issues of accessibility, customers and farmers were not interviewed directly). Because farmers did not drive the change after the vote was taken to move toward a market orientation, secondary data and notes from industry forums soliciting feedback (often highly critical) about the change programmes were used instead of interviews. Customer feedback was accessed through industry conferences (again, much of it critical), secondary publications, each Board's Web site, and trade press articles on the programmes. More than 120 documents were sited for this study. Together, these multiple sources improved the quality of the final interpretation and helped ensure triangulation.[66] The unit of analysis was each case studied. Therefore, information from each interview and the secondary sources were combined into one case manuscript. In total, this process resulted in a transcript of 103 pages (45 pages for MNZ, 58 pages for NZGIB).

The cases were analysed using Eisenhardt's method for within-case and cross-case analysis.[67] Each case first was analysed to gain a richer understanding of the processes that each underwent to move toward a market orientation. A timeline was developed for each. As each case achieved different degrees of market orientation, the cases were compared to assess similarities and differences and gain greater understanding of the processes involved. Theoretical categories were elaborated on during open and axial coding procedures.[68] Throughout the analysis, we tacked back and forth between literature on change and the data, which led to the development of a number of theoretical categories and sub-categories.[69] Such practices are consistent with other studies focused on implementing a market orientation.[70]

Throughout the study, we adopted a number of methods for improving the quality of the research. Experts were used to help select the cases; two researchers provided independent interpretations of the findings; multiple interviews were conducted; and respondents had the opportunity to provide feedback about the initial findings, all of which reinforced reliability. Colleagues performed independent coding of the transcripts, but the interviews were conducted by the same interviewer, reducing the role of bias.[71]

Findings

Background information about each case and the lead-up to implementing market-oriented change appears provided in the Appendix. We suggest readers view this Appendix before examining the findings further. Information detailing the situation prior to and after the market-oriented change for each case is identified in Tables 2.2 and 2.3. Lewin's three-stage model of change was appropriate for capturing the change processes of both cases and explains the difference in outcomes between the two cases. Each of Lewin's three-stages is addressed next.

Table 2.2   NZGIB before and after market-oriented change

Organisational Characteristics	Before	After
Organisation		
Structure	Government-mandated industry body with marketing staff and elected board. Members fund generic marketing support.	Government-mandated industry body with marketing staff and elected board. Separate brand programmes (labelled with the brand name Cervena) now franchised and run by councils of franchisees and Board (funded by franchisees).
Markets served	Primarily Germanic countries. Sell meat and leather on commodity exchanges. Antler velvet and animal parts (penises, tails, and hooves) sold unprocessed to Asian agents.	High quality targeted at up-market restaurant buyers and specialist meat sellers; lesser quality cuts sold to mass retailers as unbranded. Skins sold to international tanners that sell direct to profile brands such as BMW. Antler markets same as before, though antler powder now sold to high-energy drink manufacturers as a branded ingredient.
Market orientation		
Culture	Short-term focused: Responsibility for product ended at farm gate; commodity cycles natural; product focused; customers and network members viewed as antagonistic.	More longer-term focused: 'Pasture to plate' view of responsibility for product, though marketing activity viewed by members as a cost that should be borne by those who directly benefit; greater on-farm responsibility but still commodity cycle–focused; view network members and customers as partners.
Focus of organisation	Price, product, and production efficiency.	Product quality, support programmes, brand, and relationships.
Customer orientation	Focused on price. No orientation toward leather processors, and Asian medicine channels controlled by agents.	Support services surrounding brand directed at segments. Programmes developed in conjunction with network members. Disconnect in relation to pricing and supply stability.

Table 2.2   *Concluded*

Organisational Characteristics	Before	After
Competitor orientation	Viewed competitors as other deer-selling nations.	NZGIB now views competition as other meat producers, regardless of type. This view not shared fully within industry but is dominant.
Inter-functional coordination	Antagonism among supply chain members.	Recognition of mutual dependence and greater coordination between farms and processors, including adoption of processor-specific on-farm quality standards.
Performance outcomes		
Customer perceptions	Large supplier of high quality, low priced venison. Skins viewed as low grade.	General belief in superior product quality and latent desire to support NZGIB, but support now conditional on price and supply stability. Most (95%) skins A-grade.
Financials	In 1992, price per kilo = $NZ5.70. Total export value in 1992 = NZ$100 million, total exports = NZ$55 million.	Height of success in 2000–01: economic farm surplus for deer NZ$1,000 per hectare. 2004 figures: $NZ26 per hectare, judged unsustainable. Price for venison per kilo 2004: $NZ3.75 (down from $10 in 2001). Programmes added NZ$108 million in value 1998–2003 (total average yearly sales of approximately NZ$236 million). 2004 value: NZ$213 million.

Table 2.3   **MNZ before and after market-oriented change**

Organisational Characteristics	Before	After
Organisation		
Structure	Government-mandated industry body with marketing staff and elected board. Members fund generic marketing support.	Government-mandated industry body with marketing staff and elected board. Members fund generic marketing support.
Markets served	Unknown – product simply blended with other wools and sold in bulk.	Targets the Merino NZ ingredient brand at elite cloth processors and fashion houses. Also will provide non-branded, lesser quality wool to elite customers for secondary labels. Joint innovation with high-profile fashion brands.

Table 2.3   *Concluded*

Organisational Characteristics	Before	After
Market orientation		
Culture	Short-term focused: Responsibility for product ended at farm gate; commodity cycles natural; product focused; customers and network members unknown.	Long-term focused: Farm to fashion view of responsibility for product; brands a source of value; customers and other network members viewed with respect; understanding of mutual dependence with channel members; value focused.
Focus of organisation	Product and combined size.	Brands, customer oriented value creation processes, and relationships.
Customer orientation	None.	Relational; very customised focus; customers can request individual animals from any farm; farmers deal direct with customers.
Competitor orientation	Other wool producers.	Competing directly with all fibres.
Inter-functional coordination	None, complicated 11-member channel viewed as enemies, no interaction.	Close ties among farmers, Board, research and development, marketing, and channel members.
Performance outcomes		
Customer perceptions	Increase in price would lead fashion houses to switch to human-made fibres. Increased marketing spending and declining market share. Woolmark programme had high awareness but little value.	Overwhelming support among customers for NZ Merino, product of choice among leading processors such as Loro Piana; MNZ ingredient brand valued by fashion houses and features inside branded garments; sales of Merino products by fashion labels to end consumers at all time high.
Financials	Prices had been declining 3–6% per year over 20 years.	Difficult to assess, as prices no longer are based on auction price but are negotiated privately. Industry and farmers report prices are sustainable and industry is the healthiest it has been.

UNFREEZING

The actors in both cases had to unfreeze long-held assumptions before they could implement market-oriented change. Poor returns, declining market share, and industry rationalisations provided the rationale for each Board to change. Both Boards engaged in vigorous and acrimonious debate about the nature of their 'product' and its position in the market. The following quote, recorded at a wool industry conference in 1997, identifies the split between those pushing for change, and those opposing it:

> 'At four per cent of the world textile supply and falling, wool should be a rare highly sought-after fiber … but we have got things seriously wrong. This generation's fabric designers are unaware that wool characteristics could give them interesting variations and enable wool to set new directions in design and finish. The obvious has been completely overlooked – the natural differences which occur in wool from animal to animal, strain to strain and across strains. These natural characteristics are a fabric designer's dream, but unfortunately are not available to them.'

Management at each Board faced entrenched opposition from farmers and some elected Board members, driven by an enduring belief that both venison and wool were intrinsically commodities and that short-term price falls were just part and parcel of being in an agricultural industry. These individuals were advocates of continuing the *status quo*. For example, a speech at the 1997 Wool Conference argued:

> 'In the real world wool is struggling to attract consumer dollars. Whether some people like it or not wool is a commodity. Whether it was the finest Merino or the coarsest crossbred, wool was an industrial raw material, an ingredient product that was neither used nor ornamented in its raw state. Consumers did not buy raw wool but finished products that they wanted.'

In both cases, historical data was used to identify the long-term nature of falling market share and returns (despite investments in generic promotion, research and development, and efficiency improvements). This objective information was used by those pressuring for change to call for a radical rethinking of each Board's approach and was particularly successful in encouraging unfreezing. In particular, emphasis was placed on developing brands, delivering value to customers based on their needs, and moving away from pipeline efficiencies.

Further cultural assumptions were also challenged. In both cases, key cultural assumptions such as the nature of interaction with the environment (passive acceptance of commodity cycles versus proactively changing the market's rules), the nature of value (commodity versus brand), the nature of relating (antagonistic versus trusting), and the nature of time (future focused versus past focused) were challenged.[72] These are listed in Tables 2.3 and 2.3 for each case. Typical of the approach to unfreezing was the call for widespread changes to marketing practices, positioning, and the assumptions that underpinned these behaviours. During the unfreezing phase, both cases questioned the totality of how they conducted business and whether their current structures were to blame for poor performance.

Those advocating change in both cases sought to build coalitions and political support for the change as a means of overcoming entrenched resistance. In both cases, statistical evidence was collected to identify that the industry's current state was not just due to normal seasonal fluctuations. Also, evidence was presented to show the downward trend in real returns for commodity producers and the effect that price uncertainty had on the size of the marketplace. Those advocating change also sought powerful supporters for their cause. For example, in 1998, the then-Minister for Agriculture told a MNZ conference he was wearing a suit of 100 per cent NZ Merino wool, which would retail in Italy for

NZ$2,000, though farmers would receive only NZ$80 at auction for the wool. Such emotive examples assisted those seeking change because their message focused squarely on what was possible to achieve. During this process, information was communicated through formal conferences, newsletters, and industry journals. This information was then used by those advocating change in peer-to-peer discussions. Because each Board required a mandate from its members for change, grassroots support for change was critical.

Both cases followed the same process during the unfreezing phase – seeking to uncover and then challenge past assumptions. To do so, they built up a powerful case linking past strategies and assumptions about the product, the nature of the industry, and the adopted strategic outlook to continued declining performance. As well, they sought to communicate an image of a possible future, emphasising key aspects of market orientation, including rethinking customer and network member relations, an emphasis on customisation and branding, and seeking to add value. Both Boards also sought to overcome resistance to change through the development of a coalition of industry members, key buyers, marketers, and even government stakeholders. During this phase, the emphasis was on increasing the forces for change and decreasing resistance to the idea of change.[73] The unfreezing phase ended when there was formal agreement for change to occur, with authorisation (from a formal vote by members) given to build the basis for a new market-oriented approach. Following this, the movement phase began.

MOVEMENT

Once past assumptions had been successfully unfrozen and challenged, it was incumbent on senior leadership to develop alternative approaches (i.e., change programmes) and then move their members toward the adoption of these programmes. The first step to encourage movement toward a market orientation was to identify a strategy that fitted with the resource base of the industry and its members. Details of both strategies are provided in Tables 2.2 and 2.3. In both cases, a market-oriented vision that could be accepted by as many members as possible was necessary. This point identifies the importance of developing buy-in among influential organisational members before embarking on market-oriented change.[74]

For both cases, market intelligence during the movement phase served three roles. First, research was conducted to identify market opportunities prior to brand development. Second, research was conducted with key buyers prior to the launch of the proposed strategy. Third, results were communicated to members to gain support for the programme and continue the momentum in favour of change. For example, NZGIB conducted research on perceptions of venison in America as part of a strategy to target high-end US restaurants. Although the results reinforced traditional perceptions of venison, they also highlighted sources of positioning – country of origin and nutrition. Research also highlighted the need for a dual marketing programme targeting both end-consumers and channel buyers. As a result, a unique strategy started to take shape that involved investing the majority of marketing resources into trade programmes and using high-profile chefs and sporting stars to build consumer awareness. Through this process, research results were communicated to members through a new internal marketing programme, including the development of a Web page, newsletters, industry conferences, and regional workshops.

Also critical during this period was the reconfiguring of established systems and functions and the development of new ones to support the desired positioning. For example, NZGIB developed quality programmes to ensure deer were not stressed prior to processing (which affects the quality of the meat). These included fragmented programmes covering transport, processors, auction, and farming, which were grouped together under an all-encompassing programme called 'From Pasture to Plate.' The MNZ developed a similar programme entitled 'fleece to fashion.' Both programmes sought to identify the newfound scope of responsibility farmers had for their product according to a strategy focused on building brand value.

Despite these consistencies, one key difference that emerged at this stage surrounded the use of auctions to set prices, which would have a profound effect on the success levels of both programmes. Both cases sought to create sustainable markets and sources of value for their members. During this phase, however, MNZ brought both industry members and channel buyers together to discuss mutual problems and possible solutions. For example:

'We would work right from the farm right through the scourers, spinners, weavers, knitters, and the whole chain. What we are doing is opening up everybody's books along the pipeline to get some real transparency in the system. Everybody has shown each other what their profit margins are – we have got it down to that level of transparency.'

Emerging from these discussions was a shared understanding of the need for consistent supply and prices before any brand programme could work. This demand was especially the case because inconsistent supply (brought on by hoarding in an attempt to push up commodity prices) made planning by customers difficult, and specialist processors in the channel had very little ability to absorb upward changes in price. By way of contrast, NZGIB remained unaware of this issue during the development of its programme and thus built in measures that focused on increased auction prices as the key driver for programme success. This situation was driven partly through ignorance of the need for changes in pricing mechanisms to support the brand's position, partly due to the larger scale of the industry, and partly as a result of the multiple uses of deer.** Therefore, MNZ was more successful at developing desired new cultural assumptions in regard to channel relationships. Effectively they challenged the old adversarial view toward channel members and developed programmes that saw all channel members as necessary to the future survival and prosperity of the industry. As a result, MNZ developed a programme to encourage farmers to sign fixed-term contracts with key buyers (by 2004, 40 per cent of sales used exclusive contracts for very large buyers). In contrast, NZGIB stuck with its original plan and continued to use auctions, which ultimately would affect its ability to refreeze new assumptions, because auction-driven inconsistencies in price and supply undermined the brand promise and relationships with key buyers, who could tolerate little uncertainty.

In each case, management needed to overcome market scepticism to the programmes and challenges to their power. For example:

'It was very, very difficult for us to sell this strategy. We had some hard difficult negotiating to do with these big garment companies because they are not used to being "pushed around." The retailer really has the strength because in every market it is the retailer that has the strength

---

** Deer can be processed for meat (venison), velvet (antler powder used in traditional Asian remedies), leather, and other body parts (hooves, tails, and sexual organs also used in Asian remedies). The Cervena programme only covered the highest quality meat; other markets were required for lower quality cuts.

because at the end of the day it is a retail buyer that places the order, not the manufacturer. Obviously we have to do development work with the manufacturer because you need that 100 per cent New Zealand grown material coming through the pipeline to satisfy the retailer, but if the retail buyers specify New Zealand Merino in advance it makes what we are doing so much easier.'

During this period, the tone of communications changed from argumentative (unfreezing) to educational and from top-down to increasingly collaborative as management sought to build support for their programmes by involving key members and customers. Also, small successes, such as initial positive feedback about the emerging programme, were communicated to members to gain support for the changed approach and, importantly, securing resources (levies were compulsory but each year there was pressure to reduce them). During the movement phase, bottom-up buy-in started to occur because though the initial debates during unfreezing identified the need to 'do something different,' the exact details of the new programme needed to be established and saleable to customers and members. By involving both sets of stakeholders in the development of the programme, both Boards were able to claim legitimacy for the final strategy.

In summary, the factors identified as driving market orientation implementation changed from the unfreezing stage to the movement stage. Leaders needed to move from building a case for change to negotiating a shared mission and developing a set of tactical strategies. Also, they needed to combine their previous top-down approach with a more negotiated style to achieve buy-in from key stakeholders. Market intelligence was being gathered to identify market opportunities, with plans needed to be adapted in the face of market feedback. Communication strategies also changed, with a new emphasis on education and feedback about early successes. As such, the change was a combination of programmatic and market-back. The focus on inter-functional coordination moved from criticising past arrangements to reconfiguring old systems and building new ones. The movement stage ceased when both firms implemented their strategies by announcing their launch to their respective industries. Following this phase, refreezing needed to occur.

REFREEZING

Refreezing involves institutionalising the assumptions developed during the movement phase.[75] Tables 2.2 and 2.3 detail the different outcomes with respect to market orientation experienced by both cases. We propose that the sustained success enjoyed by MNZ and the short-term success followed by industry crisis experienced by NZGIB can be explained by the different strategies of each Board during the refreezing process. During the refreezing phase, reinforcing the new market-oriented assumptions is critical and can be done with a combination of methods, including building systems that ensure feedback between market-oriented actions and success, communicating successes, staying on message, celebrating market-oriented values through new artefacts such as stories and myths, and rewarding people for market-oriented behaviours. Both Boards adopted a number of methods for reinforcement, though the inability to build effective feedback systems saw NZGIB fail to reinforce its market-oriented values.

As a means of reinforcing market-oriented values, both Boards focused extensively on communicating short-term wins to their members (usually through newsletters) and engaged in public relations activities to raise the profile of each industry in New Zealand. During this phase, communications were celebratory (communicating short-term wins), educational (communicating new programmes and approaches), and persuasive (focusing on the need for ongoing commitments to the programme and further funding). Both Boards tried to lessen their role in the process and allow for bottom-up buy-in through market-back learning.[76] The following example was extracted from an industry workshop held by NZGIB during this period:

'It is important for farmers to try and find some sort of balance whereby the farmers, the venison processors and exporters, and the meat processors and exporters are comfortable that they can actually supply a product at a price that is profitable for everyone, rather than seeing each other as competitors and attempting to get as much margin out of each other as possible.'

The quote focuses on responding to how farmers ought to work with business buyers according to a market orientation. Because each buyer has its own specific needs, farmers were complaining of the need to adapt to different demands. The NZGIB attempted to reinforce the assumptions of a market orientation by suggesting the benefits that would come from an increased commitment to fewer buyers. This continued emphasis on the programme's goals was a hallmark of the refreezing phase, though it was constantly undermined by the lack of supportive policies ensuring feedback between performance outcomes and market-oriented actions.

The differences between the two cases approach during this period relates to the increased use of customised feedback systems that enabled adaptation to individual customers' needs by each grower with MNZ versus the use of the one-price schedule for venison farmers. Both Boards focused on quality improvement, though MNZ went one step further and built systems that enabled growers to make ongoing improvements in their product to suit the need of individual buyers (whereas NZGIB members were slow to react to customer-driven changes because they would have to change the entire generic programme). For example, MNZ made changes to the way wool was graded and sold. Previously, customers had no ability to identify whence the wool came, nor any scientific evidence for the quality of the wool. Because there was no feedback loop, farmers also had little market-based information to inform their improvements to farming practices, which could lead to improvements in wool quality. In turn, MNZ invested funds in a central grading facility called PAC. After shearing, each fleece was separately bagged and graded scientifically, providing better information to customers, because each fleece could be linked to an individual animal, farm, and customer. This information enabled growers to identify which fleece went into each bale and the price of that bale, which improved on-farm decision making and allowed growers to make improvements based on individual customer requirements.

The market intelligence provided by the PAC also enabled each farm to fine tune its approach and extract causality to identify the benefits of the programmes.[77] In the case of MNZ, the Board could illustrate a causal link between its activities and market-based success. Farmers thus could engage in a market-back learning orientation (generative learning).[78] Such a result meant MNZ not only had a market orientation but also enjoyed high quality market-oriented outcomes and drove markets through new innovations with buyers, such as co-developed products like Zealander, Opossum Wool, and Denim Wool. In contrast, NZGIB did not develop intelligence systems that could identify the benefit of this one programme to the overall industry, which is

> 'part of the fact that everyone was active in the market before we got there. If you were starting with a new market and no one actually had any market share you could probably sort out some territories. We might have factored in transparency in so that the benefits of the programme to the producers could have been seen. You might have asked the exporters to have a second (pricing) schedule for branded sales.'

Farmers voted to make the brand programme (accounting for only 10% of sales) financially independent in 2001 under the banner, 'Who Benefits, Who Pays,' whereby market-oriented programmes had to be funded by those who directly benefited. This requirement resulted in a counter-movement that undid some of the changes and limited the refreezing that could take place. Although some market-oriented objectives were achieved, they were of lower quality than those of MNZ, and by 2002, the industry as a whole entered its worst downturn in a decade (the situation had worsened by 2005), though branded venison prices have fallen less and customers still retain interest in the programme.

During this time, NZGIB communications changed from an educational tone to a more critical one, whereby Board members sought to regain the momentum by working within the current framework of auction pricing. What was needed was path-breaking change,[79] whereby the rules of the game changed radically, rather than path-driving change, whereby NZGIB continued to reinforce structural aspects of the old system that were no longer appropriate. For NZGIB, the inability to move farmers away from auctions continued to lead to market-based problems and undermined relationships with key buyers, which in turn influenced the quality of market-oriented behaviours.[80] Therefore, NZGIB leadership again had to defend its approach and attempt to drive radical changes into the method of supply in venison markets. In contrast, MNZ used the auction system to create interest in wool by running a finest wool competition, whereby the world's wool buyers bid once a year on the finest wool bale.

In summary, refreezing involved building tight inter-linkages among strategies, cultural assumptions, supportive structures, communication and leadership style, and learning style. That is, implementing a market orientation requires the development of a multi-layer (culture, systems, structures, and strategies), self-reinforcing system. Refreezing was dominated by bottom-up driven change, with top-down change taking an educational and celebratory role. During this stage, the reinforcement of market-oriented values required market information that could build causal linkages between actions and performance. Refreezing was also achieved through encouragement of customisation and innovation, driven by interactions between farmers and customers (and made possible by increasingly targeted market intelligence).

Discussion

This chapter has addressed two questions. First, we confirm that Lewin's planned change model captured the dynamics of adopting a market orientation.[81] Second, we combine planned change theories with the limited research on implementing a market orientation. We thus posit that leadership, the form and use of market intelligence, and the forms of inter-functional coordination, learning style, and challenges change across the three phases of unfreezing, movement, and refreezing. Through a comparison of two cases, we identify how the role and importance of these variables changes across the three phases of the change process (see Table 2.4). This identification receives support from real-time industry data, providing the first examples of the process of change toward market orientation. As such, this article both identifies new insights and extends extant theory by building on the results of Kennedy et al. and Narver et al.[82]

The adoption of a market orientation by these cases illustrates the complex and often politicised process involved in implementing marketing programmes. To implement a market orientation, the marketers had to overcome a lack of influence and formal power. They formed coalitions with key stakeholders, used market research to influence organisational members, and continued to build and sustain support for their market orientation throughout the change process. To date, little emphasis has been placed on marketers' political roles within firms, even though this role is necessary to implement marketing programmes. Further research should examine how marketers effectively implement strategies, given that they often have little direct authority over various functional areas that are critical to the implementation of a market orientation.

Table 2.4   Key roles, activities and challenges during market-oriented change process

	Unfreezing	Movement	Refreezing
Time of each stage	NZGIB: 1 year MNZ: 1 year	NZGIB: 2 years MNZ: 3 years	NZGIB: ongoing. MNZ: 1 year
Senior leadership role	1. Build critical mass for change. 2. Build broad-based support. 3. Appeal to hearts and minds.	1. Negotiating a shared vision. 2. Adapting market orientation to resource base. 3. Gaining buy-in from members and key stakeholders. 4. Selling the vision.	1. Communicating short-term wins. 2. Continued emphasis on programme's goals. 3. Encouraging bottom-up buy-in consistent with original vision.
Inter-functional coordination	1. Challenge the totality of current inter-functional arrangements.	1. Reconfiguring traditional arrangements to support new strategy. 2. Developing new supportive structures and systems to complement new strategy.	1. Incremental improvements. 2. Reinforcing shared-vision.

Table 2.4   *Concluded*

	Unfreezing	Movement	Refreezing
Market intelligence	1. Data identifying causal relationships between past practices and ongoing decline. 2. Data appealing to future possibilities.	1. Market-based research to identify customer/end consumer demands and perceptions of product/region of origin. 2. Research with key buyers to support proposed programmes before launch.	1. Extracting causality from data to identify benefits of programmes.
Programmatic versus market back change	1. Programmatic.	1. Programmatic and market-back.	1. Market-back and programmatic.
Tone of internal communications	1. Urgent, mix of fact and aspiration for new future.	1. Educational and informative. 2. Increasingly collaborative.	1. Reinforcing message. 2. Celebrate initial successes. 3. Educational.

LIMITATIONS AND FURTHER RESEARCH

There are several limitations to this research. First, the study of processes could be improved if it were conducted in real time and longitudinally rather than through a reliance on historical information and interviewee recall. Further research should examine the change process involved in moving to a market orientation as part of a longitudinal, participant observer study. Second, our findings have relied heavily on recall of a few organisational members, whereas additional research could benefit from interviews conducted with a wide range of stakeholders. Third, the results herein could have been improved by interviewing farmers and members critical of the programmes undertaken to uncover further cultural assumptions behind opposition to change. This effort would identify the tension involved in managing across different levels of culture, including those at organisational, functional, and individual levels. Fourth, these results focused solely on radical planned change efforts. Research could examine more evolutionary, emergent efforts, perhaps with firms closer to a market orientation. Research also should be conducted in different cultural contexts and different organisations, because these findings are focused on one country and two cases of a specific form of organisation.

Implications for managerial practice

The findings give rise to several managerial implications. First, the stage model of change identified in Table 2.4 provides the beginning of a road map for managers seeking to implement market-oriented change. Although the organisations studied underwent revolutionary change, such a process also can be adapted to firms that require more evolutionary change. For example, firms that already have a set of market-oriented values but struggle to implement them effectively in the marketplace may be able to skip the unfreezing and movement stages and address issues of refreezing, perhaps by building in feedback systems and identifying short-term wins.

Second, marketing managers seeking to implement marketing strategies in industrial firms often need to draw on a wide range of organisational members (usually outside marketing) and key customers. The cases identify several important political strategies associated with implementing marketing policies. Marketing managers should sell the benefits of change to non-marketing staff prior to undertaking a change effort. This undertaking is part of an important coalition-building process within the firm. Also, marketers need to gain top leadership support for the change and ensure this support continues over a sustained period. The stages identified in Table 2.4 will help marketers brief senior managers on their roles during the change period.

Third, marketing managers must integrate market-oriented culture, learning style, and systems and structures to ensure effective refreezing. To ensure the co-development of such an integrated system, marketers will need to give consideration to the development of educational materials, reward systems, methods of working together, and systems that provide clear feedback loops between the actions of employees and performance, and they must reward employee-driven innovation. Changes to reward systems and the encouragement of employee risk taking (to challenge past practices) will help ensure a market-back learning style that can reinforce a bottom-up commitment to market-oriented change. (The role and characteristics of such change champions, and the management of such individuals, need further research.) Such changes will require cross-functional support and again highlight the need to build collations of support with other functions to undertake successful market-oriented change efforts.

Appendix: Background Information About Each Case

THE NEW ZEALAND GAME INDUSTRY BOARD (NZGIB)

To gain market power, NZGIB formed in 1985 with the explicit aim of coordinating the growth of the deer industry in New Zealand. The Board is a statutory body and has the power to levy all industry members (farmers, processors, and exporters) to fund generic research and development, quality programmes, and industry-level marketing strategies, with a specific focus on developing export markets. Unlike other boards (statutory monopoly boards are typical of many agricultural industries in Australasia), NZGIB does not trade in product; rather, it develops generic campaigns that run in parallel with the individual branding programmes of processors. The success of NZGIB would be measured in increased market share *and* increased commodity prices. From 1985 to 1990, NZGIB was successful in creating low-price export markets for New Zealand farmed venison, eventually dominating the market (New Zealand is currently the largest deer farming nation with a total stock of 1,600,000 animals). The main market continued to be Germany, though this market was very competitive and controlled by six buyers (who colluded to set prices), and demand was driven by traditional game consumers who wanted to consume 'wild' deer during two traditional hunting festivals. Also, major buyers used venison from different countries to develop their own branded blend. As a result, they were reluctant to cooperate with NZGIB, which wanted to develop a country-of-origin programme capitalising on the positive environmental image of New Zealand among Germans. As well, the farming of venison under strict quality controls (developed by NZGIB) had led to higher quality products (wild venison naturally has large variations in quality, cut size, tenderness, and taste) and greater production efficiencies, yet prices were falling. The Chernobyl disaster, which saw radioactive fallout spread across traditional foraging grounds for European deer, prompted consumers to refuse to eat venison. Without a clear market identity, New Zealand venison was treated no differently than European products, and the world price plummeted, resulting in many farm bankruptcies, the closure of New Zealand's one specialist venison meat processing plant, and the exit of several other processors from venison production. Because of this rationalisation, NZGIB was charged with creating more sustainable markets for venison, including the development of New Zealand-owned brands (for an overview of the Cervena brand, see Beverland's account[83]) and high-value niches in new markets. As of 2004, 49 per cent of venison exports went to Germany, with the rest being sold (in declining order of sales) in Belgium, Sweden, France, the United States, Austria, Switzerland, Italy, Netherlands, the United Kingdom, and others.

MERINO NZ (MNZ)

Until the early 1990s, New Zealand wool was sold through the International Wool Secretariat (IWS), which consisted of Australian, New Zealand and South African wool farmers. The IWS had developed the Wool Mark programme as a means of stabilising prices and increasing market share. The New Zealand Wool Board was responsible for regulating the industry, and unlike NZGIB, had 'single desk seller' powers – that is, it was the only entity in New Zealand allowed to market and sell wool. Wool prices were artificially held up through stockpiling. This unsustainable situation ended in the early 1990s, when wool price guarantees were removed and returns to growers plummeted. Merino wool makes up less than 5 per cent of the total New Zealand wool clip and was usually blended in with other coarser wools or Merino from Australia or South Africa, which is not as strong or as fine (in microns) as New Zealand merino because of the different climatic conditions. The small size of the Merino industry (4% of the total clip) within the overall New Zealand wool farming community (the Wool Board was controlled by farmers who voted on strategies) meant that these growers had little voice, and their higher quality products, desired by luxury fashion houses, had little identity or value. Following New Zealand's withdrawal from the IWS in 1994, Merino growers began petitioning the Wool Board to exit and form their own industry grouping. They were successful and, in 1995, launched Merino New Zealand (without single desk-selling powers) – an industry development organisation dedicated solely to marketing New Zealand merino fibre. In 1996, following extensive market research, the New Zealand Merino brand was launched, targeted at the world's best cloth processors and leading fashion brands. Like NZGIB, MNZ is funded by a compulsory levy on farmers (but not processors) and run by a full-time marketing team and elected board. Moreover, MNZ is responsible for industry development, market development and growth, price maintenance, and the development of industry-owned brands. In contrast to NZGIB, MNZ is a niche strategist, accounting for just 2 per cent of the world's fine wool sales.

References

1. Baker, William E. and Sinkula, James M. (1999), 'Learning orientation, market orientation, and innovation: Integrating and extending models of organizational performance', *Journal of Market-Focused Management*, vol. 4, no. 3–4, pp. 295–308; Harris, Lloyd C. (2000), 'The organizational barriers to developing market orientation', *European Journal of Marketing*, vol. 34, no. 5–6, pp. 598–624; Kennedy, Karen Norman, Goolsby, Jerry R., and Arnould, Eric J. (2003), 'Implementing a customer orientation: Extension of theory and application', *Journal of Marketing*, vol. 67, no. 4, pp. 67–81; McNaughton, Rob B., Osborne, Paul, Morgan, Robert E. and Kutwaroo, G. (2001), 'Market orientation and firm value', *Journal of Marketing Management*, vol. 17, no. 5–6 pp. 521–42; Weerawardena, Jay and O'Cass, Aron (2004), 'Exploring the characteristics of the market-driven firms and antecedents to sustained competitive advantage', *Industrial Marketing Management*, vol. 33, no. 5, pp. 419–28.

2. Avlontis, George J. and Gounaris, Spiros P. (1997), 'Marketing orientation and company performance: Industrial vs. consumer goods companies', *Industrial Marketing Management*, vol. 26, no. 5, pp. 385–402; Gounaris, Spiros P. and Avlontis, George J. (2001), 'Market orientation development: A comparison of industrial vs. consumer goods companies', *Journal of Business and Industrial Marketing*, vol. 16, no. 5, pp. 354–381; Weerawardena and O'Cass, op. cit.

3. Avlontis and Gounaris, op.cit ; Gounaris and Avlontis, op. cit.

4. Narver, John C., Slater, Stanley F., and Tietje, Brian (1998), 'Creating a market orientation', *Journal of Market-Focused Management*, vol. 2, no. 1, pp. 241–55.

5. Day, George S. (1994), 'The capabilities of market-driven organizations', *Journal of Marketing*, vol. 58, no. 4, pp. 37–52; Harris, op. cit.; Jaworski, Bernard J. and Kohli, Ajay K. (1996), 'Market orientation: Review, refinement, and road-map', *Journal of Market-Focused Management*, vol. 1, no. 2, pp. 119–35; Kennedy et al., op. cit.; Narver et al., op.cit

6. Harris, op. cit., p. 619.

7. Narver et al., op. cit.

8. Kennedy et al., op. cit.

9. Narver et al., op. cit.

10. Kennedy et al., op. cit.

11. Day, op. cit.; Jaworski, Bernard J. and Kohli, Ajay K. (1993), 'Market orientation: Antecedents and consequences', *Journal of Marketing*, vol. 57, no. 3, pp. 53–70; Slater, Stanley F. and Narver, John C. (1995), 'Market orientation and the learning organization', *Journal of Marketing*, vol. 59, no. 3, pp. 63–74.

12. Kennedy et al., op. cit.

13. Lewin, Kurt (1951), *Field Theory in Social Science*, New York, Harper and Row.

14. Homburg, Christian and Pflesser, Christian (2000), 'A multiple-layer model of market oriented organizational change', *Journal of Marketing Research*, vol. 37, no. 4, pp. 449–43; Narver et al., op. cit.

15. Bisp, Søren (1999), 'Barriers to increased market-oriented activity: What the literature suggests', *Journal of Market-Focused Management*, vol. 4, no. 1, pp. 77–92; Deshpandé, Rohit and Webster, Frederick E. (1989), 'Organizational culture and marketing: Defining the research agenda', *Journal of Marketing*, vol. 53, no. 1, pp. 3–15.

16. Bisp, op. cit.

17. Narver et al., op. cit., p. 242.

18. Deshpandé and Webster, op. cit., p. 4.

19. Crocombe, Graham T., Enright, Michael J., and Porter, Michael E. (1991), *Upgrading New Zealand's Competitive Advantage*, Auckland, Oxford University Press.

20. Ibid.; Beverland, Michael B. (2005), 'Creating value for channel partners: The Cervena case', *Journal of Business and Industrial Marketing*, vol. 20, no. 3, pp. 127–35.

21. Narver et al., op. cit.

22. Ibid.; Beverland, op. cit.

23. Homburg and Pflesser, op. cit.

24. Kennedy et al., op. cit.; Narver et al., op. cit.

25. Ibid.

26. Narver et al., op. cit., p. 243.

27. Narver et al., op. cit.

28. Kennedy et al., op. cit.; Schein, Edgar H. (1992), *Organizational Culture and Leadership*, 2nd ed., San Francisco, CA, Jossey-Bass.

29. Narver et al., op. cit.; Schein, op. cit.

30. Harris, op. cit.

31. Kennedy et al., op. cit.; Narver et al., op. cit.

32. Narver et al., op. cit.

33. Van de Ven, Andrew H. and Poole, Marshall S. (1995), 'Explaining development and change in organizations', *Academy of Management Review*, vol. 20, no. 3, pp. 510–33.

34. Deshpandé, Rohit and Farley, John U. (1998), 'The market orientation construct: Correlations, culture, and comprehensiveness', *Journal of Market-Focused Management*, vol. 2, no. 3, pp. 237–39; Homburg and Pflesser, op. cit.

35. Narver et al., op. cit.

36. Lewin, op. cit.

37. Narver et al., op. cit.

38. Burnes, Bernard (2004), 'Kurt Lewin and the planned approach to change: A re-appraisal', *Journal of Management Studies*, vol. 41, no. 6, pp. 977–1002; Lewin, op. cit.; Wilson, David (1992), *A Strategy of Change: Concepts and Controversies in the Management of Change*, London, Routledge.

39. Wilson, op. cit., p. 8.

40. Burnes, op. cit.; Lewin, op. cit.

41. Narver, John C. and Slater, Stanley F. (1995), 'Market orientation and the learning organization', *Journal of Marketing*, vol. 59, no. 3, pp. 63–74.

42. Ibid.; Schein, op. cit.

43. Wilson, op. cit.

44. Bisp, op. cit.

45. Lewin, op. cit.

46. Bisp, op. cit.; Lafferty, Barbara and Hult, G.Thomas M. (2001), 'A synthesis of contemporary market orientation perspectives', *European Journal of Marketing*, vol. 35, no. 1-2, pp. 92–109; Narver et al., op. cit.

47. Lafferty and Hult, op. cit.

48. Narver et al., op. cit.

49. Burnes, op. cit.; Lewin, op. cit.

50. Becker, Jan and Homburg, Christian (1999), 'Market-oriented management: A systems-based perspective', *Journal of Market-Focused Management*, vol. 4, no. 1, pp. 17–41; Day, op. cit.; Kohli, Ajay K. and Jaworski, Bernard J. (1990), 'Market orientation: The construct, research propositions, and managerial implications', *Journal of Marketing*, vol. 54, no. 2, pp. 1–18.

51. Slater and Narver, op. cit.
52. Kennedy et al., op. cit.
53. Bisp, op. cit.
54. Baker and Sinkula, op. cit.; Lafferty and Hult, op. cit.; Weerawardena and O'Cass, op. cit.
55. Baker and Sinkula, op. cit., p. 305.
56. Bell, Simon J., Whitwell, Gregory J., and Lukas, Bryan A. (2002), 'Schools of thought in organizational learning', *Journal of the Academy of Marketing Science*, vol. 30, no. 1, pp. 70–86.
57. Baker and Sinkula, op. cit.
58. Homburg and Pflesser, op. cit.; Yakimova, Raisa and Beverland, Michael (2005), 'Organizational drivers of brand repositioning: An exploratory study', *Journal of Brand Management*, vol. 12, no. 6, pp. 445–60.
59. Narver et al., op. cit.
60. Baker and Sinkula, op.cit.
61. Eisenhardt, Kathleen M. (1989), 'Building theories from case study research', *Academy of Management Review*, vol. 14, no. 4, pp. 532–50; Matthyssens, Paul and Vandenbempt, Koen (2003), 'Cognition-in-context: Reorienting research in business market strategy', *Journal of Business and Industrial Marketing*, vol. 18, no. 6–7, pp. 595–606; Yin, Robert K. (1994), *Case Study Research: Design and Methods*, Thousand Oaks, CA, Sage Publications.
62. Lewin, Jeffrey E. and Johnston, Wesley J. (1997), 'Relationship marketing theory in practice: A case study', *Journal of Business Research*, vol. 39, no. 1, pp. 23–31.
63. Eisenhardt, op. cit.
64. Strauss, Anselm and Corbin, Juliet (1998), *Basics of Qualitative Research*, 2nd ed., Newbury Park, CA, Sage Publications.
65. Ibid.
66. Ibid.; Yin, op. cit.
67. Eisenhardt, op. cit.
68. Strauss and Corbin, op. cit.
69. Spiggle, Susan (1994), 'Analysis and interpretation of qualitative data in consumer research', *Journal of Consumer Research*, vol. 21, no. 3, pp. 491–503.
70. Kennedy et al., op. cit.
71. Lincoln, Yvonna S. and Guba, Egan (1985), *Naturalistic Inquiry*, Beverly Hills, CA, Sage Publications; Strauss and Corbin, op. cit.
72. Schein, op. cit.
73. Lewin, op. cit.
74. Wilson, op. cit.
75. Burnes, op. cit.
76. Narver et al., op. cit.
77. Kennedy et al., op. cit.
78. Baker and Sinkula, op. cit.; Narver et al., op. cit.
79. Siggelkow, Nicolaj (2001), 'Change in the presence of fit: The rise, the fall, and the renaissance of Liz Claiborne', *Academy of Management Journal*, vol. 44, No 4, pp. 838–57.
80. Baker and Sinkula, op.cit.
81. Lewin, op.cit.
82. Kennedy et al., op. cit.; Narver et al., op. cit.
83. Beverland, op.cit.

Marketing Agricultural Products

3 Moving Toward Market Orientation in Agri-food Chains: Challenges for the Feed Industry

BY STEFANIE BRÖRING*

Keywords

market orientation, agri-food chains, feed industry

Abstract

This chapter investigates the implementation of market orientation at the level of an agricultural-input factor supplier. The feed industry traditionally has adopted a medium degree of market orientation and a limited end-consumer orientation. However, it provides the building block of animal-derived foods, which are increasingly valuable consumer products. Therefore, market orientation embracing the entire food chain becomes an ever more important issue for the feed industry as an input supplier. Against this background, this chapter investigates: (1) the current challenges of the feed industry, which necessitate a more profound market orientation, (2) how the feed industry as a player on the input side of the food chain embraces the challenges of implementing a market orientation, and (3) the different influences with regard to the business model, based on different levels of vertical integration in the feed industry.

Introduction

The feed industry represents an important stepping stone in the production of animal-derived foods. This industry traditionally has been regarded as a mere supplier of commodities, without any need for a market or value chain orientation. Built on a traditional business-to-business (B2B) model, the industry perceived consumer trends

* Professor Stefanie Bröring, University of Applied Sciences Osnabrük, Oleenburger Landstr. 24, 49090 Osnabrük, Germany. E-mail: s.broering@fh-osnabruek.de.

as far away, and likewise, consumers and the public did not pay much attention to it. Only through food scandals has the public gotten more involved with the feed industry, unfortunately with negative connotations. To improve food safety and avoid feed-borne food scandals, numerous quality programmes have been implemented. However, these measures mostly have been reactive and present a production standard rather than an example of active market orientation. This approach is changing at the moment; as the praxis shows, the impact of feed on product quality can be demonstrated and result in a positive impact. For example the Dutch dairy company Campina has launched a 'healthier milk' that contains more unsaturated fatty acids because of the special diet fed the dairy cows.[1] However, proactive anticipations of trends along the food chain and their implications for feed manufacturers, or even a 'feed push' approach toward innovations at the level of the feed industry that can make an impact on the entire food value chain, are only emerging and still very rarely observed. Market orientation (MO) seems a difficult endeavour, because trends in consumer markets seem distant and thus rather difficult to absorb for a supplier of agricultural input factors – the feed industry's traditional role.

This chapter seeks to contribute to the literature pertaining to MO in food supply chains,[2] in particular by exploring the development of MO in the feed industry. Thus, this contribution delivers basic insights into the question of how to implement a MO among actors at the very front end of a food supply chain. In addition to the feed industry's position as a supplier of agricultural input factors, it can be characterised by different business models depending on the different levels of vertical integration. Hence, the challenges of building a MO may vary according to the degree of vertical integration, which influence the degree to which a partner in the very front end of the food chain has information about the back end. Different business models may require different approaches to developing a sufficient degree of MO. To gain a better understanding of why and how the feed industry needs to establish a MO, this chapter investigates current challenges to the feed industry, measures to improve MO, and the influence of the business model on the adoption of MO.

The remainder of this chapter is organised as follows: Section X.2 contains a brief literature review on MO, followed by a more detailed description of the feed industry. Section X.4 overviews current challenges for the feed industry, which exemplify the need for increased MO. The degree of MO and measures to increase MO in the feed industry then receive more extensive exploration. Finally, drawing on these findings, Section X.6 derives some conclusions, provides managerial recommendations, and also highlights areas for further research.

Market Orientation: An Overview of Existing Research

Before exploring what challenges firms in the feed industry face and how they develop to become more market oriented, a closer consideration of the definitions and characteristics of MO is necessary. This chapter follows Narver and colleagues,[3] who define MO as a general approach toward running a certain business, underscored by the company's culture. Therefore, 'market orientation is the organizational culture that most effectively and efficiently creates the necessary behaviours for the creation of superior value for customers.'[4] Literature on MO is well established,[5] initially postulated by Drucker in 1954.

However, when it comes to agricultural markets and the feed industry, a traditional B2B industry sector, the discussion of MO is rather new.[6] In addition to understanding MO as a culture, Kohli and Jaworski employ a behaviouristic approach and argue that MO is constituted by three dimensions:[7]

1. Generating market-related knowledge about customers and competitors.
2. Distribution of that knowledge inside the company.
3. The ability to react on the basis of that market knowledge and be consistent with the market concept.

General agreement in MO literature indicates that ongoing, systematic information collection about customers and competitors, cross-functional sharing of that information in the company, and rapid responsiveness to competitor actions and changing market needs are at the centre.[8] Narver and colleagues expand this definition of MO to feature pro-active MO.[9] That is, MO would be reactive only if there were no anticipation of upcoming, evolving needs. Pro-active MO is especially important for the success of new products. The relationship between market orientation and new product success seems contingent on the type of innovation.[10] In addition, extant literature argues that MO is positively influenced by supply chain management.[11] Thus, not only supply chain management itself, which refers to the way the supply chain is controlled to deliver on promises to meet customer needs,[12] but also the different relationships in the supply chain,[13] must be taken into account when analyzing MO. Strong supplier relationships positively affect the generation of market-related knowledge and more rapid responses to market information, allowing for improved customer responsiveness. Supply chain management seems especially important for long supply chains, as in the case of animal-derived foods, because value creation within the supply chain depends on how well each stage of the chain processes raw materials and information to add value for downstream customers.[14] Furthermore, MO differs with respect to the chosen strategy type[15] and degree of vertical integration,[16] because the supply chain configuration depends on the level of integration within the supply chain. According to Webster,[17] a supply chain can be characterised by different types of integration, reaching from pure transactional relationships to buyer–seller partnerships and strategic alliances to full vertical integrations. This differentiation is especially relevant for the feed industry, which consists of different strategy types, depending on the degree of vertical integration.

According to Beverland and Lindgreen,[18] moving from a commodity orientation to MO requires a company to change not only its strategic outlook and marketing practices but also its culture. Assessing MO in this sense also means that different layers of the food chain need to be tackled, including supplier markets, direct customers, and customers' customers. The last form of MO is increasingly important to the feed industry in its efforts to adopt pro-active behaviour, though it also remains unaddressed due to barriers against it. For example, path dependency and the resulting market-related absorptive capacity create major hurdles to building chain-overarching MO. As Cohen and Levinthal note,[19] it is always easier to learn about related areas. From a theoretical standpoint, the construct of absorptive capacity – the ability to recognize, value, and acquire new information to apply it to commercial ends[20] – provides the prerequisite for MO. Because the feed industry is located at the very front of the value chain, its market-related absorptive capacity to develop a MO, which also embraces trends at the consumer level, seems rather difficult.

Before exploring a practical case of MO, the next section presents an overview of the feed industry, identifies the markets the feed industry deals with, and notes the major challenges.

Feed Industry: Facts, Business Models, and Value Chain Positioning

SOME FACTS ABOUT THE FEED INDUSTRY

Global animal nutrition production has increased steadily since the mid 1980s and amounted to 637 million tons in 2006. As illustrated in Figure 3.1, the compound feed sector consists of three main sub-sectors: pig, cattle, and poultry/laying hens. Each roughly represents one-third of total production, though pig feed is the most important feed stuff. A look at the geographical distribution of feed (Figure 3.2) shows that the largest market is the United States, with a world market share of 23 per cent (175 million tons), followed by Europe (140 million tons). Since the millennium, emerging markets in Latin America, Russia, and Asia have exhibited the highest growth rates. The world's largest manufacturer, Charoen Pokphand (18 million tons), is based in Thailand.[21]

In Europe, meat and other animal-derived products represent 45 per cent of the total value of farm production, which was a market of €126.5 billion for the EU25 in 2005. In general, the market for feed stuffs depends on the market for livestock products. In 2006, the EU25 livestock farming sector produced 45 million tons of meat (21 million tons of pork, 11 million tons of poultry, and 8 million tons of beef), 131 million tons of milk, and 6 million tons of eggs. Pork meat consumption thus explains the high volume of pork

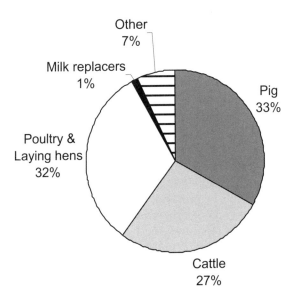

Figure 3.1   Global production of feed, 2006

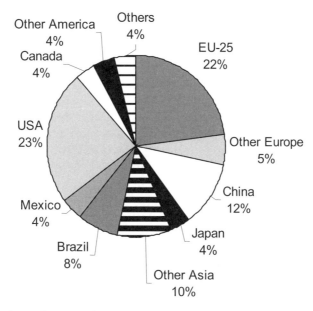

Figure 3.2 Global producers of feed, 2006

feed produced. However, owing to rising demand for poultry, the market for poultry feed shows the highest growth rates.[22]

In general, compound feeds are produced from a broad mixture of raw materials, vitamins and minerals. Owing to rising raw material costs, animal feed is an increasingly important cost factor, up to 80per cent of production costs of farm animals. Therefore, the exact calibration of energy levels of feed products by optimisation plays an increasingly important role in controlling the production costs of livestock. Feedstuffs are designed to achieve a pre-determined performance, so advanced methods formulate feeds according to the demands of the livestock farmer. For all species, the availability of carbohydrate sources such as wheat and the supply of protein crops (soybeans) are pre-requisites to ensure the production of feeds of both high quality and at competitive prices for livestock farmers.[23] In the European feed sector, protein sources such as soybeans need to be imported from overseas. Because they cannot be substituted with any other locally grown crop, the EU faces some severe challenges in the years to come.[24] From 2009 onward, new soybeans that have been genetically modified but not yet authorised by the European Food Safety Authorization (EFSA) will distort the protein supply, if there is no introduction of threshold levels that allow a certain 'pollution' level of soybeans by imported, EU-unapproved, genetically modified soybeans.

In addition to these constraints on the raw material supply side, the feed industry is implementing ever-increasing standards of quality and safety. The role of animal feed in the production of safe food is increasingly recognized worldwide.[25] Recent feed-borne scandals, such as the outbreak of bovine spongiform encephalopathy (BSE) in the United Kingdom, and other more common problems, such as salmonella and other micro-organisms, have encouraged the feed industry to take severe corrective measures and methods for their control, including the obligatory Hazard Analysis Critical Control Point (HAACP) assessment.[26] The compound feed industry is subject to a complex body of regulations, both at EU and national levels. Numerous certification schemes are in place to ensure higher degrees of feed and food safety (e.g., TrusQ, Safe Feed, QS, GMP). The Dutch Product Board Animal Feed (PDV) has introduced Good Manufacturing/ Managing Practice quality assurance (GMP) standards, which require that as of 2000, the quality of animal feed must be guaranteed. As a result, all ingredients for animal feeds must be GMP or equivalently certified.[27] Feed manufacturing not only reflects the demands of the farmer but also increasingly those of the entire food chain and society. Following the 'farm to fork' principle,[28] legislation helps improve trust in the quality of feed production for livestock and livestock products for consumers. Regulation affects production schemes and quality control systems and also controls the way feed products are marketed. For example, the 'open declaration' obliges all feed manufacturers to put all ingredients and the composition of nutrients on their labels. That requirement makes feed products easily comparable for customers and competitors alike. At the same time, health claims, comparable to the situation of foods, are prohibited, which makes product differentiation difficult to communicate.[29]

Business models and levels of vertical integration in the feed industry

To detail the role of the feed industry in the food chain, it is necessary to distinguish different strategies observable in the feed industry, because the role of the feed supplier and interfaces with other partners in the chain depend on them. According to the International Feed Industry Federation (IFIF), three generic strategies (Figure 3.3) determine production and delivery systems.

    This chapter focuses on strategy type (a), the independent feed supplier, which is still the most common form in Europe. In this business model, the feed industry is part of the entire food chain and has many interfaces with partners up and down the food chain, as well as with related industries. This type is likely characterised by a transactional supply chain configuration.[30] In type (b), the cooperative structure, the feed company is jointly

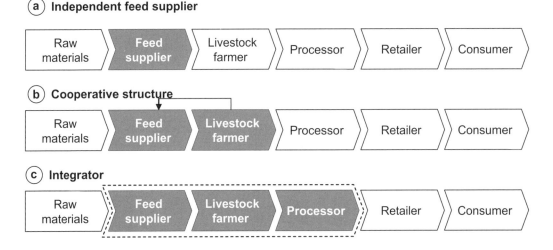

Figure 3.3   Different business models in the feed and livestock production industry

owned by the farmers. This model is quite widespread in the European feed industry. In contrast, the integrator type (c) is the dominant business model in chicken production throughout Europe.

As mentioned previously, the role of the feed industry in the chain of animal-derived foods depends on the generic business model, such as whether it acts on the basis of transactional or integrated relationships with its direct partners. However, the chain of animal-derived foods is not the only supply chain important to the feed industry. As illustrated in Figure 3.4, the feed industry is part of the interrelated chains of plant-derived foods, plant-derived fuels, and animal-derived foods. Because they all draw from the same raw materials, interdependencies are strong[31] and must be taken into account when analyzing the feed industry.[32] From this perspective, the most important partner still is the individual livestock farmer, because farming remains the major customer base of the feed industry. Depending on the individual business model, there may be relatively strong ties to raw material processors and traders, as well as with the chemical industry for the supply of premixes, minerals, and vitamins. The food industry (plant-derived food chain) can also function as a supplier to the feed industry, because by-products such as wheat bran are valuable input factors.

In addition to these direct buyer–seller relationships, the feed industry contains indirect relationships with food processors that deal with animal-derived foods. For example, the effect of certain feeding strategies on the constitution of the carcasses of pigs, cattle, or broilers and the milk fat composition of dairy products or the quality of eggs can be further investigated. Recently, more information from slaughterhouses or dairy companies gets evaluated and used to optimise a specific feeding program. Even

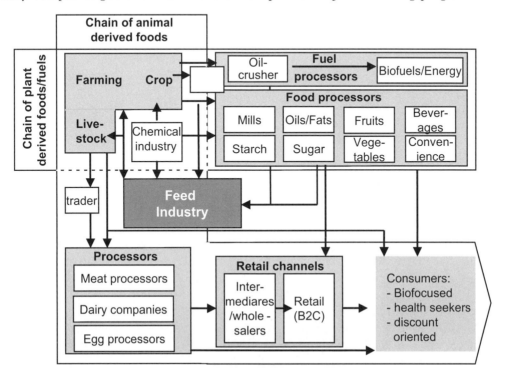

Figure 3.4   The role of the feed industry in the supply chain

though more efficient animal production remains the primary goal of innovation, there is increasing interest in carry-over effects from animal feed (e.g., vitamins, fatty acids) in consumer products.[33]

Major Challenges for the Feed Industry: Need for Market Orientation

Traditionally, the feed industry has seen its responsibility to be assuring supply with feed and, in most cases, other agricultural input factors. Therefore, its MO was limited to immediate customers, namely, livestock and crop farmers. Its MO in terms of orienting all activities with a view to the entire value chain was not necessary. The industry is thus more supplier market oriented than consumer or value chain oriented, because raw material make up 80 per cent of the costs of goods sold, and farmers are very price sensitive. In the case of easily comparable products, such as finisher diets for pigs, compound feed has similar characteristics to those of commodities. Because commodity production represents the opposite of MO,[34] the limited degree of MO can easily be explained.

However, the role of the feed industry has changed. It is part of the entire food chain but also has undergone many changes triggered by increased quality control systems. In addition, the feed sector has become more knowledge-intensive and offers more possibilities for product differentiation, resulting in a greater need for MO. Moreover, MO has become more important because consolidation processes, on both the farm level and the feed compound producer level, have led to increased competition. New forms of customer loyalty programmes, including a higher degree of services to farms and consulting offers, have become increasingly important. As Table 3.1 shows, the feed industry faces different challenges along the value chain, which can be distinguished as follows:

(A)   Raw material and feed production-related challenges.
(B)   Farm-level–related challenges.
(C)   Consumer market-related challenges.

As illustrated in Table 3.1, raw material supply and quality control play very important roles. For some species (especially broilers), access to raw materials that have not been genetically modified (GM) (e.g., soybeans, which according to legislation (EC) No 1829/2003 and (EC) No1830/2993 need not to be labelled) becomes crucial. This demand is triggered especially by retailers that want to label animal-derived products as GM-free, but the label can be issued only if the animal has been fed on a GM-free diet. This GM-free market segment has been evolving mainly in the broilers market, but it puts additional constraints on production processes and quality controls in the feed industry, because a completely separate production plant for GM-free feed would be required to reduce the risk of contamination.

Considering the overall protein supply, with respect to the European feed manufacturers, this challenge may be even more pressing, because EU-unapproved, GM soybeans will enter the EU. The first EU-unapproved GM soybean likely to be cultivated in the United States and exported to the EU is MON 89788, a replacement for the Roundup Ready soybean 40-3-2, which in 2005 was planted on approximately 60 per cent of the

Table 3.1   Challenges for the feed industry

(A) Raw material and production-related challenges	(B) Farm-level–related challenges	(C) Consumer market-related challenges
Raw material supply: • Agricultural raw materials are becoming increasingly volatile • Shortages of certain minerals (e.g., feed phosphates) • Unapproved new soybeans leading to a potential protein shortage in the EU	**Customer structure:** • Consolidation process of farms: Customer-structure becomes important (high degree of farms with potential to survive is needed) • Vertical integration plays a dominant role • Buying centres of farmers with increased bargaining power are increasingly widespread	**Control of retailers:** • Retailers are increasingly active in controlling the supply chain (e.g., labelling 'GMO-free') • In certain cases retailers even establish quality schemes for specific production programmes (e.g., obligatory feeding scheme for dairy farmers)
GMO-free products: • Access to GM-free raw materials • Involving NGOs to ensure that the feed industry follows environmental guidelines (responsible soy programme)	**Customer loyalty:** • Customer loyalty programmes from different feed suppliers • Relationship marketing becomes increasingly important	**Consumer behaviour:** • Consumers are very price sensitive but at the same time are postulating animal-welfare standards, which lead to increasing production costs
Quality control during processing and production: • Separate production facilities for each species • Within a species, separate production facilities for GM and non-GM feed • Increased safety demands create higher production costs	**Knowledge base:** • Farmers are increasingly knowledgeable, which requires a knowledgeable sales force • Prevention of diseases, the contribution of feed to animal health and welfare require feed companies to build related knowledge	**Outbound quality control:** • Quality control is increasingly important for feed production • Ensuring compliance with environmental concerns like greenhouse gas impacts of feed (e.g., reduction of methane emission from dairy cows)

global soybean area. An authorisation dossier for MON 89788 was submitted to EFSA in November 2006 and is now subject to evaluation. Thus, EU-unapproved GM soybeans will be mixed with approved soybeans and exported together before the authorisation procedure is finished at the EFSA level, which means EU livestock production will be severely challenged by a shortage of EU-approved protein sources. Following the worst-case scenario outlined by a study carried out by the EU Commission, EU pork meat production would drop up to 39 per cent in 2009 and 2010.[35] At the same time, a sharp increase of the EU price level would attract higher imports from overseas (fed EU-unapproved GM soybeans). In the long run, to avoid the negative perceptions of GM crops, the feed industry will need to adopt a non-GM certification scheme (e.g., Cert-ID's non-GMO certification) to ensure trust among consumers. This example reveals just some of the challenges the EU feed industry faces on the input side of the value chain. These challenges are rather EU-specific, because the acceptance of GM foods is generally higher among consumers in the United States, Latin America, and Asia.[36]

With regard to farm-level challenges, the feed industry is challenged by a change in its own customer structure, as farms go through consolidation processes. Competition among the remaining large farms will increase, and customer loyalty built through long-term customer relations may become less important. Furthermore, an increasing share of the market is not accessible because of the greater use of vertical integration models (especially in the broiler market), which include all steps of meat production from raw material supply and feed production to slaughtering. Hence, customer loyalty programmes must be developed for increasingly knowledgeable customers, and salespeople for the feed companies must have increased training so they can deliver knowledge about livestock production to farmers.

The consumer market-related challenges include increasing retailer control over the labelling of food products. In addition to the standardised, well-established quality control systems and labels (e.g., GMP+, QS, TrusQ), retailers and consumer food companies seek new ways to differentiate products. Usually the feed industry does not play a role in these developments. However, in some cases, product development includes the feed industry, such as when the Dutch dairy company Campina launched an innovative milk product that was rich in unsaturated fatty acids. The innovation is based on a change in cow feeding schemes. Therefore, feed companies need to foresee changes at the consumer base and then get involved in systemic innovations that involve different partners of the chain.[37]

Even though consumer behaviour and trends at consumer level seem far away from the feed company's perspective, they increasingly should be translated into feed-based innovations. In the past, opportunities for reducing production costs by enhancing the feed conversion ratio drove innovation. But in the future, certain quality attributes of animal-derived feed products and opportunities to influence these attributes through feeding schemes likely will become increasingly important. Moreover, as the public grows more concerned about the environmental impact of feed (e.g., methane emissions of cows), they have triggered new feed-related R&D programmes to address these issues. Some feed programs attempt to control environmental pollution by reducing certain constituents of excrements (RAM-reduced ammonia feeds). Furthermore, the public is increasingly interested in the role of feed with respect to animal welfare.[38] Finally, it is important for the feed industry to anticipate upcoming legislation that will regulate food safety and environmental constraints.

The challenges described in Table 3.1 indicate that the feed industry has different issues to address and increasingly is moving from a commodity to a more market-oriented industry. This development necessitates a strong orientation toward supply, toward the customer, and toward the customers´ customer markets. The feed industry is obliged to employ a MO that include s the entire food supply chain, but how can such a 'chain-overarching' MO be developed?

Adopting Market Orientation in the Feed Industry

GENERAL ASSESSMENT OF THE FEED INDUSTRY

Considering the three major challenges within the feed industry (Table 3.1), the question becomes how different firms in the feed sector might respond to them with a MO. As detailed in Section 3, the feed industry contains three different business models (see Figure 3.3), which represent responses to the challenges on the supply, farm, and consumer levels. The integrated type, which pools together everything from feed supply to slaughter and consumer products manufacturing in one company, seems to struggle with fewer difficulties anticipating changes at consumer level, because its customer interface occurs at the retail level. However, this position differs in the classical cooperative model and even more in the stand-alone, non-integrated feed supplier business model. The non-integrated feed supplier, which is the prevalent form, is challenged by its knowledge gaps in many areas related to its direct market and competitive environment. Across the entire supply chain, this business model seems to experience the most challenges with regarding to translating trends at the end consumer level back into feed developments. The following case study explores how a family-run business that is not integrated but instead concentrates on feed supply has dealt with current challenges and developed itself to a market-oriented company.

ASSESSING MARKET ORIENTATION AT THE BRÖRING GROUP

The BRÖRING Group is a family-held feed producer that was founded in 1891 as a local feed mill and grain trader in Dinklage in northern Germany. Despite a traditional commodity-based product production approach, the company produces 1.2 million tons of feed per year, based on a detailed consulting and service concept that includes farmers but also increasingly the entire chain. The company has undergone a significant change from just feed supply to the supply of both feed and knowledge about animal nutrition, housing, animal health, and environmental measures.

Market-related knowledge results from strong customer relationships. Furthermore, information about legal developments can be absorbed by playing an active role in the German feed producers association (DVT) (see Figure 3.5). Because relationships with downstream partners in non-integrated feed companies are not as tightly coupled as they would be in vertically integrated hierarchies, buyer–seller relationships were of tremendous importance for BRÖRING if it hoped to be market-oriented and generate relevant information. On the raw material supply side, supplier relationships generate relevant market information and translate it into feed calibrations. Therefore, a close collaboration among the purchasing, production and sales department was crucial and the basis for market-oriented feed production – especially for feed products that offer fewer opportunities for product differentiation and thus are relatively comparable to the farmer, because pricing possibilities in the market are determined by raw material prices. In terms of adopting a MO for the immediate market, the company's strong buyer–seller relationships and efficient internal communication processes were key, in line with Kohli and Jaworski's MO dimensions.[39]

For the immediate customer base of BRÖRING, the livestock farmers, the company developed strong market capabilities by implementing a customer-relationship management programme and a wide array of services. Starting as a rather reactive commodity supply

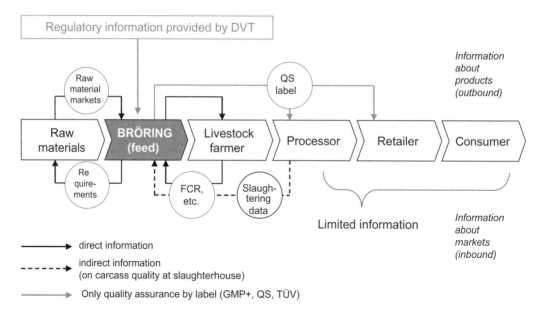

Figure 3.5 Market information from the feed industry's perspective

organisation, the entire company underwent a change to become more service-oriented. For example, in addition to its core business, it offers consultancy services for farmers, such as hygiene programmes, advice in livestock housing systems, individually calibrated feed compositions that fit other raw materials a farmer may have, knowledge transfer in piglet and sow nutrition, and so forth. This service also encompasses joint applied research with customers to calibrate the feed conversion ratio (FCR) and evaluate the efficiency of new feed products. In terms of Kohli and Jaworski's definition of MO,[40] BRÖRING has developed the ability to react to its market knowledge. This capability is especially important in its efforts to address challenges at farm level, because consultancy to an ever-shrinking customer base (due to consolidation) can encourage long-term customer relationships and customer loyalty among the A-customers who offer the greatest future potential.

However, at the level of the market for animal-derived foods, there has been very limited development of the three dimensions of MO, and access to consumer market-related knowledge is rather indirect and constrained. The reason likely pertains to missing market-related absorptive capacity.[41] Therefore, not only market knowledge generation but also its distribution in the company and the ability to react on it remains rather delayed. To develop a system of 'consumer-responsive agriculture,' the feed suppliers that are not vertically integrated need to establish close links with downstream partners. In recognising this challenge, BRÖRING joined an integration system that produces feed for broilers. The feed mill is jointly owned by the food processor, slaughterhouse, broiler farms, and BRÖRING. In this consortium, information from retailers is more accessible, because the food processor shares it with the consortium. This situation differs entirely from pig production, which shows little tendency to integrate. Input suppliers like BRÖRING still receive rather weak signals from the consumer. The vast potential for product differentiation induced by different feeding schemes – and the possible role of the feed industry in adjusting feeding strategies to customers' needs – is not yet in place.

Conclusions

The feed industry, a partner at the very front of the entire food chain, is increasingly challenged by not only raw material shortages but also increasing expectations of food safety from retailers and consumers. In turn, MO has become crucial for this industry and encompasses three levels: raw materials (supplier market), farmers (direct customers), and end-consumers (customers' customer market). As discussed in Section 4, the feed industry faces different challenges related to these market levels (see Table 3.1). To respond to supplier market-related challenges, such as raw material shortages, companies require close buyer–seller relationships for all purchasing processes. As Martin and Grbac state, stronger supplier relationships positively affect MO, and this claim holds in the feed industry too. The numerous certification schemes foster reciprocal investments between the feed industry and its raw material suppliers. Likewise, the feed industry has improved its MO tremendously, from 'tossing products over the fence' to delivering highly specialised, customised feed solutions for individual farms, as BRÖRING is doing. In this case, MO consists of a high degree of services and sophisticated forms of customer relationship management, which allow for information gathering, distribution of information inside the company, and response capabilities. The three characteristics of MO, as described by Kohli and Jaworski,[43] thus are present in feed industries and their immediate customer base.

However, at the third level of MO, which features the consumer level and an entire value chain orientation, the feed industry is not yet very advanced, and consumer trends still seem far away from daily business. In this respect, MO differs according to the business model of the feed company (integrators vs. single partners in the chain). The flow of information may be easier in integrated forms. The supply chain configuration in agri-chains plays an important role for developing MO. For a feed company that acts separately as a partner in the chain and focuses on relations with farmers, the MO of the entire chain is a huge challenge. Chain MO also seems especially crucial for an independent, not integrated feed producer that, in contrast with an integrated system, does not automatically have access to relevant information from the consumer market. A feed manufacturer that is not integrated therefore should be especially pro-active in its MO. The more partners there are in the downstream chain, the more important it is for a firm to move from a market to a chain orientation.

Regarding the traditional business model of a feed company supplying bulk animal feed for livestock production, no direct link appears with consumer goods companies, retailers, and the consumer. These gaps help explain why consumers have only learned about this industry recently through bad news such as food scandals. Nevertheless, feed and food-related safety crises (especially BSE and dioxin) offered particularly important impulses to enhance existing quality programs. The integration of the HACCP and GMP+ standards and upstream extension of the quality assurance to all suppliers of feed ingredients has resulted.[44] But quality assurance is not MO, because quality and safety are basic pre-requisites for successful marketing of any kind of product. The next step for the feed industry thus requires moving from quality assurance toward consumer responsiveness. The ways to enforce this move in practice and underline it with theory create interesting questions for agri-business research (as suggested by the newly founded Homer Nowlin Chair of Consumer Responsive Agriculture at Michigan State University).

The role of the feed industry, located as it is at the very front of the chain, seems like an interesting topic to explore further.

References

1. Bröring, S. (2008), 'How systemic innovations require alterations along the entire value chain-the case of animal derived functional foods', *Journal of Chain and Network Science*, vol. 8, no. 2, pp. 107–19.
2. Beverland, M.B. (2005), 'Creating value for channel partners: The Cervena case', *Journal of Business and Industrial Marketing*, vol. 20, no. 3, pp. 127–35; Manson, K., Doyle, P. and Wong, V. (2006), 'Market orientation and quasi-integration: Adding value through relationships', *Industrial Marketing Management*, vol. 35, pp. 140–55.
3. Narver, J.C., Slater, S.F. and Douglas, L.M. (2004), 'Responsive and proactive market orientation and new product success', *Journal of Product Innovation Management*, vol. 21, no. 5, pp. 334–47.
4. Ibid., p. 242.
5. Drucker, P. (1954), *The practise of management*, New York; Levitt, T. (1960), 'Marketing myopia', *Harvard Business Review*, vol. 38, no. 4, pp. 45–57; McNamara, C.P. (1972), 'The present status of the marketing concept', *Harvard Business Review*, vol. 36, no.1; Shapiro, B.P. (1988), 'What the hell is market orientation?', *Harvard Business Review*, vol. 66, no. 6, pp. 119–26; Kohli, A.K. and Jaworski, B.J. (1990), 'Market orientation: the construct, research propositions, and managerial implications', *Journal of Marketing*, vol. 54, no. 2, pp. 1–18; Day, G.S. (1999), 'The capabilities of market-driven organizations', *Journal of Marketing*, vol. 58, no. 4, pp. 37–53.
6. Beverland, M.B. and Lindgreen, A. (2007), 'Implementing market orientation in industrial firms: a multiple case study', *Industrial Marketing Management*, vol. 36, no. 4, pp. 430–42.
7. Kohli and Jaworski, op. cit.
8. Jaworski, B.J. and Kohli, A.K. (1996), 'Market orientation: review, refinement and roadmap', *Journal of Market Focused Management*, vol. 1, no. 2, pp. 119–36.
9. Narver, J.C., Slater, S.F, and Tietje, B. (1998), 'Creating a market orientation', *Journal of Market-Focused Management*, vol. 2, no. 1, pp. 241–55.
10. Lukas, B.A. and Ferrel, O.C. (2000), 'The effect of market orientation on product innovation', *Journal of the Academy of Marketing Science*, vol. 28, no. 2, pp. 239–47.
11. Martin, J.H. and Grbac, B. (2003), 'Using supply chain management to leverage a firm's market orientation', *Industrial Marketing Management*, vol. 32, pp. 25–38; Hsieh, Y.-C., Chiu, H.-C., and Hsu, Y.-C. (2008), 'Supplier market orientation and accommodation of the customer in different relationship phases', *Industrial Marketing Management*, vol. 37, pp. 380–93.
12. Lee, H.L., Padmanabhan, V., and Whang, S. (1997), 'The bullwhip effect in supply chains', *Sloan Management Review*, vol. 38, no. 3, pp. 93–101.
13. Hingley, M. (2001), 'Relationship management in the supply chain', *International Journal of Logistics Management*, vol. 12, no.2, pp. 57–71.
14. Manson et al., op. cit.
15. Matsuno, K. and Mentzer, J.T. (2000), 'The effects of strategy type of the market orientation performance relationship', *Journal of Marketing*, vol. 64, no. 4, pp. 1–17.
16. Manson et al., op. cit.
17. Webster Jr., F.E. (1992), 'The changing role of marketing in the corporation', *Journal of Marketing*, vol. 56, no. 4, 20–38.

18. Beverland and Lindgreen, op. cit.

19. Cohen, Wesley M. and Levinthal, Daniel A. (1990). 'Absorptive capacity: a new perspective on learning and innovation', *Administrative Science Quarterly*, vol. 5, no. 3, pp. 128–52.

20. Ibid.

21. FEFAC, Federation Europeenne des Fabricants d'Alimentation composes pour animeau (2007), *Statistical Yearbook 2006*, Brussels.

22. Ibid.

23. FAO, WHO (2002), 'Protein sources for the animal feed industry', *Expert Consultation and Workshop*, 29 April–3 May 2002, Bangkok.

24. Krüsken, B. (2008), 'Crash test dummies.', *Feed Magazine*, vol. 7–8, pp. 6–7.

25. FAO, WHO (2007), 'Animal feed impact on food safety', *Report of the FAO/WHO Expert Meeting*, FAO Headquarters, 8–12 October 2007, Rome.

26. Feil, A. and Jansen, H.-D. (2003), 'HACCP in der Futtermittelindustrie–Ein Leitfaden für die Nutzung', *Feed Magazine*, vol. 6, pp. 178–88.

27. Product Board Animal Feed (2006), General introduction to the GMP⁺ Certification Scheme in the Animal Feed Sector, The Hague, Netherlands.

28. European Commission, DG Press and Communication (2004), *From Farm to Fork: Safe Food for Europe's Consumers*, Brussels.

29. Bröring, op. cit.

30. Webster, op. cit.

31. Lenk, F., Bröring, S., Herzog, P., and Leker, J. (2007), 'On the usage of agricultural raw materials–energy or food? An assessment from an economics perspective', *Biotechnology Journal*, vol. 2, no. 12 (Special Issue on "Energy Production"), pp. 1497–1504.

32. Poignée, O., Hannus, T., Jahn, V., and Schiefer, G. (2005), 'Informationssystem QM-G–Schienennetz zur Gewährleistung der Rückverfolgbarkeit und Qualitätssicherung in der Futtermittelwirtschaft', in Schiefer, G. (Ed.), *Rückverfolgbarkeit und Qualitätsmanagement in der Getreide- und Futtermittelwirtschaft*, pp. 1–20.

33. Bröring, op. cit.

34. Narver and Slater, op. cit.

35. European Commission, DG Agriculture and Rural Development (2007), Economic Impact of unapproved GMOs on EU Feed Imports and Livestock Production, Brussels.

36. Huffmann, W.E. (2003), 'Consumers' acceptance of (and resistance to) genetically modified foods in high-income countries: effects of labels and information in an uncertain environment', *American Journal of Agricultural Economics*, vol. 85, no. 5, pp. 1112–1118.

37. Bröring, op. cit.

38. Makking, C. (2008), 'Dierenwelzijnindicatoren nodig voor classificatie', *De Molenaar*, vol. 111, no. 7, pp. 28–33.

39. Kohli and Jaworski, op. cit.

40. Ibid.

41. Cohen and Levinthal, op. cit.

42. Martin and Grbac, op. cit.

43. Kohli and Jaworski, op. cit.

44. Den Hartong, J. (2003) 'Feed for food: HACCP in the animal feed industry', *Food Control*, vol. 14, no. 2, pp. 95–9.

4 Business-to-Business Brand Orientation

BY MICHAEL B. BEVERLAND[*]

Keywords

branding, internal marketing, implementation

Abstract

This chapter proposes that managers attend to both the external identity of the brand and firm-level supportive practices that will enable staff to reinforce the brand promise. Drawing on the experience of Merino NZ, I identify six components of a brand orientation that can guide managers of other cooperatives. Thus, in this chapter I will identify the importance of brand orientation for agri-business, explore how Merino NZ implemented a brand strategy, identify six components of brand orientation, and suggest ways managers can enhance brand value.

Why Brands Matter in Business Markets

Interest in the strategic value of brands in business markets has grown in recent years as practitioners rethink ways to compete and build a sustainable competitive position in industrial markets.[1] Recent figures identify that almost 21 per cent of North American business marketers are focusing primarily on building brand awareness, up from 17.5 per cent in 2005.[2] Examples of successful branding strategies in business-to-business include Intel, Caterpillar, Bluescope Steel, UPS, IBM, De Beers, McKinsey and Company, and TNS Research. This emerging stream of research has identified that branding by industrial firms provides the basis for differentiation, results in price premiums, has a positive influence on buyer choice (for both new buys and rebuys), reinforces a positive reputation including perceptions of product and service quality, and assists salespeople in accessing customers and selling to them.

[*] Professor Michael B. Beverland, School of Economics, Finance and Marketing, RMIT University, GPO Box 2476V, Melbourne, Victoria 3001, Australia. E-mail: mbeverland@yahoo.com. Telephone: + 61 416 102 492.

Furthermore, research reveals that corporate reputation (built as a result of branding, among other things) is second only to price in influencing perceptions of value in business markets.[3] When brand equity is high, customers are often more prepared to pay a price premium for the product and more likely to engage in favourable word-of-mouth communications regarding the firm and its brands.[4] As well, building brands has often gone hand-in-hand with business-to-business firms' drive to reposition away from price competition or escape commodity cycles. For example, De Beers has developed a brand programme targeting business buyers to combat the threat of new entrants and synthetic diamonds. Likewise, Intel has sought to build equity through branding to position itself as the chip of choice among computer manufacturers and thus capture a greater share of the motherboard. In services, IBM (along with other business service providers) has attempted to brand itself as a 'solutions provider' to extract greater value from clients and position away from the image of a low-cost, information technology seller.

Despite the positive effects of branding in business markets, and the preceding examples of firms attempting to build a branded position, research provides little information about how industrial firms develop a brand orientation. Brand orientation is defined as 'an approach in which the processes of the organisation revolve around the creation, development and protection of brand identity in an ongoing interaction with target customers, with the aim of achieving lasting competitive advantages in the form of brand.'[5] Brand orientation goes beyond market orientation (which focuses primarily on adapting to the market, an approach that is believed to be problematic for brands that must attain a high degree of stability) and suggests that firms adopting a branding strategy must effectively 'live their brands' by developing a brand-supportive dominant logic. Such an approach is important, because a long-term approach to managing brands is fundamental to maintaining brand equity.[6]

Research supports the need to address this question, given the identified importance of brand-supportive processes and cultures to ongoing brand leadership[7] and the benefits gained from a logic whereby managers 'direct their actions and practices toward the development, acquisition and leveraging of branded products and services.'[8] Although research has examined some tactical aspects of brand orientation,[9] the strategic firm-level requirements of such an orientation remain under-researched. In practice, a brand strategy requires significant restructuring of the firm, resulting in the brand influencing all external communications and actions by the firm – as reinforced by Caterpillar's recent reorganisation, in which marketing staff have the final say on all marketing communications across the diversified group.

Recent theorising about corporate branding strategies also supports the need for cultural and structural support for brands (similar to a dominant logic or brand orientation),[10] as does research on branding in business markets,[11] though it remains silent on the important practical issue of how firms develop such an orientation and the nature of this orientation. I explore this issue in this chapter, with reference to a longitudinal case study – Merino NZ.

Turning Wool into a Brand: Background on Merino NZ

Merino NZ (MNZ), founded in 1995, represents all farmers of merino wool in New Zealand (700 growers in 2006) and is charged with industry development and building intangible capabilities (such as brands) for the New Zealand industry. A niche strategist,

MNZ accounts for just 2 per cent of the world's fine wool sales. Although it does not trade directly in product, it is the main front for the industry and plays a central role in liaising between sellers and buyers as part of developing a sustainable industry (MNZ has no statutory monopoly powers to force farmers to belong). In 1996, following extensive market research, the New Zealand Merino brand launched, targeted at the world's best cloth processors and leading fashion brands.

The 'Merino NZ' brand programme was officially launched globally to wool buyers in 1996. The development of this programme had not been without challenges, because the final programme fundamentally questioned 150 years of traditional practices. Many debates took place behind the scenes regarding whether to build brand equity with end consumers or business buyers and whether wool could ever be differentiated. On this last point, the debates were quite fierce, as many in the industry believed money would be better spent improving efficiency and production levels to increase returns and gain market power. Their view was that wool was intrinsically a commodity, and marketing could do little to change that fact.

In contrast, others believed that a continuance of past practices was the cause of their problems rather than a sound strategy. They argued that increasing efficiency only resulted in lower returns, and resource limitations made it difficult to increase production to the levels of large producing nations such as Australia and South Africa. Also, gaining greater power over wool buyers was based on a myopic view of the competitive marketplace because end users choose from many raw materials to produce final garments. In the past, price rises resulting from increased market power merely saw end buyers select alternatives such as cotton or invest in human-made materials. Finally, they argued that New Zealand Merino wool had a large number of product differences (resulting from different climactic conditions and investments in production technologies and farm practices) that gave them the ability to target manufacturers with greater customisation. Focusing a campaign on consumers that sought to raise awareness about the benefits of wool and its natural qualities would result in little long-term value if key business buyers did not support the programme. As a result, the brand programme was perceived as a new experiment for the industry, such that early successes would reinforce a commitment to industry-level change.

Prior to the brand launch, MNZ had undertaken significant market research among key buyers and fashion houses (i.e., lead users). Given the small production base, the MNZ brand was positioned as an up-market ingredient brand. Thus, MNZ would aim to co-brand with select buyers that had the desired up-market positioning to increase the equity of its own brand. Because there were only a few key wool buyers, brand-marketing programmes would need to be adapted to the needs of each buyer. Also, MNZ would need to develop programmes that supported other key buyers without an up-market position, because relatively little wool is needed to produce up-market men's suit, so other markets needed to be found and developed to ensure sustainability.

Results to date have been impressive. The brand is now the first choice for many leading fashion houses and for the wool processor Loro Piana. The product is a leading ingredient brand, valued by fashion houses and often featured in branded garments. Industry representatives, independent reports, and farmers confirm that prices are sustainable and that the industry is healthier than it has ever been. With regard to financial performance, on-farm returns have increased from $14 per hectare to $100 per hectare (this figure is important, because it affects the capital value of the farm and provides the measure farmers use to capture the benefits of brand investments). Although 45 per cent of growers use fixed-term contracts (the highest quality suppliers), auction-based prices for MNZ branded wool have risen from 840 cents per kilo to 1170 cents per kilo in the past three years (they were steadily improving prior to that), whereas the commodity price for non-branded wool has declined (400 cents/kilo to 350 cents/kilo for strong wool and 800 to 480 cents/kilo for medium strength wool over the same period). Although contract prices are difficult to gauge, MNZ reports suggest that sellers have gained a 50 per cent price premium over 10 years.

Business-to-Business Brand Orientation

I identified six components of a business-to-business brand orientation from the sampled case: a defined brand essence, top management support for the brand, a brand equity management system, a strategic approach to brand fortification, brand architecture, and a brand-oriented culture.

BRAND ESSENCE

A clearly defined brand position drives the strategic and tactical actions of MNZ, including partner choice, co-branding, public relations activities, and joint marketing activities. The firm's corporate brand is positioned as a high-end (top 5 per cent of fashion buyers), global, environmentally conscious, high-quality ingredient brand from New Zealand. The brand's role is to provide a point of difference for New Zealand Merino farmers that can be owned (i.e., protected by law) and used to deliver sustainable returns to the cooperative's membership. Urde[12] argues that a brand's essence (or mission) must address questions such as: 'Why does this brand exist? What does the brand stand for? Who is the brand? How is the goal of the brand to be achieved?' The MNZ brand answers these questions and guides the entire marketing programme in an attempt to convey a consistent brand image to consumers to maintain and enhance the existing position of the brand. For example:

> 'In the textile trade everyone knows who the top brands are and we want to be associated with them. In Japan, the wealthiest consumers go after the old recognized European brands, so that's what we unashamedly go for.' (marketing manager, MNZ)

This passage identifies the relationship between the brand's essence and customer selection in key markets. The choice of these partners both reflects and reinforces the brand's essence, thus helping MNZ achieve its desired position. Because MNZ actively selects a small group of customers to work with, MNZ limits its short-term growth to reinforce the brand position and build long-term value. Such a move lies in direct contrast with its commodity past, when undifferentiated products sold to all comers (Merino wool previously was part of the 'Wool Mark' programme, a quality mark that was widely circulated and provided little differentiation). To reinforce the brand's position further, MNZ has developed co-branding and joint promotional strategies, adapting to the needs of each target customers. For example:

> 'We recognize that at that top end of the market the production and manufacturing companies are different, so we do special things with each of them. Their "face" to the market is completely different from ours so we have to do unique things for each of them.' (CEO, MNZ)

This passage identifies an important challenge for business brand marketers: the need to adapt the marketing programme to each individual customer[13] and maintain consistency in the brand identity and image. To manage this challenge, MNZ deliberately developed an ingredient brand strategy to be targeted primarily at business customers (public relations activities that target lead buyers also help drive demand for garments made with MNZ-branded fibre among consumers). As a result, they add significant value to the customer without diluting their position, because the basic components of the ingredient brand programme remain the same, even though the materials are customised around the customer brand's position and imagery. The brand's logo ('Merino NZ') is sewn into the final garment (produced by business customers such as fashion houses) to indicate country of origin, quality, and exclusivity. One way of achieving this image is to invest in training materials for the customers' staff to reinforce the brand's essence. For example, MNZ produces CD-ROMs for use in staff training sessions to inform them about the product, its origin, and the brand values.

Two other strategies used by MNZ also reinforce the brand's essence. First, to overcome resource limitations (i.e., small promotional budget), MNZ developed an annual 'Fleece of the Year' competition. This competition involved scientifically grading various farmers' wool to identify the finest wool, then auctioning off a single bale of this wool to the highest bidder. The resulting price would gain widespread publicity and drive consumer and fashion house demand for the cloth (suits made out this material were specially branded with 'Fleece of the Year' promotional material). Such a strategy also reinforced the exclusivity of the brand, as only 200 men's suits could be produced from this bale. Gifts of cloth made from this bale were given to high-profile heads of state (e.g., Presidents Clinton and Bush) on their visits to New Zealand, further reinforcing perceptions of quality, exclusivity, and building in unique brand associations. For example:

'We conducted a special competition/auction for the growers. We invited buyers from Europe and Asia to buy the finest bale and that got us very good press coverage. I think it's good for the growers too. Only a very small part of a grower's total clip is of this ultra fine quality (15 to 16 microns), but everybody wants to be the winner so this drives change in the industry. It drives people to make improvements in their farms and systems. The manufacturer gets his money back because he markets it very exclusively and each of his suits will have a special label and special marking on it.' (CEO, MNZ)

Second, joint marketing activities built excitement about the brand, raised awareness, added value to customer relationships, and reinforced the brand's essence. For example, MNZ's 1997 advertising campaign featured Miss World 1997, who modelled a range of 100 per cent Merino garments made by the UK fashion house John Smedley, against the backdrop of the South Island (reinforcing the country of origin). The company also sponsored a fashion show in London as part of its international marketing programme, attracting more than 100 fashion buyers and media. In 1998, Merino NZ launched 'The Merino Trail,' a tour aimed at business customers (the programme consisted of visits to 45 South Island high country farmers and 2 North Island farmers), and helped build emotional bonds between buyers and the brand. This strategy also reinforced the difference between MNZ (which identified individual farmers and the link between different climatic and farming conditions and unique wool characteristics) and other commodity sellers that sold large volumes of undifferentiated products.

TOP MANAGEMENT SUPPORT FOR THE BRAND

Top management plays an important role in the strategic management of a brand. The CEO and management team are responsible for continually driving organisation-wide support[14] and demonstrating passion for the brand. To build a credible brand and gain customer support, MNZ management needed to break the traditional practice of growers selling their wool on commodity exchanges and attempting to manipulate supply to push up prices. Research revealed that buyers required greater certainty of price and supply and that they operated on slim margins that could not easily be passed on downstream. As a result, MNZ sought to convince growers to sign fixed price and supply contracts with their buyers for five years. Estimated efficiency gains from one contract would increase profits by 10 per cent, and growers supplying the contract would receive prices 40 per cent above the season's average wool price. Because such contracts represented path-breaking change, even customers needed convincing that the contracts would be beneficial; one customer who signed a supply contract agreement with MNZ saw it as a 'gamble' initially but quickly became convinced of the benefits:

One of the benefits to growers is that it's a known price, so they aren't so much at the whim of the auction systems. It's a very bankable item for them as they can foresee what income they will get from the portion of their clip that is contracted. More importantly is allows growers to build an on-going relationship with the company.[15]

The involvement of senior management in marketing communications and relationship management (among other things; see the other sections of findings) helped demonstrate real commitment to the brand. The Merino Trail programme, Fleece of the Year competition, and other activities, such as working individually with farmers to form relationships and contracts, also helped ensure buy-in from the farmers. The combination of top-down support and bottom-up buy-in are essential for achieving lasting, market-oriented change.[16] Also, by working closely with downstream customers, MNZ gained customer support for the brand programme, thus ensuring greater loyalty and cooperation, factors that increase brand equity.[17] This equity was essential for creating a brand-oriented culture and ensuring ongoing support for the programme.

Finally, MNZ's management team supported the brand in other ways: aligning hiring goals with the brand's values, promoting long-term brand objectives at trade shows, and orienting reward systems around the brand (by tying farmer returns directly to increases in brand value). This involvement in the day-to-day activities of brand management by senior management and their ongoing passion helps ensure continued support for the programme and resources (farmers pay a compulsory levy on their sales to MNZ but ultimately vote on the amount of levy and MNZ's budget). However, displaying passion for the brand was not enough by itself; senior management also needed to demonstrate the effectiveness of their brand management programmes through the relationships between marketing activities and brand equity.

BRAND EQUITY MANAGEMENT SYSTEM

As part of a brand's long-term strategic focus, brand equity must be constantly monitored. By monitoring brand equity, it is possible to make tactical adjustments to the brand positioning programme to maintain the strategic thrust of the brand.[18] Monitoring brand equity involves building systems that ensure customer feedback about the brand and its relative position gets regularly collected and used to inform changes in the marketing programme. MNZ developed a brand equity management system that covered key aspects of relative brand position (formal tracking studies and qualitative feedback from buyers and industry commentators), product and relationship quality, and the links between marketing expenditures and brand equity increases and improved returns to farmers (income and capital value of the farm).

Research conducted by MNZ flowed back to individual sellers and customers to assist with relationship management and problem solving. For example:

'A key strategy of ours has been to bring the growers together with processors and buyers. No one (growers and buyers) knew that they could actually lay down their own specifications, to ensure their machines run at the absolute efficiency because before they were in the hands of middlemen who made the product to a certain price.' (CEO, MNZ)

As this passage identifies, research (in this case, both formal and informal) can enhance the consumer–brand relationship and adapt the marketing programme to the needs of customers. Much of the brand equity measurement system was focused on on-going adaptations to the product, delivery, and quality, which is not unusual for an ingredient brand. In relation to problems of product adaptation, research revealed that MNZ needed to change fundamentally how it produced and graded wool. The fineness of wool depended on animal stress levels, continued access to feed, and climate on each farm. The traditional means of producing wool involved shearing the sheep and bundling all the fleeces into large bales that were sent to a packinghouse. Here the wool was graded by 'feel' or hand by skilled 'wool classers.' The farmers then received payment for their wool at market prices. Owing to the small production of most farms, each farm's wool clip was blended with other wools and mixed with wool from other species to produce a final bale (Merino wool could add softness to the wool of other breeds). As such, the individual characteristics of each farm's wool were lost, and the ability to adapt the product to the needs of key buyers was limited.

The inability to trace a link between a farm and the raw materials also meant customers could not adopt preferences for certain suppliers or make individualised recommendations for improvements. Likewise, the lack of measurement systems for quality and the inability to track supply back to an individual animal meant farmers made culling decisions subjectively (usually by the way the animal 'looked') but could not make customer-driven improvements in practices.

In response, MNZ undertook radical changes in production, measurement systems, and basic research and development. It developed a product advancement centre (PAC) whose main job (other than R&D) was to measure scientifically the actual fineness of each individual fleece. To support the brand programme, allow for greater customisation and adaptation, set new industry standards, and drive future innovations, MNZ decided to bag each fleece separately after shearing and tag that bag to indicate the animal and farm of origin. This fleece's fineness was then scientifically graded, and the tag allowed customers to track the origin of the wool through the demand chain.

This shift had multiple benefits. First, customers could now trace their raw materials back to individual farms and even animals, allowing them to request supplies from preferred farms. Second, the same customer could provide feedback to individual farmers, allowing them to make specific on-farm improvements or investments. Third, end-users could build greater authenticity into their final brands by identifying a unique farm of origin in their promotional materials. Fourth, it provided farmers with scientific information that would allow them to make better culling decisions.

Finally, incremental improvements pertained to product quality. For example, based on customer feedback, MNZ developed a packaging programme that ensured all wool was packed in packaging made from natural fibres. Previously, wool had been packed in polyurethane, which often got mixed in with the wool. Because polyurethane cannot be dyed, the resulting cloth often had small specks or streaks in it. Such an action resulted in a new industry standard and further reinforcement of MNZ's 'tier 1' supplier status.

These measurement systems served to express the brand's essence and enabled management to retain supplier and buyer interest in the brand. This information was collected and communicated to members, some customers, and MNZ staff in the form of a brand equity report. Consistent with recommended best practice in this area, the report contained the results of regular tracking studies and six monthly brand audits (an assessment of key sources of brand equity from perspectives of both the firm and the consumer). These audits enabled brand managers to 'keep their fingers on the pulse,' ensuring more proactive and responsive management of the brand.[19] Depending on the outcome of the brand audit and the alignment of consumer and firm brand meaning, the brand's position may be manipulated by adding new associations, strengthening existing associations, or weakening existing associations. Some of the programmes, such as the Fleece of the Year competition, Merino Trail, and quality adjustments, resulted from tracking studies and audits. The audits enabled on-going improvements to the brand and assisted in strengthening or fortifying the brand's position.

STRATEGIC APPROACH TO BRAND FORTIFICATION

Brand fortification refers to the use of an existing brand name to introduce new products (through line or brand extensions) that enhance the equity of the parent brand.[20] Fortification enhances brand equity through either the reinforcement of existing sources of equity and/or the addition of new associations to the brand.[21] Although fortification may be used to enhance and maintain equity, it is important not to extend too far and dilute the equity of the parent brand. If new products are introduced under an existing brand name and no effort is made to provide marketing support for the parent brand, the brand equity of the parent brand may be damaged. Therefore, brand and line extensions should be carefully thought out in terms of the brand's overall strategic direction. Keller advises that the overall knowledge buyers have of the brand be taken into consideration in evaluating extension potential.[22]

The MNZ brand was a corporate brand with several brand extensions. Some of these extensions deliberately reinforced the brand position. For example, the following passage identifies the motivation for developing the Zealander brand extension for the lead buyer and cloth maker (which supplies cloth to fashion houses) Loro Piana.

'One of the most prestigious vertical spinner and weavers in the world is a company called Loro Piana. We approached Loro Piana early on because they produced fabrics for 180 years. We went to them with our story and told them what we were trying to achieve and so on, and they decided to do some trials and make 100 per cent New Zealand Merino woven fabrics. That partnership resulted in a product called 'Zealander®', which is a brand name we gave to

a range of 100 per cent New Zealand Merino, and as a result they purchase somewhere in the order of 4–5 per cent of our total clip. The thing about Loro Piana is by getting with them very early, they are opinion leaders, if they do something others follow, and so by them getting into bed with us, and being dedicated to New Zealand Merino it means that other textile companies start sitting up and taking notice of New Zealand Merino.' (Marketing manager, MNZ)

The brand essence thus drove partner selection and determined the choice to develop an exclusive brand extension with this critical lead user. This extension provides the buyer with a further point of exclusivity that it can use with its downstream customers (e.g., fashion houses). As well, Loro Piana, which has a larger marketing budget, can promote this product with advertising, thus overcoming the resource limitations faced by MNZ. The extension helps fortify the brand by reinforcing relationship quality, quality associations, exclusivity, and country-of-origin image and by raising awareness among important fashion houses and their consumers. Several other extensions had this same effect.

The strategic approach to brand fortification was further reinforced by MNZ's solution to another key problem. MNZ represents all Merino growers and must therefore create viable markets for all wool, regardless of the level of fineness. Due to the relationship among climate, feed availability and quality, and fleece quality, not all farmers can deliver the finest quality wool. Therefore, MNZ developed separate brand programmes, often partnering with well-known brands to develop new branded products lines (e.g., blends between wool and fur, denim and wool). For example, Just Jeans (an Australian retail chain of mid-range fashion shops aimed at younger consumers) worked with MNZ to develop *Denimwool*, a blend of cotton and 19.5–micron New Zealand Merino wool.

Key buyers often have their own brand extensions that require raw materials, but they do not fit the desired position of MNZ. In these cases, MNZ adds value to the relationship if it can provide a total solution to the customer. For example, one Japanese buyer with two ranges (an upper range of 15,000 suits and a lower brand that sells in excess of 20,000 suits a season) was allowed to use the MNZ brand in the upper range but not in the lower range, even though this product is made of 100 per cent New Zealand Merino wool. However, MNZ developed joint promotional material with this customer for both lines to assist it in the marketplace. All of this effort helps the company increase its customer base (through mutual adaptation), without damaging its perceived positioning.

In both cases, these extensions and strategies do not undermine the MNZ brand's essence, but they do meet the growth needs of the firm and reinforce customers' perceptions of MNZ as an innovative market leader that solves key problems. This strategy, plus the role that brand essence plays in customer selection and marketing promotions, reinforces both a strategic approach to fortification and the claim that MNZ leverages its brand architecture.

BRAND ARCHITECTURE

These last two sections primarily draw on the data presented previously. Brand architecture is an important component of MNZ's strategy, and its overall brand orientation, focused on adopting an appropriate brand structure (or architecture), is important to the enhancement of the brand's strategic focus and ongoing leadership. Within a brand's architecture, it is necessary to specify the roles of each brand to realise the relationship among the brands in a brand portfolio. These role relationships help guide the brand portfolio structure of an organisation and maintain the strategic focus for any given brand.[23]

Previous sections identified how MNZ uses the corporate brand, brand extensions, co-brands, and joint marketing to ensure its ongoing leadership (e.g., providing a total solution involving the corporate brand and non-branded ingredients, supported with joint marketing materials focusing on country of origin) and the development of extensions such as Denimwool and Zealander. Laforet and Saunders[24] argue that though firms tend to favour one brand structure over another, companies largely use a hybrid approach to organise their brands, as confirmed by MNZ's use of corporate brands, endorsed brands, co-brands, and a house of brands.[25] A hybrid approach appears appropriate for many industrial firms, which often deal with few customers and must adapt their marketing approach to meet each customer's complex needs.

BRAND-ORIENTED CULTURE

These five factors likely reinforce a brand-oriented culture. For example, the brand's essence reflects statements and beliefs about what is important and valuable, two essential elements of a culture. Top management effectively models desired behaviour and therefore plays a critical role in reflecting and enforcing cultural norms. The alignment of resources, strategies and tactics, and measurement systems is essential for reinforcing commitment to cultural values. A brand-focused culture is fundamental for maintaining a strategic brand focus. Finally, brand orientation requires an alignment between culture and structure, as addressed by the brand's architecture, supportive systems, and clearly brand management defined roles.

A brand-oriented culture also is demonstrated by the constant improvements and adaptations undertaken by MNZ to enhance brand equity and customer satisfaction. The case description identified the importance of forming relationships with key customers and then reinforcing those relationships for the overall success of the brand programme. Such capability is underpinned by a more fundamental ability to work with customers. The development of relationship formation and management capabilities that are network focused supports the brand programme because it allows greater awareness of and buy-in to the programme across a complicated 11-member supply chain (all of whom are necessary to transform raw wool into cloth and end products).

The case description also identified customer-driven changes in the process, packaging and product that enhanced the brand and built a superior market position (i.e., competitive advantage). MNZ invested significantly in basic research and development to ensure more breakthrough process and product innovations and thus enhance the brand position and ensure ongoing leadership. Such innovative capabilities were network focused, in that they sought ways to improve quality throughout the entire process. MNZ called this programme 'From Paddock to Garment.' Such a formalisation of core values in strategic documents that drive action reflects the wider brand-oriented culture.

Discussion

These findings contribute in a number of ways. First, to my knowledge, this study is the first investigation of strategic brand orientation in business-to-business markets.

As identified in the introduction, the benefits of branding are becoming apparent in industrial markets, yet less is known about how such firms capture brand benefits over the long term. I identify six interlocking components of business-to-business brand orientation that I posit can assist business marketers in growing and maintaining brand equity in the long run. Consistent with research on market and brand orientation, I suggest that structural, systemic and cultural support for the brand at the firm level will result in improved tactical outcomes.[26] Further research is needed to support this prediction.

The findings also support Urde's proposition that brand orientation must go beyond market orientation, because successful brands involve driving the market and stability of image.[27] The six components identified in this case include aspects of market orientation (brand tracking and auditing and a brand-oriented culture), as well as elements that marry the needs of the market with the firm's resources (brand essence and strategic brand fortification). This approach is likely appropriate for leader or niche brands but may be less so for follower brands. Again, research is needed to test the boundary conditions of this proposed model.

This chapter also adds to knowledge about brand orientation. Research into brand orientation has tended to favour a focus on the behavioural aspects of this construct, which usually considers the benefits of an integrated tactical brand programme for consumer brands.[28] This preference parallels research into market orientation that focuses only on the benefits of customer and competitor orientations and cross-functional coordination rather than the cultural aspects of being customer oriented. Considering the emphasis placed on strategic management of the brand, brand supportive systems, and capabilities by leading brand researchers, examining both aspects of a brand orientation provides an important extension to understanding of how to implement brand strategies in business (and consumer) markets.

Implications for Managers

The findings have several implications for managers. First, they suggest business-to-business brands benefit from supportive systems, structures, and cultures. Marketing managers in such firms must gain support for organisation-wide initiatives to support and protect the brand. Caterpillar is one example of such an approach, where a senior manager has final say about all brand initiatives within the global group. Crucial to gaining this support will be the development of effective measures of the return on brand investments. Second, the findings suggest that for firms embarking on branding strategies, the degree of change required will go beyond the tactical level and thus must involve change at the firm level. Third, the findings suggest that brand orientation in business-to-business firms involves managing network relationships, because suppliers and customers may need to cooperate to deliver brand benefits. Fourth, the results suggest that brand essence in an industrial context (depending on network position and market conditions) must allow for both stability of image and flexibility of delivery, because many business customers demand customisation. This need in turn suggests that best practices from the business-to-consumer realm may not transfer completely to the business-to-business world.

References

1. Bendixen, M., Bukasa, K., and Abratt, R. (2004), 'Brand equity in the business-to-business market', *Industrial Marketing Management*, vol. 33, no. 5, pp. 371–380; Beverland, M.B., Napoli, J., and Lindgreen, A. (2007), 'Global industrial brands: A framework and exploratory examination', *Industrial Marketing Management*, vol. 36, no. 8, pp. 1082–1093; Kotler, P. and Pfoertsch, W. (2007), 'Being known or being one of many: The need for brand management for business-to-business (B2B) companies', *Journal of Business and Industrial Marketing*, vol. 22, no. 6, pp. 357–62; Lamons, B. (2005) *The Case for B2B Branding*, Thomson: Sydney; Low, J. and Blois, K. (2002), 'The evolution of generic brands in industrial markets: The challenges to owners of brand equity', *Industrial Market Management*, vol. 31, no. 5, pp. 385–92; Mitchell, P., King, J., and Reast, J. (2001), 'Brand values related to industrial products', *Industrial Marketing Management*, vol. 30, no. 5, pp. 415–25; Walley, K., Custance, P., Taylor, S., Lindgreen, A., and Hingley, M. (2007), 'The importance of brand in the industrial purchase decision: A case study of the UK tractor market', *Journal of Business and Industrial Marketing*, vol. 22, no. 6, pp. 383–93.

2. *Marketing News* (2006), 'Marketing fact book', July 15, pp. 27–37.

3. Mudambi, S. (2002), 'Branding importance in business-to-business markets: Three buyer clusters', *Industrial Marketing Management*, vol. 31, no. 6, pp. 525–33.

4. Bendixen et al., op. cit.; Beverland et al., op. cit.

5. Urde, M. (1999), 'Brand orientation: A mind set for building brands into strategic resources', *Journal of Marketing Management*, vol. 15, no. 1–3, pp. 117–33; see pp. 117–18.

6. Aaker, D.A. and Joachimsthaler, E. (2000) *Brand Leadership*, The Free Press: New York.

7. Aaker and Joachimsthaler, op. cit.; Beverland et al., op. cit.; Hankinson, P. and Hankinson, G. (1999), 'Managing successful brands: An empirical study which compares the corporate culture of companies managing the world's top 100 brands with those managing outsider brand', *Journal of Marketing Management*, vol. 15, pp. 135–55.

8. Napoli, J. (2006), 'The impact of nonprofit brand orientation on organisational performance', *Journal of Marketing Management*, vol. 22, pp. 673–94.

9. Ewing, M.T. and Napoli, J. (2005), 'Developing and validating a multidimensional nonprofit brand orientation scale', *Journal of Business Research*, vol. 58, no. 6, pp. 841–53; Napoli, op. cit.

10. Balmer, J.M.T. (2001), 'Corporate identity, corporate branding and corporate marketing', *European Journal of Marketing*, vol. 35, no. 3/4, pp. 248–91; Brown, T.J., Dacin, P.A., Pratt, M.G., and Whetten, D.A. (2006), 'Identity, intended image, construed image, and reputation: An interdisciplinary framework and suggested terminology', *Journal of Academy of Marketing Science*, vol. 34, no. 2, pp. 99–106; Hatch, M.J and Schultz, M. (1997), 'Relations between organizational culture, identity and image', *European Journal of Marketing*, vol. 31, no. 5/6, pp. 356–65; Urde, M. (2003), 'Core value-based corporate brand building', *European Journal of Marketing*, vol. 37, no. 7/8, pp. 1017–1040; Van Riel, C.B.M. and Balmer, J.M.T. (1997), 'Corporate identity: The concept, its measurement and management', *European Journal of Marketing*, vol. 31, no. 5/6, pp. 340–55.

11. Beverland et al., op. cit.

12. Urde (1999), op. cit.

13. Ford, D. and Associates (2002) *The Business Marketing Course: Managing in Complex Networks*, John Wiley and Sons: Chichester.

14. De Chernatony, L. (2001) *From Brand Vision to Brand Evaluation: Strategically Building and Sustaining Brands*, Butterworth-Heinemann: Jordan Hill, Oxford.

15. *The New Zealand Farmer* (1007), July 24, p. 6.

16. Beverland, M.B. and Lindgreen, A. (2007), 'Implementing market orientation in industrial firms: A multiple case study', *Industrial Marketing Management*, vol. 36, no. 4, pp. 430–42.

17. Keller, K.L. (2003) *Strategic Brand Management: Building, Measuring and Managing Brand Equity*, Prentice-Hall: Englewood Cliffs, NJ.

18. Napoli, op. cit.

19. Keller (2003), op. cit.

20. Keller, K.L. (1999), 'Managing brands for the long run: Brand reinforcement and revitalization strategies', *California Management Review*, vol. 41, no. 3, pp. 102–24.

21. Park, W.C., Jaworski, B.J., and MacInnis, D.J. (1986), 'Strategic brand concept-image management', *Journal of Marketing*, vol. 50 Oct, pp. 135–45.

22. Keller (2003), op. cit.

23. Aaker and Joachimsthaler, op. cit.

24. Laforet, S. and Saunders, J. (1994), 'Managing brand portfolios: How the leaders do it', *Journal of Advertising Research*, vol. 34, Sept/Oct, pp. 64–76.

25. Keller (2003), op. cit.

26. Aaker and Joachimsthaler, op. cit.; Beverland and Lindgreen, op. cit.; Beverland et al., op. cit.; Napoli, op. cit.

27. Urde (1999), op. cit.

28. Ewing and Napoli, op. cit.; Napoli, op. cit.

5 *Improving Market Orientation in the Scottish Beef Supply Chain Through Performance-related Communications: The Case of the McIntosh Donald Beef Producer Club and Qboxanalysis*

BY PHILIP LEAT,[*] CESAR REVOREDO-GIHA,[†] AND BEATA KUPIEC-TEAHAN[‡]

Keywords

Scottish agriculture, performance-related communications, beef supply chain

Abstract

The reform of the Common Agricultural Policy in 2003, and particularly the introduction of the Single Payment Scheme, was intended to make farmers more market-oriented and competitive. In this context, the purpose of this chapter is to discuss whether

[*]   Mr. Philip Leat, Food Marketing Research, Land Economy and Environment Research Group, Scottish Agricultural College (SAC), Ferguson Building, Craibstone Estate, Aberdeen AB21 9YA, UK. E-mail: philip.leat@sac.ac.uk. Telephone: + 44 1224 711048.

[†]   Dr Cesar Revoredo-Giha, Food Marketing Research, Land Economy and Environment Research Group, Scottish Agricultural College (SAC), King's Buildings, West Mains Road, Edinburgh EH9 3JG, UK. E-mail: cesar.revoredo@sac.ac.uk. Telephone: + 44 131 535 4344.

[‡]   Dr Beata Kupiec-Teahan, Food Marketing Research, Land Economy and Environment Research Group, Scottish Agricultural College (SAC), King's Buildings, West Mains Road, Edinburgh EH9 3JG, UK. E-mail: beata.kupiec@sac.ac.uk. Telephone: + 44 131 535 4186.

performance-related communication strategies offer the possibility of improving farmers' market orientation. This possibility is studied through the case of McIntosh Donald, a beef processor located in the north-east of Scotland and a major red meat supplier for Tesco, and its encounters with Qboxanalysis, a performance-related communication system that was introduced by the processor to its Beef Producer Club members in March 2005. Results indicate that the enhanced communication strategy has the potential to increase farmers' performance and market orientation, not only through the use of the Qboxanalysis system but also through the Producer Club's wider activities. However, additional efforts are required to engage producers that are less proactive.

In this chapter, we will first present a brief background of the marketing of cattle in Scotland and the challenges presented by the 2003 reform of the Common Agricultural Policy. Second, we discuss whether performance-related communication strategies offer the possibility of improving farmers' market orientation. Third, we study the case of Qboxanalysis, a performance-related communication system introduced by McIntosh Donald, a Scottish beef processor, with the assistance of Tesco. Fourth, we present conclusions summarising the benefits and challenges of the performance-related communication system for different supply chain stakeholders.

Introduction and Background to the Research

Since it came into force in 1962, the European Union's Common Agricultural Policy has supported increases in farm production, thereby steadily generating overproduction, agricultural support budget pressures, accusations of excessive market protection and distortion, and concerns about the environmental impact of agricultural intensification. All these pressures have contributed to growing support for fundamental reform of the Common Agricultural Policy.

The new agricultural policy measures adopted by the European Union's farm ministers in 2003 seek to reform the Common Agricultural Policy in ways that encourage EU farmers and businesses to become more market-oriented, competitive and sustainable, both economically and environmentally. The main element of this reform has been the introduction of farm support that is decoupled from production, through the Single Payment Scheme, which commenced in 2005–06. In the case of cattle production in Scotland, the Single Payment Scheme has replaced production-linked subsidies supporting both beef cows and cattle during their production.

Another source of pressure for the beef supply chain to adopt a more market-oriented strategy comes from the presence of imported beef (the United Kingdom was 78 per cent self-sufficient for beef in 2006[1]), which in the case of further trade liberalisation would become a more serious competitor for the domestic industry. This presence may be particularly threatening because consumer loyalty to stores (i.e., large multiple retailers, which are also importers of beef) usually is higher than that to brands or products,[2] even those with strong regional or local identities. Furthermore, in the absence of local produce, consumers tend to choose a substitute product (i.e., one that is not produced locally) rather than postpone the purchase or look for the product in an alternative outlet.

Policy and market forces, over time, will force farmers to adopt more market-focused strategies to survive within the new market environment. The main marketing channels now available to producers of finished beef cattle in Scotland are to market directly to an abattoir on a 'deadweight' basis, with the value of the sale determined by the deadweight of the animal's carcase, its fatness, and conformation (shape of the carcase), or to sell through an auction market with open-bidding and the value determined by the animal's liveweight. Other less common options involve selling through a dealer, buying agent or marketing group. Deadweight selling, in which factors in addition to weight determine carcase value, is clearly a more market-focused approach than live weight sales.

An indication that farmers are pursuing more market-focused strategies is provided by Revoredo-Giha and Leat in a study of how Scottish cattle producers plan to cope with the Common Agricultural Policy's 2003 reforms.[3] They find that 37 per cent of farmers were planning to take measures to improve the quality of their production. Whilst this is an appropriate response that would allow farmers to achieve a better alignment of production with market requirements with respect to quality and at the same time secure quality-related price premiums, it is not an easy task to accomplish for many producers. Many farmers are not actively part of a supply chain within which information about quality requirements and the rewards for quality improvement are readily communicated. As shown by FOODCOMM[4] and Leat and Revoredo-Giha,[5] farmers as a whole are the most difficult component of the beef supply chain to draw into integrated supply chain activities. Many sell their store animals and finished livestock through auction markets, without a clear view of the final customers and their requirements, and they lack the possibility to benchmark their production – in terms of physical performance or quality – against that of other producers.

Finally, research shows that supply chain relationships in the beef chain can be significantly improved when chain arrangements offer the opportunity for commercial rewards to chain participants,[6] including farmers, and that good communication is necessary for the development of sustainable chain relationships.

The described context, though challenging, may open opportunities for farmers to develop alternative and improved marketing channels, possibly by establishing new forms of partnerships with processors. Thus, the purpose of this chapter is to discuss whether performance-related communication strategies have the possibility of improving farmers' market orientation. To study this potential, the chapter uses the case of McIntosh Donald, a beef processor located in the north-east of Scotland and a major red meat supplier for Tesco. It introduced Qboxanalysis, a performance-related communication system, to its Beef Producer Club (also referred to as the Tesco Beef Producer Club) members in March 2005.[7]

Literature Review

Prior to the Common Agricultural Policy reform of 2003, production-related support tended to reduce incentives for primary producers to become more proactive in the operation of their supply chains. This point is particularly important, because an efficient and effective collaborative supply chain can provide a critical competitive advantage.[8]

This section provides a brief overview, based on prior literature, of some of the main issues surrounding the use of performance-related communication – in particular, its impact on increasing the market orientation of the supply chain (as market information flows along the chain) and its effects on supply chain cohesion through improved communication and increased trust between farmer and processor.

The flow of information from consumers to farmers in a market-oriented food supply chain can be represented as in Figure 5.1. A fundamental prerequisite of good marketing performance within a free market environment is awareness of the customers and their needs. Harmsen et al. note that market orientation involves a focus on, and responsiveness to, customers and competitors as part of an external orientation.[9] Within the context of supply chains and their performance, this awareness should be extended to embrace the needs of other chain participants as well. Such awareness invariably involves information sharing.[10]

Because the flow of information is an important component in the operation of the supply chain, it is not surprising that communication has emerged as an important factor in achieving successful inter-firm cooperation.[12] Communication allows chain participants to learn about, and react to, changes in the requirements and expectations of other chain participants and assist superior chain performance, which can be enabled by modern information technologies. Furthermore, enhanced transparency, through an information-sharing mechanism linking supply chain partners, is one of the most critical drivers of supply chain success.[13] Increasingly, communication of comparative performance information, which enables benchmarking, splay a role in furthering enterprise and chain performance.

With respect to information transmission, information technology has an important role in improving communication and cohesion within the supply chain. It should be noted that information technology selection and usage is a key strategic consideration in efficient consumer response, where information is transmitted from consumers to all parties in the supply chain. Enabling technologies, such as electronic data interchange, create the basis for data transmission among chain partners. By storing data about customers, stocks, sales, competitors, and so forth in a centralised location, it is possible to divide it into use-oriented and decision-oriented forms. Overall, the use of modern and harmonised information technology can give supply chain partners an information advantage that can lead to a significant competitive advantage.[14]

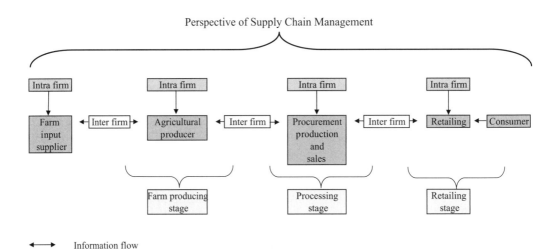

Figure 5.1   Flow of information in supply chain management (SCM)

Source: Based on Dangelmaier et al.[11]

Communication in supply chains can be influenced by many factors. Most of them can be allocated to one of the following groups: (1) communication behaviour, (2) information quality, or (3) communication tools. These factors, along with some of the key influences on them, appear in Figure 5.2.

Several elements in Figure 5.2 are also part of the system's performance-related communication. As a particularly important aspect, the system can enable decision-making by reducing uncertainty.[15] Moreover, Farace et al. define information in terms of the reduction of uncertainty.[16] The greater the uncertainty, the greater the need for information, which is particularly important for beef producers because they can use information over time to improve the quality of their finished animals.

Information quality is another important aspect of communication quality. Low and Mohr[17] use relevance, accuracy, reliability and timeliness to characterise the quality of marketing information, drawing on work by O'Reilly.[18] Relevance means that only useful and significant information for the decision process, or for achieving specific objectives, will be transmitted. Accuracy refers to the clear and precise formulation and transmission of information. Reliability concerns the trustworthiness of the information. Timeliness adds a temporal dimension: up-to-date information allows the receiver to react appropriately.

The indicators of relevance and accuracy do not refer solely to information quality but also to information quantity. Both should be appropriate to the situation. Previous research has shown that managers tend to believe that more information is better[19] and that a lack of information is connected to poor decisions. However, information overload can occur when communication costs decrease as a result of new technologies, and information may be transmitted without processing. Thus, important information can get lost in reams of irrelevant messages. It is essential that the transmission of information is undertaken in the appropriate quantity and in a way that the user can apply to management decisions.

Regarding the effects that performance-related communication has on supply chain cohesion, results from FOODCOMM[20] show that the most important contributor to good business relationships is effective communication, which comprises adequate communication frequency and high information quality. Within the same project, we have identified that communication quality is a significant determinant of sustainable supply chain relationships within the UK beef supply chain.

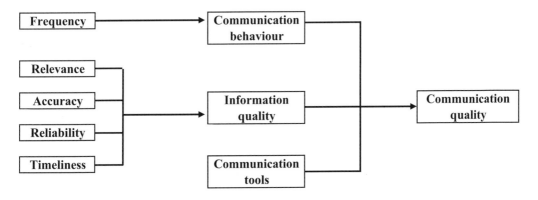

Figure 5.2   Factors influencing communication

Source: Based on FOODCOMM.

Finally, the appropriate and successful functioning of the communication system may enhance trust and satisfaction amongst businesses in the supply chain and therefore their commitment. These three values, as many studies have shown,[21] are important components of the quality of relationships within the supply chain.

Methodology

The information used in this case study arises from an EU Sixth Framework research project known as FOODCOMM.[22] It was gathered through a series of in-depth interviews conducted during August and September 2007. The 11 interviews collected information from the main stakeholders within the supply chain, including farmers (4 persons), representatives of the processor (2) and retailer (1), beef production advisers (2), and developers of the technology (2).

Some of the interviews were conducted on a face-to-face basis, whereas others were conducted over the telephone with further information exchanged by e-mail. In addition, secondary information was collected from Internet sources and written material.

Information on the marketing environment faced by the Scottish beef industry was drawn both from the FOODCOMM project and a Scottish Executive Environment and Rural Affairs Department-funded project about the implications of the Common Agricultural Policy (IMCAPT).[23] In addition, the case study has been enhanced by additional information supplied by Innovent Technologies Ltd.

The topics explored in the interviews were as follows:

*   the marketing environment context of the Scottish beef industry
*   the history of Qboxanalysis and its development
*   the aims of the initiating stakeholders
*   the information transmitted
*   the costs and mechanics of its operation
*   the potential and actual benefits of the system for the respective stakeholders
*   the level of uptake by farmers
*   the implications of the system for beef production management, supply chain operation, performance, and relationships.

Case Study

This section is structured as follows: first, we present the origins of Qboxanalysis and the businesses involved, and second, we fully describe the Qboxanalysis system, including the operation of the system, the information provided, the costs of the system, and its uptake by farmers.

THE ORIGINS OF QBOXANALYSIS AND THE BUSINESSES INVOLVED

The Qboxanalysis system originally was developed by Willie Thomson, Technical Director of Harbro Ltd., a progressive livestock feed company based in the north-east of Scotland. The development of Qboxanalysis started in 2002 with the aim of providing pig farmers with a management tool that would help them ensure more of their pigs hit target specifications with respect to weights and carcase probe values (a determinant of carcase quality). The system not only enables the user to investigate how to improve pig enterprise performance but also notes the financial consequences of the changes. In essence it provides a fact-based analysis of the pig enterprise and allows a proficient manager or adviser to plan performance improvements.

The property rights for Qboxanalysis belong to Innovent Technology, a software company closely related to Harbro Ltd. that is also based in the north-east of Scotland. Innovent has extensive experience developing and operating a range of Web-based systems for enterprise monitoring and planning in the farming sector, all under the Qbox name.

Qboxanalysis for beef cattle was developed during 2003 and 2004 and made available to the McIntosh Donald Producer Club members in March 2005. McIntosh Donald is part of the Grampian Country Food Group and is a major slaughterer and processor of beef cattle in the north-east of Scotland. In total, it slaughters approximately 80,000 cattle per year, some 15 per cent of the Scottish kill, and counts Tesco as one of its major customers. The company includes an established network of more than 1,000 beef cattle producers who supply beef cattle for slaughter on a deadweight basis and who operate within the structure of the Producer Club. This club has a series of activities, aimed at strengthening the flow of information and relationships among the retailer, processor, and farmer suppliers.

At the outset, McIntosh Donald recognised that its slaughtering and processing operation generated a considerable amount of information about the quality and life-time performance of the cattle it was procuring. It also noted that communicating this information back to farmer suppliers, in an appropriate format, would provide farmers with the opportunity to better identify the performance of their cattle, both in their own right and relative to the cattle of other producers supplying the factory.

From the interviews, it was clear that through Qboxanalysis, McIntosh Donald aimed to provide information to farmers that would enable them to make better on-farm management decisions with respect to their beef production. The company also hoped that the Qboxanalysis system would help strengthen farmer–processor relationships; in the longer run, the company could benefit because more cattle would meet their target specifications in terms of weight, fatness, and conformation. Figure 5.3 presents McIntosh Donald's preferred cattle specifications, which range from E to R in terms of conformation and 3 to 4L in terms of fatness.

KEY POINTS:

- McIntosh Donald, with the support of Tesco, has an existing Producer Club within which it can communicate with farmers on matters of mutual interest.
- Qboxanalysis provides comparative information about the on-farm performance of beef cattle.
- Qboxanalysis aims to facilitate improved on-farm management decisions and improve the overall quality of beef cattle production, in line with market requirements.
- An informal system of horizontal collaboration assists vertical integration with respect to communication and production improvements.

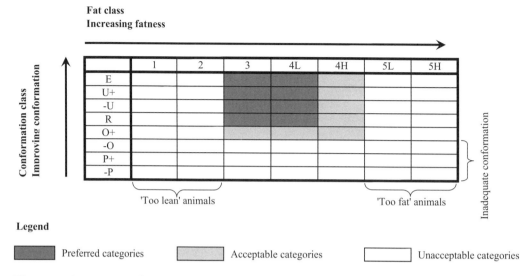

Figure 5.3   McIntosh Donald target specifications for fatness and conformation

THE QBOXANALYSIS SYSTEM

The operation of Qboxanalysis

McIntosh Donald supplies basic data to Innovent Technology on a weekly basis. Innovent sends out an e-mail each Friday to all registered Qboxanalysis suppliers of cattle for that particular week. The farmer clicks on the Qboxanalysis link in the e-mail and enters the system with a specific username and password. The system is totally confidential; a user only has access to his or her own results and comparable figures for the complete McIntosh Donald cattle intake.

The information is provided on a 7-day and 13-week basis, so the user can compare the results of his or her cattle with all others slaughtered in that week or examine the comparative performance of his or her cattle over the past quarter. This provision of quarterly data recognises different production methods and feeding performance across the farming season, thus enabling 'like compared with like' (e.g., performance of winter-fed cattle in one year with that of another). The system also provides the opportunity for a 365-day summary, and if a Qboxanalysis registered farmer does not submit cattle to McIntosh Donald for a prolonged period, he or she receives a periodic reminder about the availability of the 365-day summary.

KEY POINT:

- Detailed comparative information is available on a frequent and regular basis.

The information provided by Qboxanalysis

McIntosh Donald gathers the data for Qboxanalysis by scanning cattle passports (which list the farmer and age, sex and breed of the animal) and uploading the data from the

Hellenic abattoir data system, which carries all the slaughter data for the plant. Thus there are no significant additional costs in gathering the information.

The data in the Qboxanalysis system distinguish between Scotch steers, Scotch heifers and Scotch young bulls and include the following indicators: the numbers of cattle, average weights, conformation score (using the EUROP classification but translated onto a scale of 1–8 to assist comparison), fat score (using the 1–5H classification but expressed on a scale of 1–7), age at slaughter; average value, deadweight gain per day, estimated liveweight gain, and percentage with fluke.

KEY POINT:

- An extensive range of comparative data is collected at no additional cost to the processor.

Thereafter, a Whole Life Margin Monitor expresses the margin on the animal slaughtered, given a range of feed costs per head per day. A Whole Life Breed Performance Monitor also compares, for steers and heifers separately, the average daily liveweight gain for each of the main crosses, including data about the individual producer and the processor. The breeds covered are Charolais Limousin, Simmental, and Belgian Blue crosses.

Figure 5.4 shows the information available to a farmer on the Qboxanalysis system (similar data are also provided for the 365-day period). This producer has supplied 16 steers and 12 heifers in the previous 7 days and 189 steers and 222 heifers in the past 13 weeks. Over the 13 weeks, his steers have been on average 7 kg heavier than the plant average, and his heifers are about 16 kg heavier.

The steers, despite the heavier weight, were 33 days younger than the plant average at slaughter (717 versus 750 days), and the heifers were 15 days younger (701 versus 716). The deadweight and liveweight gain were marginally better than average. In terms of carcase value, these steers were £23 better than average and the heifers £30 better.

The Whole Life Margin Monitor shows the margin that would have been earned on animals submitted in the last 7 days for a range of feeding costs. If the producer had feeding costs of 90 pence per day, he would have earned a margin of £161 on his steers as opposed to the plant average of £108 (an additional 49 per cent).

KEY POINTS:

- The physical performance of a producer's cattle, across a range of criteria, are clearly compared with the average results for cattle going through the plant.
- The financial consequences of cattle performance are indicated in terms of both carcase value and net margin.

The costs of the system

The initial development costs of the system were borne by Innovent Technology and Harbro. When McIntosh Donald adopted the system, it made an initial payment of £10,000, with a further £10,000 coming from Tesco. Tesco is a major customer of McIntosh

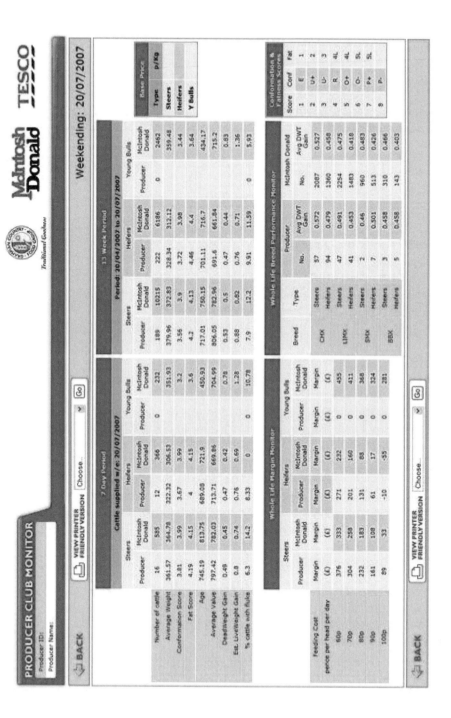

Figure 5.4 Sample producer's steers and heifers compared with plant average

Source: © Innovent technology 2005.

Note: The Grampian Food Group logo is reproduced with the permission of VION Food Group Ltd.

Donald and wished to support the processor in furthering its communication with farmers and thereby improve the quality of cattle being supplied to the processor, as well as to assist farmers in improving the performance of their beef enterprises. These payments enabled the system to be installed and made operational. In addition, the ongoing annual operational cost of £10,000 is borne by McIntosh Donald, which at current slaughtering levels works out to approximately 12.5 pence per head. As noted previously, there are no additional data collection costs for Qboxanalysis within the abattoir.

At the present time, there is no direct cost to producers who are members of the McIntosh Donald Producers Club. The system is available to any of the Club members who wish to register with Qboxanalysis. All it requires is an e-mail address and access to the Internet. Thereafter, the only cost is the time spent accessing and studying the information.

KEY POINTS:

- Qboxanalysis is free to the farmer user.
- The establishment costs have been paid by the processor and retailer.
- Following the initial establishment costs, the annual running cost to the processor is very modest, at approximately 12.5 pence per carcase.
- The retailer aims to assist farmers in improving the quality of cattle supplied and the performance of their beef enterprises.

Uptake of the system by farmers

As at August 2007, the system included 429 registered farmer users, of whom 100–150 were regular users (i.e., delivering cattle for slaughter and logging onto Qboxanalysis). This number of registrations is very close to the number of McIntosh Donald cattle suppliers who have an e-mail address. The farmers delivered approximately 15,000 (19%) of the 80,000 cattle supplied to the factory annually.

Those who register to receive Qboxanalysis are already connected to the Internet for other reasons, rather than getting connected to access Qboxanalysis. The use of the system is also constrained because currently the data presented are of most relevance to a farmer who both breeds and finishes his or her own cattle, ready for slaughter. Much of the performance information relates to the whole life of the cattle concerned – age at slaughter, weight gain per day over the life of the animal, margin over the whole life, and so forth.

The appeal of Qboxanalysis to farmers who are beef finishers, that is, those who buy 'store' animals that others have bred and then feed them through to slaughter, will be greatly enhanced when it carries a module that reports on performance over the finishing period. For this capability to be achieved, purchase data would have to be entered into the system, including the holding of birth, weight at purchase and time of purchase. This system could provide finishers with accurate data about the performance of cattle during the time on their farm. The introduction of a finishing module occurred in early 2008, and it has the potential to identify the source of cattle most likely to achieve the performance level desired by the finisher. Moreover, a by-product of this development is the ability to compare the performance of store calves prior to purchase, which could provide useful information about the influence of different feeding regimes in the cattle's early life. With the introduction of a Suckler Herd Monitor, the breeder potentially could receive information about how store animals performed through to slaughter, which ultimately could influence decisions about the genetic qualities and management of suckler cows and bulls.

The type of information that might be made available to a breeder through the Suckler Herd Monitor is presented in Figure 5.5. In this case, the breeder's Simmental Cross calves have finished at an average daily liveweight gain of 1.17 kg, compared with the factory average of 0.85 kg. This difference has a dramatic effect on the margin of the animal, taking it up to £183 compared with a loss of £9 for the average animal with a lower live weight gain.

KEY POINTS:

- Greater Internet connectivity levels by farmers will enable Qboxanalysis-type communication.
- The power, appeal, and uptake of the system could be considerably enhanced by a finishing period module and Suckler Herd Monitor.
- These modules ultimately will enable the better selection and management of breeding stock, improved management of young stock, and better store stock selection decisions by finishers.

BENEFITS OF QBOXANALYSIS

This section explores the benefits of Qboxanalysis for farmers, processors and retailers, emphasising those elements that increase the cohesion of the supply chain and farmers' market orientation.

Benefits for producers and farmers' use of the system

From the farmer's perspective, the data from Qboxanalysis:

- highly accurate,
- quick and easy to access at no cost,
- straightforward in their analysis of carcase classification and value achieved,
- health check reports for fluke, and
- summaries of trend data over time.

The case study reveals three broad types of farmers registered with Qboxanalysis. First, there are those who are registered with the system but who infrequently log on or make use of it. This group may be as many as 65 per cent of those registered with the system. Second, there are farmers who log on to the system and use it to provide confirmation that their beef production enterprise is operating satisfactorily. Such farmers are generally operating at average or above-average levels of performance. This relatively passive usage is in itself beneficial, in that it reassures those with basically sound beef husbandry practices. Moreover, in time such users may become more proactive in developing their beef production based on Qboxanalysis information. Third, a smaller group logs on to

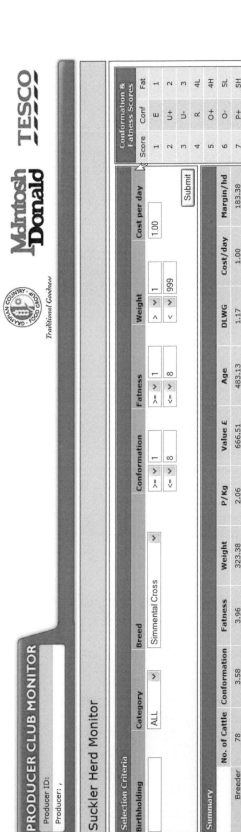

Figure 5.5 Sample feedback to the rearer on how cattle have performed to finishing

Source: © Innovent technology 2005.

Note: The Grampian Food Group logo is reproduced with the permission of VION Food Group Ltd.

the system regularly (when they put cattle away for slaughter) and use the information gained to influence their enterprise management practices and decisions. For example, such producers may engage in:

- weighing animals at a younger age and batching them according to weights rather than age
- weighing cattle more regularly and being more selective about which animals are put away for slaughter
- gaining a better understanding of the relationship between the liveweight of animals and their deadweight
- changing the bull that is put onto the suckler cows
- confirming the quality of a particular source of store cattle
- reviewing feeding rations to try to achieve better weight gain and earlier finishing
- treating cattle for fluke when they come onto the farm
- reviewing the grazing used by stock when fluke problems have arisen
- putting animals that are not ideal for McIntosh Donald to another market (e.g., through the livestock market).

Those who use the system regularly find that the information is easy to understand, once they have familiarised themselves with how the data are presented.

The second and third groups of producers, namely, the reassured and active users, together may represent 35 per cent of registered users and 10–15 per cent of McIntosh Donald's cattle suppliers.

To make full use of Qboxanalysis, a farmer must be motivated to improve the performance of finished cattle and have the capabilities to decide what farm-related changes need to be made to the cattle production system (e.g., genetics/source of stock, feeding systems, improved animal health and welfare). As an example of what may be achieved, consider a breeder-finisher who has steadily studied and responded to the Qboxanalysis data for his cattle for more than three years. He has experienced a 32-day reduction in days to slaughter (486 to 454) and an improvement in deadweight gain of 0.05 kg per day (from 0.73 to 0.78). At the same time, the change in the value of his carcases has matched that of the plant average.

When a farmer requires help evaluating the practical implications of the Qboxanalysis data, a farm or feed adviser would seem the obvious source of assistance.

KEY POINTS:

- Qboxanalysis information may provide evidence of potential improvements in beef enterprise management.
- The reassured and active users of Qboxanalysis represent up to 35 per cent of registered users and 10–15 per cent of McIntosh Donald's cattle suppliers.
- Active use of the system by a farmer depends on the existing performance of his or her finished cattle, the scope for improvement, and his or her motivation and capabilities.
- Some farmers need advisory assistance to get the best out of the information provided.

Benefits for the processor

Qboxanalysis has many benefits for McIntosh Donald. First, it enables the company to take a proactive approach toward farmer suppliers in helping them achieve better efficiency and reduced costs on the farm, as well as producing carcases that better match market needs. In this sense, it can help farmers achieve better carcase values and higher net margins, while also strengthening its image and relationship with farmers. Second, it provides an accurate analysis of producer performance and can improve supply chain management by providing clear evidence of beef production trends. Longer term, a greater proportion of beef carcases may hit the processor's ideal carcase specification. The company accepts carcases of 250–410 kg without penalty, but to remain competitive, a narrower range of 270–380 kg would better meet customers' needs, and 350 kg is ideal. In mid-2007, the factory average was close to this level. As far as fatness is concerned, 4L is ideal, but in mid-2007, some 25 per cent of animals were fatter at 4H. Overfat animals cost the company about £24 per carcase in lost revenue, as an overfat animal yields 2 per cent less saleable meat. Fat, which is effectively purchased at £2.10 to £2.20 per kg, must be trimmed and sold for 6 pence per kg. There is thus considerable financial benefit to be achieved from the production of cattle of ideal fatness (4L) and conformation (R or better). In 2007, overfat animals, representing a total of 20 per cent of supply, could cost the processor more than £380,000. Third, the system may provide further opportunities for payment notification, e-mail marketing, and the promotion of Producer Club initiatives.

KEY POINTS:

- Qboxanalysis enables the processor to help beef cattle suppliers with production and marketing decisions and thus can strengthen communications and relationships with producers, as well as the company's image.
- It can also improve the quality of overall cattle intake and provide major cost savings.
- It can assist in analysing cattle supplies and suppliers over time.

Benefits for the retailer

Tesco sources beef and lamb from some 10,000 farmers and endeavours to operate and support supply chains within which all parties derive benefits. Its Producer Clubs, operated by its main suppliers, are major channels of two-way communication with farmers.

In supporting the introduction of Qboxanalysis, Tesco is seeking to raise awareness about production efficiency and encourage performance comparisons amongst producers, geographical areas, and breeds/crosses. It particularly wishes to support those producers that are more progressive in seeking production improvement and greater market orientation. It perceives the system's value in its basis in hard current and historical facts and the foundation it provides for discussing or considering production improvements. It also draws farmers' attention to the potential benefits of IT systems and provides an incentive for them to 'get connected.'

As far as the processor is concerned, Tesco perceives the potential benefits of more high-quality animals entering the supply chain with less waste (fewer animals outside the ideal specification), as well as earlier finishing, which should give rise to better eating quality. A further benefit is that the system helps Tesco identify progressive producers with whom it can engage for the future development of the industry.

Tesco is very satisfied with the system's performance to date. It perceives value in future developments (i.e., finishing module and Suckler Herd Monitor) and in the wider adoption of the system (another supplier, also part of the Grampian Country Food Group to which McIntosh Donald belongs, is looking to introduce it). Tesco recognises that for some farmers the information may appear quite complex and that its presentation may need further development to facilitate its interpretation.

KEY POINTS:

- Tesco regards Qboxanalysis as a means to help raise awareness about production efficiency by encouraging performance comparisons amongst producers.
- It sees the potential benefit of improved quality over time in the intake of cattle by the processor.
- The system helps it identify more progressive farmers with whom it can engage on future industry development.
- Tesco is very satisfied with the system and welcomes its wider adoption and further development.

THE INFLUENCE OF QBOXANALYSIS ON FARMER–PROCESSOR RELATIONSHIPS AND FARMERS' MARKET ORIENTATION

At present Qboxanalysis appears to be of benefit to those farmers with a strong commercial orientation toward their farming activities and marketing of their cattle. As market pressures further impinge on the sector, perhaps through greater import penetration and reduction in the Single Payment Scheme, active interest in the system may increase. Further planned developments in the system, specifically those that will assist beef finishers and breeders, will also widen its appeal to beef producers.

The farmers who are actively using the system are mainly those who engage regularly in other Producer Club activities. Consequently, any influence that Qboxanalysis may have on processor–farmer relationships is difficult to disentangle from the relationship influences of the wider Producer Club activities.

First, significant numbers of farmer suppliers have a sound relationship with the processor, particularly those who readily engage over marketing and livestock suitability issues. Relationships are frequently based on the personal interactions and bonds that exist between the two parties. The company's staff (field and procurement) are readily described as 'very professional,' 'friendly,' 'always helpful,' 'responsive to queries,' and so on. The factual information provided by Qboxanalysis provides a sound basis for discussions between the farmer and field staff about how to achieve improvements in the beef enterprise, which thereby enriches the relationship further.

Second, farmers who are commercially oriented and seeking improvements in their farm enterprises appreciate what the Producer Club is doing to make farmers aware of their own performance, market developments, and challenges facing the factory and industry, as well as to bring producers together to encourage interactions both amongst themselves and with the factory and retailer. Such farmers express a strong level of satisfaction with McIntosh Donald. However, Qboxanalysis provides information of potential commercial value that is not readily available from other marketing sources (though it may be complemented by other farm record data systems), and as such, it strengthens the trading relationship, and satisfaction with it, for farmers who see the value in Qboxanalysis data.

Third, there appears to be limited impact on the level of trust in the processor, though exceptions exist. Trust appears to be influenced by price and the state of the market, as well as by personal relationships with key company personnel. However, Qboxanalysis adds greater transparency to farm enterprise performance and provides guidance about how, in the medium to longer term, a producer might improve the market performance of his or her animals.

KEY POINTS:

- The system provides the most benefit to commercially and market-oriented farmers.
- The system reinforces the good relationship that many farmers have with the processor (there is a strong personal element to this relationship).
- For some, the value of the data strengthens the relationship with the processor.
- There is little impact on the trust that farmers have in the processor (it is already reasonable).
- Market-oriented users of the system appreciate its value (actual and potential) and derive satisfaction from the comparisons it provides and its indications of how to improve things.

Conclusions

The key points arising from the case study are as follows:

- McIntosh Donald, with the support of Tesco, has an existing Beef Producer Club within which it communicates with farmers on matters of mutual interest;
- Qboxanalysis provides comparative information about the on-farm performance of beef cattle;
- Qboxanalysis aims to facilitate improved on-farm management decisions and improve the overall quality of beef cattle production, in line with market requirements;
- An informal system of horizontal collaboration assists vertical integration with respect to communication and production improvement.

THE NATURE OF QBOXANALYSIS

- Detailed comparative information is made available on a frequent and regular basis.
- An extensive range of comparative data is collected at no additional cost to the processor.
- The physical performance of a producer's cattle, across a range of criteria, is clearly compared with the average results for cattle going through the plant.
- The financial consequences of cattle performance are indicated in terms of both carcase value and net margin.
- Qboxanalysis is free to the farmer user. The establishment costs have been paid by the processor and retailer.

- Following the initial establishment costs, the annual running cost to the processor is very modest at approximately 12.5 pence per carcass.
- The retailer aims to assist farmers in improving the quality of cattle supplied and the performance of their beef enterprises.

UPTAKE OF QBOXANALYSIS

- Greater Internet connectivity levels by farmers will enable Qboxanalysis-type communication.
- The power, appeal and uptake of the system will be considerably enhanced by a finishing period module and Suckler Herd Monitor.
- These monitors ultimately will enable the better selection and management of breeding stock, improved management of young stock and better store stock selection decisions by finishers.

BENEFITS AND USE OF THE SYSTEM

Farmers

- Qboxanalysis information may provide evidence for potential improvements in beef enterprise management.
- The reassured and active users of Qboxanalysis may represent up to 35 per cent of registered users and 10–15 per cent of McIntosh Donald's cattle suppliers.
- The active use of the system by a farmer depends on the existing performance of his or her finished cattle, the scope for improvement, and his or her motivation and capabilities.
- Some farmers will need advisory assistance to get the best out of the information provided.

Processor

- Qboxanalysis enables the processor to help beef cattle suppliers in their production and marketing decisions and further strengthens communications and relationships with producers, as well as the company's image.
- Qboxanalysis can improve the quality of the processor's overall cattle intake and provide major cost savings.
- Qboxanalysis can assist the processor in analysing cattle supplies and suppliers over time.

Retailer

- Tesco regards Qboxanalysis as a means to help raise awareness about production efficiency by encouraging performance comparisons amongst producers.
- Tesco sees the potential benefits of improved quality over time in the intake of cattle by the processor.
- The system helps Tesco identify more progressive farmers with whom it can engage in future industry development.

- Tesco welcomes further development of the system.
- Tesco is very satisfied with the system and welcomes its wider adoption and further development.

Qboxanalysis effect on farmer–processor relationships and farmers' market orientation

- The system provides the most benefit to commercially and market-oriented farmers.
- The system reinforces the good relationship that many farmers have with the processor (there is a strong personal element to this relationship).
- For some, the value of the data strengthens the relationship with the processor.
- There is little impact on the trust that farmers have in the processor (it is already reasonable).
- Market-oriented users of the system appreciate its value (actual and potential) and derive satisfaction from the comparisons it gives and the indications of how to improve things.

In summary, McIntosh Donald's decision to introduce Qboxanalysis has offered potential benefits to all parties in the supply chain. It represents an effort to improve the on-farm performance and market orientation of beef farmers through performance-related communications. It also has the potential to improve the performance of the processing and supply chain operation by reducing the number of animals that fall outside the ideal specification for the processor and its main customer, thus reducing waste and saving costs. Benefits also accrue to the retailer (which helped pay for the installation of the system), not least of which is the opportunity to be proactive in assisting the performance of farmers and the whole supply chain. Furthermore, Qboxanalysis offers a way to improve business relationships along the supply chain, thereby complementing the other chain integration activities of the McIntosh Donald Beef Producers' Club.

Acknowledgements

This case study derives from the FOODCOMM project, *Key Factors Influencing Economic Relationships and Communication in European Food Chains* (FOODCOMM, SSPE-CT-2005–006458), which was funded by the European Commission as part of the Sixth Framework Programme. It also derives from Scottish government–funded research into 'Sustainable Farming Systems.'

We thank the members of the FOODCOMM project and the persons interviewed in the case study. We particularly thank Mr. Alan McNaughton and Mr. Eric Buchan, Managing Director and Procurement Director, respectively, of McIntosh Donald; Mr. Michael Martin, Chairman, McIntosh Donald Producer Club; Mr. Willie Thomson, Harbro Feeds Ltd. and Innovent Technologies Ltd.; and Ms. Alice Pattinson, Producer Club Manager, Tesco Stores Ltd., all of whom made especially valuable comments and amendments to a preliminary version of the case study. Any remaining errors are solely the responsibility of the authors.

References

1. Meat and Livestock Commission (MLC) (2007), *A Pocketful of Meat Facts 2007*, MLC, Milton Keynes, UK.
2. Mintel (2006), *Attitudes Towards Ethical Foods*, Mintel Group, London.
3. Revoredo-Giha, C. and Leat, P. (2007), *Red Meat Producers' Preferences for Strategies to Cope with the CAP Reform in Scotland*, AA211 Special Study Report, Scottish Executive Environment and Rural Affairs Department (SEERAD), Edinburgh.
4. FOODCOMM Project (2006), *Key Factors Influencing Economic Relationships and Communication in European Food Chains*, Workpackage Report 2: Review of Food Chain Systems, European Commission-funded FP6 research project. Available at http://www.foodcomm.eu. Accessed: December, 2007.
5. Leat, P. and Revoredo-Giha, C. (2007), 'Building collaborative agri-food supply chains: The challenge of relationship development in the Scottish red meat chain', *British Food Journal*, vol. 110, no. 4/5, pp. 395–411.
6. FOODCOMM Project (2007), *Key Factors Influencing Economic Relationships and Communication in European Food Chains*, Workpackage Report 4: Analysis of survey data. Available at http://www.foodcomm.eu.
7. An account of the McIntosh Donald Producers' Club, also known as the Tesco Producers' Club, can be found in Fearne, A. (1998), 'The evolution of partnerships in the meat supply chain: insights from the British Beef Industry', *Supply Chain Management: An International Journal*, vol. 3, no. 4, pp. 214–31.
8. See, for example, Dyer, J.H. and Singh, H. (1998), 'The relational view: Co-operative strategy and sources of inter-organisational competitive advantage', *Academy of Management Review*, vol. 23, no. 4, pp. 660–79; Sahay, B.S. (2003), 'Supply chain collaboration: The key to value creation', *Work Study*, vol. 52, no. 2, pp. 76–83; Power, D. (2005), 'Supply chain management integration and implementation: A literature review', *Supply Chain Management: An International Journal*, vol. 10, no.4, pp. 252–63.
9. Harmsen, H., Grunert, K.G., and Declerck, F. (2000), 'Why did we make that cheese? An empirically based framework for understanding what drives innovation activity', *R&D Management*, vol. 30, no. 2, pp. 151–66.
10. Peterson, J., Cornwell, F., and Pearson, C.J. (2000), 'Chain stocktake of some Australian agricultural and fishing industries', Bureau of Rural Sciences, Canberra. Available at http://affashop.gov.au/PdfFiles/PC12761.pdf. Accessed: December, 2007.
11. Dangelmaier, W., Pape, U., and Rüther, M. (2001), 'Supply Chain Management bei werksübergreifender Frachtkostenoptimierung', *WISU*, vol. 10, pp. 1368–82.
12. See, for example, Bleeke, J. and Ernst, D. (1993), *Collaborating to Compete*, John Wiley and Sons, New York; Mohr, J.J., Fisher, R.J., and Nevin, J.R. (1996), 'Collaborative communication in interfirm relationships: Moderating effects of integration and control', *Journal of Marketing*, vol. 60, N. 3, pp. 103–115; Tuten, T.L. and Urban, D.J. (2001), 'An expanded model of business-to-business partnership formation and success', *Industrial Marketing Management*, vol. 30, no. 2, pp. 149–64.
13. Min, H. and Zhou, G. (2002), 'Supply chain modelling: Past, present and future', *Computers and Industrial Engineering*, vol. 43, pp. 231–49.
14. Mau, M. (2000), *Supply Chain Management*. John Wiley and Sons Ltd, Frankfurt.
15. Duncan, T. and Moriarty, S.E. (1998), 'A communication-based marketing model for managing relationships', *Journal of Marketing*, vol. 62, pp. 1–13.

16. Farace, R.V., Monge, P.R., and Russel, H.M. (1977), *Communicating and Organizing*. Addison-Wesley, Reading, MA.

17. Low, G.S. and Mohr, J.J. (2001), 'Factors affecting the use of information in the evaluation of marketing communications productivity', *Journal of the Academy of Marketing Science*, vol. 29, no. 1, pp. 70–88.

18. O'Reilly, C.A. (1982), 'Variations in decision makers' use of information sources: The impact of quality and accessibility of information', *Academy of Management Journal*, vol. 25, no. 4, pp. 756–71.

19. O'Reilly, C.A. (1980), 'Individuals and information overload in organizations: Is more necessarily better?', *Academy of Management Journal*, vol. 23, no. 4, pp. 684–96.

20. Fischer, C., Hartmann, M., Bavorova, M., Hockmann, H., Suvanto, H., Viitaharju, L., Leat, P., Revoredo-Giha, C., Henchion, M., McGee, C., Dybowski, G., and Kobuszynska, M. (2008), 'Business relationships and B2B communication in selected European agri-food chains–first empirical evidence', *International Food and Agribusiness Management Review*, vol. 11, no. 2, pp. 73–100.

21. Lagace, R.R., Dahlstrom, R., and Gassenheimer, J.B. (1991), 'The relevance of ethical salesperson behavior on relationship quality: the pharmaceutical industry', *Journal of Personal Selling and Sales Management*, vol. 4, no. 1, pp. 39–47; Moorman, C., Zaltman, G., and Deshpande, R. (1992), 'Relationships between providers and users of market research: The dynamics of trust within and between organizations', *Journal of Marketing Research*, vol. 29, no. 3, pp. 314–28; Wray, B., Palmer, A., and Bejou, D. (1994), 'Using neural network analysis to evaluate buyer–seller relationships', *European Journal of Marketing*, vol. 28, no. 1, pp. 32–48; Storbacka, K., Strandvik, T., and Grönroos, C. (1994), 'Managing customer relationships for profit: The dynamics of relationship quality', *International Journal of Service Industry Management*, vol. 5, no. 5, pp. 21–38; Bejou D., Wray B., and Ingram T.N. (1996), 'Determinants of relationship quality: An artificial neural network analysis', *Journal of Business Research*, vol. 36, no. 2, pp. 137–43.; Lewin, J.E. and Johnston, W.J. (1997), 'Relationship marketing theory in practice: A case study', *Journal of Business Research*, vol. 39, no. 1, pp. 23–31; Hennig-Thurau, T. and Klee, A. (1997), 'The impact of customer satisfaction and relationship quality on customer retention: A critical reassessment and model development', *Psychology and Marketing*, vol. 14, no. 8, pp. 737–64; Boles, J.S., Barksdale, H.C., and Johnson J.T. (1997), 'Business relationships: An examination of the effects of buyer-salesperson relationships on customer retention and willingness to refer and recommend', *Journal of Business and Industrial Marketing*, vol. 12, no. 3–4, pp. 253–64; Dorsch, M.J., Swanson, S.R., and Kelley, S.W. (1998), 'The role of relationship quality in the stratification of vendors as perceived by customers', *Journal of the Academy of Marketing Science*, vol. 26, no. 2, pp. 128–42; Rosen, D.E. and Suprenant, C. (1998), 'Evaluating relationships: Are satisfaction and quality enough?', *International Journal of Service Industry Management*, vol. 9, no. 2, pp. 103–125; Lang, B. and Colgate, M. (2003), 'Relationship quality, on-line banking and the information technology gap', *International Journal of Bank Marketing*, vol. 21, no. 1, pp. 29–37; Bennet, R. and Barkensjo, A. (2005), 'Relationship quality, relationship marketing, and client perceptions of the levels of service quality of charitable organisations', *International Journal of Service Industry Management*, vol. 16, no. 1, pp. 81–106.

22. FOODCOMM Project (2005), *Key Factors Influencing Economic Relationships and Communication in European Food Chains*, SSPE-CT-2005-006458.

23. Scottish Agricultural College (SAC) (2006), *Implications of the CAP Reform (IMCAPT)*, SAC, Edinburgh.

6 *Production and Marketing Innovation in the Argentine Beef Sector: The Prinex Case*

BY HERNÁN PALAU,* SEBASTIÁN SENESI,† AND FERNANDO VILELLA‡

Keywords

institutional innovation, organisational innovation, technological innovation, competitiveness, networks

Abstract

In this chapter, we introduce the Prinex case, a network of cattle producers that implemented beef origin and quality assurance systems and achieved success in the most demanding niche markets in Europe, the United States, Brazil and Chile. After we describe the Argentine beef agri-business, we outline Prinex history, the way it coordinated transactions, its marketing strategies, and its constraints and limitations. We will describe the beef international agri-business and Argentinean beef sector in particular; define strategic and tactical actions for marketing beef in international markets; and identify constraints and limitations for exporting and marketing beef in Argentina.

International Beef Agri-business

In recent years, several agri-food crises have broken out in various countries of Europe, the United States, Japan, Canada and Argentina. They were caused by diseases transmitted by

---

*   Mr. Hernán Palau, Research Area Coordinator, Food and Agribusiness Programme, School of Agronomy-UBA. Av. San Martín 4453 (CP: 1417), Buenos Aires, Argentina. E-mail: hpalau@agro.uba.ar. Telephone: + 5411 4524 2490.

†   Mr. Sebastián Senesi, Sub-Director Food and Agribusiness Programme, School of Agronomy-UBA. Av. San Martín 4453 (CP: 1417), Buenos Aires, Argentina. E-mail: ssenesi@agro.uba.ar. Telephone: + 5411 4524 2490.

‡   Mr. Fernando Vilella, Director Food and Agribusiness Programme, School of Agronomy-UBA. Av. San Martín 4453 (CP: 1417), Buenos Aires, Argentina. E-mail: vilella@agro.uba.ar. Telephone: + 5411 4524 2490.

food. Consumers consequently are much more concerned now about what they eat.[1] The disease that had the greatest impact was BSE (bovine spongiform encephalopathy, or 'mad cow disease'), which is said to have killed more than 10 people in the United Kingdom in 1996.[2] In addition, BSE was found in Canadian and US herds in 2002 and 2003, which led to the reduction of North America's share of beef exports. Other cases in Europe and the United States involved meat contaminated with *Campilobacter sp.*, *Escherichia coli*, *Listeria monocytogenes*, and *Salmonella sp.*,[3] as well as foot-and-mouth disease.

In response to such events, certain consumer segments began to demand food origin and quality assurances,[4] though other consumers were only concerned about taste aspects, such as good flavour or meat palatability. These changes in the agri-business environment offer good opportunities to farmers who might add value to their products through differentiation strategies, adapted to meet higher consumer demands.

To respond to the new situation, the industry, the public sector and non-governmental organisations (NGOs) have designed and implemented food safety and quality standards – traceability systems – in line with consumer expectations. For example, in Argentina, two resolutions (15/2003 and 391/2003) were designed and implemented to assure the origin (traceability) of beef to be exported to the EU.

In Argentina, 80 per cent (2.5 million tons) of the total beef production is commercialised in the domestic market. Argentine consumers demand a good price–quality relationship. Consumers generally are not worried about food safety;[5] quality means tenderness and juiciness, not food safety. Argentine beef enjoys a good reputation among consumers in many developed countries, so the remaining 20 per cent (500,000 tons) of the production is exported, but mostly as a commodity. The most important market is Europe, which imports almost 50 per cent of total Argentine beef exports. The players align their transactions with a commodity strategy: low frequency of transaction, low asset specificity, and so on. Importers have great bargaining power and sometimes eliminate the possibility of contracts with small or medium-sized companies.

However, food crises have created niche markets of consumers in Europe and other countries (e.g., Chile, Brazil, United States) who demand specialty foods – in this case, beef. They demand not only food safety but also tenderness, ethics in terms of the way the animals are raised and fed (e.g., grassland), organic offerings and traceability. These strategies are not applied in Argentina for either domestic or foreign markets because of the predominance of local consumption (80 per cent of total production) and the lack of institutional or organisational incentives to invest in such systems. Nevertheless, 26 companies (supply chain networks of farmers, which export 2,000 tons of beef) have implemented vertical coordination, from the farm to the supermarkets, and adopted a quality-driven strategy focused on the end consumer. Prinex provides an example of this type of company.

Research Area and Principal Research Questions

The international beef market is booming. Consumers in developed markets demand higher-quality food (beef). Even though the Argentine beef agri-business is oriented toward domestic markets, many companies try to develop their foreign markets, especially those that can coordinate the chain 'from the farm to the table.'

This chapter adopts the new institutional economics paradigm, integrated with a marketing strategies bibliography. Traditionally, economic studies employ the point of view of 'neoclassical economics.' This paradigm attempts to explain the functioning of the economic system according to the following assumptions: an infinite number of buyers and sellers, transparency in transactions, complete contracts, homogeneous products, complete information, factor mobility, freedom to enter and exit, and prices based on supply and demand. According to Hoff and colleagues, 'neoclassical theory explains the economic system adequately when markets 'function' reasonably well, but fails when there are missing markets and price is not all that is needed to adjust and carry out transactions.'[6]

In contrast, the new institutional economics can study a sector or system, such as the beef system described subsequently. The main difference between neoclassical theory and the new institutional economics is that the latter analyses institutions, because the legal structure enforced by laws, contracts and property rights is relevant to the economy.[7] The new institutional economics also poses two propositions: (1) 'institutions do matter' and (2) 'the determinants of institutions are susceptible to analysis by means of the tools of economic theory.'[8]

According to the new institutional economics, institutions are partly responsible for transaction costs and play an important role in the development of trade in goods and services.[9] North states that when transaction costs are high, institutions are important because they affect production and transaction costs.[10]

The new institutional economics focuses on the historical process of institutional change, the economics of property rights, and the transaction cost economic theory of the firm.[11] The great finding of this new theory is transaction costs, which are the *ex ante* and *ex post* costs of a transaction. They are the not-always-visible costs that result from negotiating, planning and carrying out a transaction or those derived from a poor negotiation, contract adjustment, and/or contract safeguards due to errors, omissions or unexpected modifications.[12] They are the 'costs of operating an economic system.'[13]

Ordóñez proposes an interesting addition to the approach of the new institutional economics.[14] At present, international commerce, and particularly the global food business, is in a state of constant growth. Every day, consumers demand more developed products that bear more information. In the current paradigm of food and agri-business, 'the systemic character constitutes one of the fundamental approaches based on the set of intra-systemic and inter-systemic interactions among different actors, not only local or regional but also global. The interactions articulate adaptation of the agrifood business systems including RandD, agriculture, industry, and distribution until consumer demand is satisfied.'[15] This complex model of relationships between agents of agri-food systems, plus international rules to trade in goods and each country's legislation, render the analysis of each agri-food system highly complex. Yet constant changes in global trends of food consumption drive the need to study how businesses respond to changes and satisfy consumer needs,[16] using theoretical frameworks related to marketing and market studies. Moreover, organising and aligning transactions to satisfy intermediate and final clients at regional, national and international levels merits an in-depth analysis. The fusion of the new paradigm of the new institutional economics and the marketing theoretical framework can help illuminate these changes.

The changes in food and agri-business have led firms to join strategic alliances through horizontal and vertical coordination, including mergers with or acquisitions of smaller firms.[17] Consequently, the spot market is no longer the most effective option to maintain competitiveness, reduce transaction costs, and promote continuous improvement of differentiated products, because it only permits trade through coordinated actions (i.e., relational-specific contracts)[18] or vertical integration, such that each party emerges as a unit within the governance structure. These steps are what actual food and agri-business global companies are doing today.

In following this argument and trying to join the new institutional economics with marketing orientation theory, we present the Prinex case study as a means to examine what constitutes best practices in adopting a market orientation. The Prinex history, the way it has coordinated its transactions, its marketing strategies and its constraints and limitations all will be described subsequently.

Methodology

This chapter follows a case study methodology, in line with the phenomenological epistemology proposed by Peterson,[19] because case studies can help clarify the new agri-business paradigm. The objective of this study is to understand the phenomenon within a complex socioeconomic reality and develop non-quantifiable theoretical models, adjusted to the context through induction.

The research is applied and descriptive,[20] and it attempts to explain the determinants of the market orientation of Prinex, a small enterprise in the Argentine beef agri-business. Methodologically, this chapter takes a macro-level and micro-level approach. The study of the sector (macro-level) is based on primary and secondary information sources. In 2007, we conducted interviews with experts in the sector (ten producers, three industrialists, two chamber representatives, three public sector representatives) to identify the sector's institutional, organisational, and technological environments. The face-to-face interviews followed a pre-established questionnaire, but extra questions were posed during the meeting. A bibliographical search and the authors' knowledge of the sector complemented this macro-level approach.

The micro-level approach (Prinex case study) consisted of contacting the CEO of Prinex (twice) and two company employees (three times) through in-depth interviews, also in 2007. Prinex was the first small or medium-sized enterprise to achieve and position its trademark in European markets, making it a leading case. This performance has been described in a previous case study presented by Ordóñez in 1998 at the IAMA Conference,[21] which was instrumental for the current study.

Also, we used an participator-observation method in the agri-industrial chains for the preceding six years, considering research reports and case studies of research groups (Food and Agribusiness Programme-PAA-School of Agronomy-UBA, PROSAP, PENSA, Global Food Network), aimed at characterising the key market-oriented elements and advancing understanding of the development of the sector, following the new institutional economics theory.

The Argentine Beef Sector

Argentine livestock amounts to 55 million head, and beef consumption is 70 kg per inhabitant per year, which represents an increase since the 2002 devaluation of the peso, when consumption was 55 kg/inhabitant/year. The extraction rate is 26 per cent (14 million animals slaughtered). The industry produces an average of 3 million tons, and 24 million cows breed 15 million male and female calves. Argentina's foot-and-mouth disease status is 'free, following vaccination.'

Total Argentine beef exports fluctuate between 300,000 and 700,000 tons. During the 1990s, exports were approximately 400,000 tons. After the peso devaluation in 2002, exports increased to more than 750,000 tons in 2005. This increase was accompanied by an increase in meat production, the price of livestock, and the price of beef for consumers. Exports had never before exceeded 25 per cent of the total production of Argentine beef.

Argentinean slaughterhouses are highly dependant on the Hilton quota, which was assigned to beef producing countries by the European Community in 1979 to introduce high-quality beef in European countries at a lower tariff. Introducing beef outside the Hilton quota earns only 12.8 per cent of the CIF price, plus €3,040 per ton of tariff, instead of 20 per cent of CIF price. Argentina has the highest volume Hilton quota among those subjected to it, at 28,000 tons per year; countries like Brazil, Uruguay, Australia and the United States do not exceed 10,000 tons/year. These volumes enter the European market at much higher prices than those of beef entering outside the quota. For example, prices per ton of beef within the Hilton quota were US$14,400 in March 2008, but they did exceed US$8,000 (March 2008) outside the quota. Before 1994, the Hilton quota was used only by meat processing plants that had access, owing to institutional resolutions. Since then, groups of livestock producers have gained greater access to a percentage of the quota (5 per cent and growing), with strict control parameters (brand development, quality protocols, and so on). This development has been very important to improve the profitability of beef producers and represents an important institutional innovation that leads to new organisational forms.

In examining the institutional environment of the beef sector, we note poor compliance with hygiene and tax regulations. The cattle and beef typing system is poor, value added taxes (VAT) on meat and agricultural products are inconsistent, and there is scant hygiene and tax control. As a result, unfair competition results between companies that export (with higher tax and sanitary control) and slaughterhouses that sell their beef to the domestic markets. Traceability is officially regulated; producers must register their animals for sale in both domestic and foreign markets. However, the local consumption culture does not reward products with certified origin and quality with higher prices, and traceability is not fully implemented in the domestic market.

The organisational environment, in turn, appears disarticulated and atomised. For example, in 2007, the top five meat processing plants that slaughtered represented only 9 per cent of the total slaughter. Currently, there are almost 500 meat processing plants and almost 800 slaughterers (a slaughterer is a person or legal entity that buys livestock, slaughters in plants that belong to others, and sells the carcases to butchers' shops; they often are associated with quick capital rotations, low structure, and informal commerce with very low added-value). In the export channels, concentration is higher (100 plants provide all Argentine beef exports, and 50 per cent of exports are in the hands of 15 companies). Noting beef consumption in the local market, the strategy of most actors is to reduce costs to reduce their prices. Atomisation, low co-ordination, and opportunistic behaviour by agents lead to high transaction costs, increasing the utilisation of the 'spot-market' as a governance structure and offering low incentives for more investments (e.g., in quality standards).

Meat processing plants can be divided into export plants and local consumption plants, according to their health and fiscal requirements. Thus, processing plants for the domestic market have no access to exports and only market their products as carcase meat with very little added value. Export processing plants have higher requirements and more development of brands, origin and quality certifications, which means they find it economically almost impossible to sell beef in the local market due to the lack of demand for such quality. This situation deepens the double standards for hygiene, tax and quality, and harms export processing plants, which face major restrictions on their exporting from both the external market and the government.

In terms of the technological environment, the traditional commercialisation system consists of selling half-carcases with bone (70 per cent of the total traded in the domestic market). Exports, in contrast, are totally segmented: boxed beef distributes the cuts where they are really in demand. It is difficult to add value to the product when marketing half-carcases, because the cuts cannot be identified, marketing strategies cannot be implemented, and many cuts get sold in places where consumers are not willing to pay higher prices.[22] Moreover, Argentina is only in an initial stage of implementing differentiation strategies. In the past ten years, firms that exported to developed countries started to develop fattening, slaughtering, boning, traceability, logistics, and distribution protocols for their products. The drivers for these new systems have been more demanding global consumers and institutional demands (especially in EU countries).

Therefore, the situation for the international beef business between 2002 and 2007 can be summarised as follows:

- greater world demand for beef, resulting from the appearance of new actors
- BSE-free status for countries like Argentina and Uruguay, which results in better prices
- Appearance of BSE in Canada and the United States
- Uruguay's entrance into the US market
- Avian flu, which decreased poultry and increased beef consumption
- Brazil ranked as the main exporter of beef worldwide.

In this situation, Argentine exports increased by approximately 100 per cent in volume, and prices went from an average of US$1,500 to US$3,500 at the end of 2007 (133 per cent increase).

Beef consumption in the domestic market also increased after the peso devaluation in 2002. The increase of the international price of beef has made it possible to increase the buying price of steers for the export meat processing industry. However, it also has generated an increase in the price of animals in the internal market. This price increase has generated higher prices of beef at the counter, which has had a great impact on inflation.

To reduce the prices of beef at butchers' shops and supermarkets, the government implemented a series of measures, including export rights, a minimum slaughter weight, percentage increases in export rights, restriction of beef exports, export quotas, maximum sale prices for retail sales (13 popular cuts), and so on. The result has been a reduction in beef exports, production and product quality disincentives (e.g., certified or branded meats), lower levels of investment in technology, and great institutional uncertainty.

Prinex Case Study

At the end of the 1980s, the business scene in Argentina was extremely turbulent. Inflation and hyperinflation distorted the business picture. Different government measures relating to fiscal, financial, exchange and monetary policies altered the value of the currency and permanently interfered with the movement of relative prices, abruptly modifying the profitability of business transactions in both internal and external markets.

In the agri-business chain, agricultural producers were the most affected because they were discriminated against by the economic policy of the export substitution model. In their daily business practices, the bargaining power of the agri-industrial and exporting sectors were greater. Such behaviours determined the appropriation of scarce available surpluses in the agri-business chain. The agricultural sector, provider of the raw materials, was left in an extremely weak state and finally had to face a deficit.

At the beginning of the 1990s, a group of 11 traditional producers formed Prinex. They had cattle stock of 125,000 animals on 160,000 hectares. However, they first founded Fexport S.A., with the objective of selling grains and oil crops, produced by the farms of the founding members, directly to external markets. This export strategy aimed to overcome several limiting factors that seriously affected the safety and profitability of the agricultural business: lack of market risk coverage, lack of financing at competitive rates, and high transaction costs of traditional commercialisation in the domestic market. Fexport's experience in the export of commodities – 10,000 tons in the first 1987–88 growing season – quickly reached 100,000 tons in the 1992–13 growing season and up to 150,000 tons in 1996, with a turnover of US$13,000,000.[23]

This successful process led to the founding of Prinex in 1989, with the object of adding value to the grains and de-commodifying through industrialisation. Their experience in logistics and commercial practices from the grain export business and the opportunity to access the idle installed capacity of the oil industry to use 'á façon' – leasing facilities of an industrial plant with idle capacity – led to the new enterprise. Prinex exported its first oil and sunflower pellets in 1989, using Cargill's industrial facilities in Necochea, 'á façon'. Oil and pellet invoicing amounted to $6,000,000 in the third year of operations, on 9,200 tons of oil and 10,500 tons of sunflower pellets.

Fexport's grain export activity and Prinex's oil and by-product export activity thus solved part of the problem. Meanwhile, livestock production, accounting for 40 per cent of the producers' shipments, did not participate in the new mechanisms and followed the so-called traditional business model.

High internal market instability, a lack of market coverage, and poor access to financing prompted Fexport's export activity in 1987. Likewise, the critical situation associated with raising Argentine cattle led the group, starting in 1992, to study the possibility of entering the beef business using the Prinex structure. During that year and early 1993, the group studied different options. The first solutions seemed as complex and confusing as the problematic conditions of the market that were forcing the change.

Regarding the origins of Prinex:

'At the worst times, when the crisis looks threatening and there is chaos, the system becomes fractured and empty spaces generate. Within these empty spaces, the opportunity arises to stop doing more of the same. The opportunity opens up for novel ways out. In a crisis, to continue doing the same thing means to continue falling. In order to get ahead, all that is left is to do *new business or do the old business in new ways… The new process should combine horizontal associations to gain scale in the provision of raw materials and different vertical integration activities to add value.' (Luis Piñero Pacheco, former CEO of Prinex)*

Prinex therefore became a homogeneous network of producers with credible commitments. To add value to the animals (and consequently to the beef they wanted to sell), they developed quality and traceability protocols to satisfy global demand for beef with certified origin and quality. However, to guarantee the provision of animals and offer the participants security, they also developed a series of contracts at horizontal and vertical levels[24] They went on to organise a network of certified producers, coordinating the business of specialty meats for export. Vertical coordination involved contracts with meat slaughters and packers to use their idle capacity by processing meat 'à façon', as well as contracts with supermarkets to provide cuts in response to specific demand. The horizontal coordination entailed contracts with farmers to gain scale economies. Finally, they were the first group of farmers to build new ways to produce, process, and distribute beef following a strict protocol of quality and traceability. This protocol defined animal parameters, geographical origins, management and feeding, age and weight before slaughter, grading, slaughtering, boning and delivery procedures.

The design was driven by a strong capacity for association and trust. Prinex was the first collective association of Argentine livestock producers, focussed around a common objective.

PRINEX'S PERFORMANCE

Prinex offers an excellent example of how companies try to function in a globalised world by meeting special consumer requirements. The strategies and performance were tailored to consumer demand, which is why Prinex developed strong contracts with different agents across the value chain. By following the protocols, it was able to build its own competitive advantage: the capacity to generate significant volumes of beef of the highest quality and climb the value chain. Prinex's business strategy is a strategy of market differentiation and segmentation, using the 'Novillo Pampeano' trademark – of known origin and quality – distributing customised cuts to high-end shops (ABC1 consumers), and de-commodifying beef.[25]

Prinex took advantage of the institutional innovation whereby producers and groups of producers could export 1,400 tons within the Hilton quota as of 1994. Prinex's Hilton quota was 120 tons this year. Although there have been no significant changes in quota assignments, there were some changes to its export performance within the quota of the associated meat processing plants (see Table 6.1). Prinex always exported more than it was assigned, except during 2000–2002, when the borders of the main Argentine beef-buying countries were closed due to recurrence of foot-and-mouth disease.

But Prinex not only exported inside the Hilton quota. Exporting through this quota represents much more income than exporting outside the quota (in terms of net income). However, to position its brand ('Novillo Pampeano'), Prinex needed to market additional amounts of beef. For example, in the 2004–05 growing season, Prinex exported more than 4,000 tons of beef under the 'Novillo Pampeano' label to 18 markets abroad. The company turnover for that growing season was US$11 million. Steers keep coming from producers associated to certified production. In all, more than 100 producers sell steers to Prinex,

Table 6.1   Evolution of the Hilton quota of Prinex alone and in association (1994–2005)

Period	Tons		Surplus	Per cent
	Assigned	Exported		
1994/95	120	326	206	172
1995/96	100	231	131	131
1996/97	130	249	119	92
1997/98	140	173	33	23
1998/99	140	102	0	0
1999/00	120	128	8	6
2000/01	118	0	0	0
2001/02	91	0	0	0
2002/03	122	285	163	133
2003/04	88	138	50	56
2004/05	105	465	360	328
Total	**1269**	**2097**	**1075**	**65**
Average	115.36	190.63	97.72	

totalling more than 14,000 head of cattle slaughtered each year. 'Novillo Pampeano' also gained a position as the most important Argentine beef brand in Europe.

MARKETING STRATEGY

Through institutional, organisational and technological innovations (multi-dimensional co-innovation[26]), Prinex managed to enter the most select market segments (ABC1) of more highly developed countries and thus obtained higher prices. The beef's origin – in 'las pampas argentinas' (the Argentine pampas), highly recognised in Europe – has been very important for positioning the brand and gaining shelf space in supermarkets.

Prinex originally obtained an additional 20 per cent mark up in its final prices compared with the average Hilton quota selling price. In Spain, Prinex even sold for more than 45 per cent more than the price of its Argentine competitors. Because of its capabilities in animal traceability, boxed beef, and full compliance with the protocol,

'Prinex is able to sell cap of rump to the best Brazilian customer, tenderloin to Corte Inglés in Spain, and silverside fillets to Chile. The highest prices customers are willing to pay for each cut go directly to the farmers. That is why the Prinex motto is: A cut for every customer and the proceeds for the farmer who delivered quality' (Carlos Odriozola, CEO of Prinex).

Prinex considers each of its stakeholders a client. Producers must have a share in Prinex's business for the company to gain their loyalty and induce them to deliver the animals in time, according to the quality parameters specified by the protocol. Therefore, producers receive bonuses based on the quality of the animals they deliver. Meat processing plants are guaranteed a certain volume of animals per week, which contributes to full use of the facilities and the optimisation of fixed costs. In addition, Prinex often incorporates new technologies in processing plants that may be used later to its advantage. Finally, supermarkets may adjust the prices that Prinex sets, based on the reputation and position of the brand according to the consumer. In summary, it is a win–win situation for all the actors in the system, adding value to the product before it reaches the end consumer.

Case Restrictions

The livestock and meats sector in Argentina undergoes constant change and faces great uncertainty, especially at the institutional level. For example, changes in export taxes are extreme. In 2004, export taxes were 5 per cent of the price; in 2005, they were 10 per cent, and in 2006, they reached 20 per cent. No one knows what they will be in the future. The lack of definition of VAT reimbursements also means that though most export companies suffer a delay of VAT reimbursement of approximately 2 months, some cases indicate they have waited up to 11 months. Commercialising beef in the domestic market remains a complicated business; there are different shades of grey when it comes to hygiene and tax issues, and highly unfair competition occurs between those that export and those that sell in the local market, mostly due to the different controls and demands. It is impossible to estimate a future buying price for the animals, because there is no futures market for livestock. Finally, local consumers are not willing to pay premium prices for quality certified beef.

The Argentine business paradigm suffers strong path dependency on the domestic market, and Prinex, as an export company, has undergone from many constraints and limitations to stay in business. The main restrictions relate to external markets. For example, European countries impose heavy taxes on Argentine beef that enter outside the Hilton quota (approximately 50 per cent). Foot-and-mouth disease, recurrent in Argentina, and the health threats related to this disease constitute another important restriction, because it prevents Argentine beef's access to high purchase-value markets (e.g., Japan). The frequency of the recurrence of the disease creates a poor image for the country and thus the closing of some markets.

Argentina also imposes strong restrictions on Prinex at the domestic level. The financial cost of paying taxes and reimbursements, tax evasion, the arbitrary assignation of the Hilton quota, and the high cost of implementation owing to the technology associated with processes and products (i.e., high level of investment in specific assets and poor access to affordable loans) are clear examples that limit the development of the business.

Regarding financial costs, the situation can be summarised as follows: for a steer, Prinex pays its associate producers a price plus VAT (10.5 per cent), and it must pay a 15 per cent export tariff. However, the VAT is an internal tax, which the state must reimburse in the short term to avoid financial imbalances; it also must reimburse 5 per cent as a promotion to beef exports. Many times the VAT reimbursement and other repayments do not occur in due time and form, which leaves the state-owing Prinex 15.5 per cent – an amount that often represents the company's profit margin. For example, in 1998, Prinex suffered a delay of 11 months for its VAT reimbursement, which created a heavy financial burden of approximately US$1,500,000. These delays are a consequence of the high levels of fiscal evasion in the system, which harms those companies that operate formally.

The Hilton quota and its allocation system also hurts such businesses. Without a Hilton quota, companies would wind up working almost at the limit of profitability. As Carlos Odriozola, the CEO of Prinex, recognises,

'What is exported to Europe outside the Hilton quota is more for positioning than for profit. When you sell outside Hilton you operate at the limit of the business margins and sometimes even at a loss. When you sell outside Hilton to Europe, most of the time the margin is below 15.5 per cent and that goes against you financially; you are paying livestock plus VAT, you are not receiving the reimbursements in due time and form, and these rise to 15.5 per cent.'

Therefore, the amount of Hilton quota a company receives is an important limiting factor, because companies can generate more profits through the Hilton quota and increase the number of producers, animals and exported beef. This situation explains why Prinex associates that have processing plants 'sell' their quota to Prinex to market their beef under the 'Novillo Pampeano' denomination. This sale provides benefits for both Prinex and the processing plant and ultimately builds trust between the parties.

Brand development, investment in fixed assets for business development, and strict control of quality protocols are highly specific assets. Poor access to credit for agricultural producers and industrialists also makes investment difficult. Therefore, the number of producers who associate together to gain scale is limited, and it is hard to expand the markets. In general, these projects find a safe number of producers and adjust their sales volume to the supply of beef. In this way, they reach a stable balance between production and supply of beef, and new markets seldom open.

Finally, the most significant issue for Prinex has been the policy of export restrictions or closures based on export taxes. The company takes a position in the market, makes contracts with suppliers and buyers, and invests in promotion and positioning in those markets where it sells products. In the face of export restrictions, even if the trademark is well positioned, it is very difficult to sustain the business, and the company must restrict its activity to limited exports, according to the specific openings in the market.

Conclusions

Prinex offers a leading case that describes the development of a beef trademark, marketed from Argentina to export markets, with all the constraints faced by small and medium companies. The 'Novillo Pampeano' brand provides a pull force, from consumer to farmers. The network of farmers, slaughterhouses, supermarkets and brokers provides a basis for quality assurance and customer satisfaction. End consumers, satisfied with the offered quality and services (information), are willing to pay extra, but the power of Prinex rests with how this information gets translated to appeal to all the participants in the network.

This case confirms the high cost of implementing new organisational designs for beef in Argentina, owing to both tangible and intangible assets. These assets include new machinery for beef cutting and slaughtering, training of personnel, new types of packaging and packing machinery, logistics investments, cooling capacities, costumer demand knowledge and just-in-time (JIT) organisation. These assets are specific, which makes the transaction through market governance structures, which require contracts or hierarchies in some cases, more difficult.[27] The institutional environment constrains their ability achieve credits, and Prinex partners must in invest in all assets on their own.

Generally speaking, this kind of project finds an established market (niche), with established volumes, resulting in an established number of producers. Investments and uncertainty in this sector limit the possibility of incorporating new members. Moreover, actual agri-business (crops) in Argentina are more profitable than the beef sector, and producers are daily exiting the business. Should Prinex invest in new markets and products, new machinery and know-how, and more beef production, or should it continue with low-scale production and pursue niche markets? The institutional environment suggests the latter choice.

Acknowledgments

The authors thank the Prinex CEO and Prinex staff for their invaluable contributions to this document. We appreciate all the knowledge transferred by Héctor Ordóñez.

References

1. Gellynck, X. and Verbeke, W. (2001), 'Consumer perception of traceability in the meat chain', *Agrarwirtschaft*, vol. 50, no. 6, pp. 368–74.
2. *The Economist* (1998a), 'The science of BSE: Bungled', March 14, pp. 21–3; *The Economist* (1998b), 'The science of BSE: Birth of a disaster', March 14, p. 22.
3. Licking, E. and Carey, J. (1999), 'How to head off the next tainted-food disaster?' *BusinessWeek*, March 1, p. 34; Schaffner, D., Schroder, W., and Earle, M. (1998), *Food Marketing: An International Perspective*, McGraw-Hill, New York.
4. Gilg, A.W. and Battershill, M. (1998), 'Quality farm food in Europe: A possible alternative to industrialised food market and to current agri-environmental policies: Lessons from France', *Food Policy*, vol. 23, no. 1, pp. 25–40.
5. Palau, H. (2005), 'Agronegocios de Ganados y Carnes en la Argentina: restricciones y limitaciones al diseño e implementación de sistemas de aseguramiento de origen y calidad. Estudio de caso múltiple', Master's thesis, Food and Agribusiness, School of Agronomy, UBA.
6. Hoff, K., Braverman, A., and Stiglitz, J. (1993), *The Economics of Rural Organization. Theory, Practice and Policy*, Oxford University Press, Oxford.
7. Ibid.
8. Matthews, R.C.O. (1986), 'The economics of institutions and the sources of economic growth', *Economic Journal*, vol. 96, pp. 903–18.
9. Coase, R. (1937), 'The nature of the firm', *Economica*, vol. 4; Coase, R. (1960), 'The problem of the social cost', *Journal of Law and Economics*, vol. 3.

10. North, D.C. (1990), *Institutions, Institutional Change and Economic Performance*, Cambridge University Press, Cambridge; North, D.C. (1994), 'Economic performance through time', *American Economic Review*, vol. 84, no. 3, pp. 359–68.

11. North, 1990, op. cit.; Demzsetz, H. (1967), 'Toward a theory of property rights', *American Economic Review*, vol. 57, pp. 347–59; Williamson, O. (1985), *The Economic Institutions of Capitalism*, The Free Press, New York.

12. Williamson, O. (1993), 'Transaction cost economics and organizational theory', *Journal of Industrial and Corporate Change*, vol. 2, pp. 107–156.

13. Williamson, 1985, op. cit.

14. Ordóñez, H. (1999), 'Nueva economía y negocios agroalimentarios', *Programmea de Agronegocios y Alimentos. School of Agronomy, UBA*.

15. Ibid.

16. Palau, H., Glade, S., Otaño, C., and Ordóñez, H. (2006), 'New market segments and consumer strategies in Argentina: Pre and post devaluation scenarios', *paper presented at the 16th International Food and Agribusiness Management Association (IAMA), World Food and Agribusiness Symposium*, Buenos Aires, Argentina, June 10–13.

17. Kherallah, M. and Kirsten, J. (2001), 'The new institutional economics. Application for agricultural policy research in developing countries', *Markets and Structural Studies Division, International Food Policy Research Institute*. Available at http://www.ifpri.org/divs/mtid/dp/papers/mssdp41.pdf. Accessed March 14, 2009.

18. Streeter, D., Sonka, S., and Hudson, M. (1991), 'Information technology, coordination and competitiveness in the food and agribusiness sector', *American Journal of Agricultural Economics*, vol. 73, no. 5, pp. 1465–71.

19. Yin, R.K. (1989), *Case Study Research: Design and Methods*, Sage, Newbery Park, CA; Peterson, H.C. (1997), 'The epistemology of agribusiness: Peers, methods and rigor', *Agribusiness Research Forum*.

20. Gil, A.C. (1994), 'Métodos e técnicas de pesquisa social', *Ed. Atlas*, São Paulo, Brazil.

21. Ordóñez, H. (1998), 'Alternative chain management in beef agribusiness. The PRINEX case', *VIII IAMA World Congress Food and Agribusiness*, Punta del Este, Uruguay, June.

22. Ordóñez, H. (2001), 'El sistema media res, la causa de la crisis ganadera', *BAE Rural*, June 22.

23. Ordóñez, H. (1998), op. cit.

24. Ordóñez, H. (1998), op. cit.

25. Ordóñez, H. (1998), op. cit.

26. Ordóñez, H. (1998), op. cit.

27. Williamson, O. (1985), op. cit.

7 *Agricultural Cooperatives and Market Orientation: A Challenging Combination?*

BY JOS BIJMAN*

Keywords

agricultural cooperative, market orientation, restructuring

Abstract

In this chapter, I argue that farmer-owned cooperatives have gone through major restructuring processes during the past 15–20 years. These changes may relate directly to the need for cooperatives to become more market-oriented. Increased attention to the demands of consumers and retailers has led to strategic reorientation, as well as organisational restructuring. As a result, the relationship between farmers and the marketing/processing cooperative has been re-engineered. In addition, new types of producer-owned firms have been established in north-western Europe. I will in this chapter describe the role of producer-owned cooperatives in marketing farm products; suggest several propositions about the relationship between market orientation and restructuring of cooperative firms; present detailed information about restructuring processes among agricultural cooperatives; and conclude that both restructured cooperatives and new producer organisations are suitable and viable business forms for marketing agricultural products.

Introduction

In 1989 an influential report was published in the Netherlands arguing that the companies processing and marketing farm products were not sufficiently market-oriented.[1] They

* Dr Jos Bijman, Wageningen University, Department of Business Administration, Hollandseweg 1, 6706KN Wageningen, the Netherlands. E-mail: jos.bijman@wur.nl. Telephone: + 31 317 483831.

were too much focused on bulk production, too little on consumer responsiveness, and were not sufficiently responsive to (changing) consumer preferences and shifting market conditions. The final conclusion of this report was that Dutch agriculture and agri-business were losing their competitiveness, in both domestic and foreign markets. The arguments and recommendations then were again published for an international audience and in an academic journal.[2]

The Van der Stee report had a major impact on Dutch agri-business. Not only the processing and marketing firms but also farmers and even the Ministry of Agriculture took these warnings seriously. Individually and collectively – agriculture and agribusiness has always been a tight network of public and private actors – firms started to develop new strategies. As a large number of these agri-business firms were (and are) cooperatives, in which farmers maintain control, changing the strategy was only possible as result of close consultation among producers, their representatives on boards of directors, and managers.

In other countries, cooperative and non-cooperative food companies also were advised to change their strategies and structures in response to changing market conditions in domestic and foreign markets. In Ireland, the Irish Cooperative Organisation Society, during the late 1980s, urged the Irish dairy industry to build a small number of large firms with the scale, cost efficiency and resources to develop new products and compete in foreign markets.[3]

Whether due to expert reports or internal arguments, it is clear that major changes in strategies and structures have taken place among agricultural cooperatives in north-west Europe during the past 20 years. Throughout the twentieth century, cooperatives have been stable organisations, with a strategy that was clear to both internal and external stakeholders. The main forms of restructuring were the many mergers of local cooperatives into larger organisations, but these mergers did not have a major effect on form or function. Since the early 1990s, however, the inertia of the cooperatives was no longer an asset. In a rapidly changing market environment (partly due to changing public policies), cooperatives were forced to respond to the challenges of a much more competitive market.

This chapter describes and explains the changes among agricultural cooperatives in north-west Europe. Are the changes specific to particular sectors or countries, or are they part of a larger development? In addition, how have the special characteristics of farmer-owned cooperatives, which make them different from investors-owned or family-owned companies, affected the strategic reorientation?

This chapter is based on a literature review, as well as unpublished information gathered through many interviews and discussions with managers, farmers, board members and experts. Given the background of the author, relatively more examples will come from Dutch cooperatives. However, findings and examples from other countries will show that the trends described and analysed in this chapter are not confined to the Netherlands.

The structure of this chapter is as follows: The next section gives an outline of the basic characteristics of farmer-owned cooperatives. Section 3 presents figures about the importance of cooperatives in marketing farm products. Section 4 details the main propositions about the relationship between cooperative restructuring and market orientation. Section 5 contains empirical information about the restructuring processes among marketing cooperatives in north-western Europe. Finally, Section 6 concludes with a brief discussion of new challenges for market-oriented cooperatives.

A Primer on Agricultural Cooperatives

An agricultural cooperative is a firm owned jointly by farmers. The main purpose of the cooperative is to support farmers in their individual farming activities. Such services can be purchasing or producing inputs (e.g., fertilizers, animal feed), provision of credit, marketing farm products, processing farm products, provision of insurance and providing technical assistance. These activities have a minimum efficient scale that goes well beyond the individual farm. By establishing a jointly owned firm that carries out these activities, farmers benefit from economies of scale. As they are the owners of the firm, transaction costs in the producer–processor relationship are low. In addition, by purchasing and selling collectively, farmers have established countervailing power with regard to sellers and buyers that are much larger and therefore have more market power. This chapter focuses on cooperatives that provide processing and marketing services.

A commonly used definition of a cooperative comes from the International Cooperative Alliance (ICA), an international interest group for cooperatives: 'A cooperative is an autonomous association of persons united voluntarily to meet their common economic, social, and cultural needs and aspirations through a jointly owned and democratically-controlled enterprise' (see www.ica.coop.org). This definition indicates one of the most interesting and at the same time most challenging characteristics of the cooperative: its double nature.[4] A cooperative is both an association (i.e., society) of members and an enterprise (the cooperative firm) in which economic activities are conducted. The association is the locus of collective action, while in the firm, the economic benefits for the members are obtained. In practice, there is no clear-cut distinction between these two organisational elements, certainly not in the mind of the members, but it may be helpful from an analytical point of view to make this separation. Societal trends have differential impacts on the association and the enterprise.

Conceptualising the cooperative firm as a service provider that works for the benefit of the members does not mean that it does not have interests in itself. Indeed, shifting to a more market-oriented strategy implies that the cooperative firm no longer positions itself as a purely dependent firm but increasingly emphasises its own role and responsibilities with regard to retailers and consumers. One of the recommendations of the Van der Stee report was that the new market conditions require the cooperative firm to become more entrepreneurial.[5]

What impact does the development toward a more market-oriented and entrepreneurial cooperative firm have on the relationship with the members of the cooperative? To answer this question, it is helpful to distinguish three elements of the member–cooperative relationship. Much of the literature on agricultural cooperatives uses a three-fold relationship: member-benefit, member-control and member-ownership.[6] Member-benefit means that the members are the primary stakeholders to benefit from the cooperative, and this benefit is obtained by (individually) using the services of the cooperative. Member-control indicates that the rights to decide which strategies and policies the co-op will follow ultimately lies with the member-producers. In practice a board of directors, chosen by and from the members, makes the strategic decisions. Member-ownership signifies that the equity capital in the cooperative is provided by the members. Using the terminology of the property rights theory, income rights are held individually (through transactions between the individual member and the cooperative), whereas decision rights are held collectively. As I argue subsequently, these particular organisational characteristics have consequences for obtaining additional equity capital, (speed of) decision making, innovation, diversification, growth and internationalisation.

Cooperatives have been established to support the economic well-being of their member-producers. This general purpose, however, can be translated into various functions that the cooperative might perform. Historically, cooperatives have focused on a limited number of objectives, such as (1) overcoming market failures, especially when access is constrained or when markets do not exist; (2) gaining economies of scale; (3) strengthening bargaining power in the relationship with customers or suppliers; (4) sharing risks, such as market risks or those risks related to natural conditions; (5) reducing transaction costs, such as by making the market more transparent; and (6) innovating, which often requires investments and entails risks that cannot be borne by individual producers. These were the main arguments used by farmers that have set up cooperatives in the past 100 years. They may still apply today, in both the developed and the developing world.

Globally, there is a revival of attention to cooperatives as a tool to strengthen the market access for and competitiveness of farmers. Now known under the name 'producer organisations,' farmer-owned cooperatives and associations are considered institutional innovations that can help small-scale farmers improve their access to more demanding markets.[7] Particularly for farmers of high-value products, such as fruit and vegetables sold to supermarkets or organics and fair trade products sold to international customers, the producer organisation is an appropriate organisational structure to support efficient transaction between farmers and their customers by improving, maintaining, and guaranteeing food quality and safety. The role of producer organisations in development was has been emphasised in the World Bank's World Development Report 2008.[8]

The Role of Cooperatives in Marketing Farm Products

Throughout the twentieth century, cooperatives have been important vehicles for marketing agricultural products. In fact, the history of most existing cooperatives dates back to the nineteenth century. In the Netherlands, cooperative legislation was introduced in 1876, and since then, many cooperatives have been established. The largest number of new establishments appeared in the period 1890–1914, a time of economic growth in Western Europe.[9]

Information on the 1995 market shares of agricultural cooperatives in the European Union was published by Van Bekkum and Van Dijk.[10] These authors discuss the main trends among agricultural cooperatives, as well as data on the importance of cooperatives on a country-by-country basis. More recent figures (see Table 7.1) have been published by Juliá-Igual and Meliá.[11] Although marketing co-ops can be found in all European countries and for all agricultural products, they are particularly strong in north-western Europe and for commodities like milk, meat, fruits and vegetables, and cereals. In addition, cooperatives have substantial market shares in the production of animal feeds and the retailing of fertilizers, seeds and farm equipment. Finally, in many countries, the provision of rural credit is strongly dominated by cooperative banks. The importance of cooperatives in the marketing of farm products has not diminished over the years. For those countries for which longitudinal data are available, the trend in market share by cooperatives is

Table 7.1   Market share (in %) of cooperatives in marketing farm products, 2003

	Milk	Fruit and vegetables	Meat	Cereals
Austria	94	n.a.	20	60
Belgium	50	80	25	40
Denmark	97	30	90	80
Finland	97	n.a.	74	n.a.
France	37	35–50*	34*	74
Germany	68	45	35	n.a.
Ireland	97	n.a.	70	69*
Netherlands	85	60	35*	n.a.
Sweden	90	n.a.	30	70
UK	55**	25–40**	10–25**	25**

* 1998 data.

** 1999 data.

Source: Based on COGECA data, Juliá-Igual and Meliá.

stable or even upward. Table 7.1 also shows that cooperatives are particularly dominant in the processing and marketing of milk. Among the 20 largest agricultural cooperatives in Europe, 8 are dairy co-ops.[12] Outside Europe, cooperatives also dominate the dairy industry, as in New Zealand and the United States.[13]

A special type of marketing cooperative that has been particularly important in the Netherlands is the auction. The grower-owned auction is the dominant sales organisation for ornamentals (cut flowers, potted plants) in the Netherlands, with almost 90 per cent of domestic ornamentals (total production value of €3.9 billion) sold through one of the four auctions. The flower auction was originally established to lower transaction costs (particularly information costs) in a non-transparent market with many different buyers and sellers and a large number of perishable products. Although information costs have been greatly reduced, due to modern information and communication technology, the market has not become less complex. The largest auction, FloraHolland, daily completes approximately 80,000 transactions with one of the 10,000 suppliers and one of the 5,000 buyers. The auction provides the growers with a low-cost marketing tool, allowing them to specialise in production activities. The auction basically is an organised spot market in which, because the auction is grower-owned, the interests of the growers are protected.

In contrast to the continuing importance of the flower auction, the auction for fruit and vegetables is no longer an important marketing tool in the Dutch horticultural industry. Until the early 1990s, for fresh produce, the auction was the dominant sales channel. However, exactly because of the need to become more market-oriented, producers have transformed the traditional auction cooperatives into marketing cooperatives that now use brokerage as the dominant price-determination method and perform logistic and trade functions that are necessary when acting as a preferred supplier to major retailers. The demise of the traditional vegetable auction was accompanied by the rise of many new cooperatives in the fresh produce industry.[14] These new cooperatives (or producer organisations) are more market-oriented and stress the importance of product innovation, of close collaboration with customers, of high food quality and safety standards, and of environmental friendly production.

Propositions

In the past two decades, market conditions for agricultural products, particularly for food products, have changed dramatically, not only because of changes in public regulation (e.g., restructuring of European agricultural policies, more strict food safety legislation), but also because of changing consumer preferences and alterations in market structures, particularly in food retail. Consumers, compared with 20 years ago, demand higher quality, more variety, year-round availability and safety guarantees. The last has become particularly crucial after major outbreaks of animal diseases such as foot-and-mouth disease, swine fever and avian flu. In reaction to consumer concerns, governments have made food legislation stricter and enhanced safety controls, and industry has responded by introducing safety control systems such as HACCP and tracking and tracing.

Food retail has become very concentrated in the past 15 years. Oligopoly market structures have strengthened the bargaining power of supermarkets. In addition, the introduction of private labels, private food quality standards (e.g., BRC, Eurep-GAP), and preferred supplier purchasing systems have affected the structure of the agri-food supply chains. Suppliers, whether they are wholesalers or manufacturing companies, have responded by strengthening their bargaining power, such as through mergers and acquisitions, both domestically and internationally.

In response to these changing market conditions, cooperatives were pressed to shift from a producer (or product) orientation toward a market orientation. Following Narver and Slater, we define market orientation as the combination of a competitor orientation, a customer orientation and an inter-functional coordination.[15] Competitor orientation means gathering information about what competitors are doing and using this information in strategic decision making. Customer orientation implies acquiring information about the demands and requirements of customers and using that information throughout the firm. Inter-functional coordination means that different departments of the firm collaborate to create customer value.

The issue of inter-functional coordination is particularly challenging in a cooperative, because the production function and the marketing function are not executed by one firm; production is carried out by the members, while marketing is conducted by the cooperative firm. When increasing market (or customer) orientation requires closer coordination between production and marketing, more centralised decision making is required.[16] This shift implies that the cooperative firm will obtain more control over the transactions with members (e.g., determining minimum quality, type of packaging, modes of transport). Therefore the following propositions:

P1   Market orientation will lead to a shift in control from the members (represented by the board of directors) to the management of the cooperative firm.

Becoming market-oriented also requires obtaining marketing knowledge, not only by setting up or strengthening a specific marketing department but also by having marketing knowledge in the main decision-making bodies of the firm. Given the important position of the board of directors in a cooperative firm, the board might seek to include marketing expertise, and therefore:

P2   Market orientation will lead to incorporating more marketing expertise in the management team, the board of directors, and the board of supervisors of the cooperative firm.

Market orientation also means that firms behave more strategically; that is, they consider their choices and activities in competition with other firms. A major tool in maintaining or even improving competitiveness is innovation. Market orientation may lead to a continuous product innovation effort, not only to create additional value for customers but also to maintain and strengthen the firm's reputation. Therefore:

P3   Market orientation will lead to greater innovativeness in cooperative firms.

Market orientation further might require a larger scale of operation, for several reasons. Innovation is an expensive and risky endeavour. It is common to use equity capital to finance R and D. In cooperatives, equity capital is provided by the members, who must make a trade-off between more investment in their own farm and more investment in the jointly owned firm. Farmers may not be able or willing to invest additional capital in the cooperative firm, particularly if this money is used for risky investments. Thus, cooperatives that follow a strategy of product innovation will seek additional sources of equity capital. Therefore:

P4   Market orientation will lead to a restructuring of the financing of the cooperative firm.

Finally, innovation may not be the only reason for pursuing firm growth. Another reason is the company size needed to establish consumer brands. Developing and maintaining consumer brands requires substantial and often sunk investments. Moreover, becoming a preferred supplier of large retail companies and supplying all products in a particular category (e.g., all fruit and vegetables, all dairy products) requires a large scale of operation. Therefore:

P5   Market orientation will lead to firm growth, such as through mergers, acquisitions, and internationalisation.

Restructuring of agricultural cooperatives

Agricultural cooperatives have responded to these challenges in various ways. Some responses have been studied extensively and mapped quantitatively; others have received less scholarly attention but can be deduced from more qualitative sources, such as case studies and professional media. Cooperatives and their members have revised their strategies and structures in the following directions: (1) growth through mergers, acquisitions, and internationalisation; (2) new financial structures; (3) restructuring corporate governance; and (4) establishing new producer organisations. The following sections briefly discuss each of these issues.

GROWTH THROUGH MERGERS, ACQUISITIONS, AND INTERNATIONALISATION

The traditional growth strategy of agricultural cooperatives has been horizontal integration, mainly through mergers. Cooperatives were often price takers and did not consider themselves competitors. As cooperatives received constant urging from their members to improve efficiency and provide services at the lowest cost, mergers among cooperatives were a well-known and relatively smooth process in the agricultural industry. This strategy of continuously seeking economies of scale in operations worked particularly well in situations of market protection, such as the agricultural policies of the European Union.[17]

However, starting in the early 1990s, cooperatives increasingly felt that this continuous quest for economies of scale was not sufficient to improve or even sustain competitiveness. Changes in consumer preferences, changes in the Common Agricultural Policy of the EU and changes in retail concentration led cooperatives, similar to other food companies, to pursue other growth strategies. The main arguments to defend a proposed merger shifted from economies of scale in operations to countervailing power and scale for product development and brand building. For instance, when four dairy cooperatives in the northern part of the Netherlands merged into Friesland Dairy Foods, in 1998, one of the main arguments they offered was the need to establish a countervailing power against the large domestic retailers.[18]

The 2008 merger between the two largest Dutch dairy cooperatives into Friesland Campina, which became the largest dairy cooperative in Europe (with an expected turnover of more than €9 billion), was justified through several arguments. First, in Europe, the pressure on profit margins is rising due to the consolidation and growth of private labels and discounters. Second, outside of Europe, supermarkets are increasing their share of food retail. Third, large non-co-op dairy firms, like Danone, Nestlé, and Lactalis, are expanding their worldwide positions.[19] As these arguments show, operational motives (economies of scale) have been replaced by strategic motives, such as horizontal competition with other food companies and vertical competition with retailers.

Competing with non-cooperative companies in the market for final food products (i.e., consumer products) requires a scale that often exceeds national borders. Competitors are huge multinational food companies, like Nestlé, Danone and Unilever. Because the degree of specialisation possible is determined mainly by the size of the market, companies from small countries are forced to internationalise their activities to become cost efficient. Particularly for cooperatives from small European countries like the Netherlands, Belgium, Ireland and Scandinavian countries, the domestic market provides only limited opportunities for growth.

Internationalisation of cooperatives can proceed along different lines, involving more or less embeddedness in foreign economies. Just like any other firm, co-ops can acquire foreign companies or newly establish foreign subsidiaries. But more interesting is that cooperatives may have members in different countries. For a long time, foreign membership was rare, but since the 1990s, that has gradually been changing, particularly for cooperatives from small countries. Examples include the 2000 merger of the dairy cooperatives MD Foods (Denmark) and Arla (Sweden) into Arla Foods and the 2001 merger of the dairy cooperatives Milchwerke Köln/Wuppertal (Germany) and De Verbroedering (Belgium) into Campina (the Netherlands). Both Arla Foods and Campina were truly international cooperatives. In the winter of 2004–05, they entered into merger talks, though this merger was never realised owing to their insurmountable differences in financial structure and their inability to come to an agreement about the new leadership of the company.[20]

The internationalisation amongst dairy cooperatives also has been influenced by changing EU dairy policies (direct results of the 1994 GATT Uruguay Round agreements). The resulting decrease in subsidised exports and enhanced access to internal markets increased competition for European dairy producers. The cooperatives were used to following a low cost strategy, focusing on processing large quantities of milk, which meant they were more affected by these changes than their non-cooperative competitors, which had always put more emphasis on developing and selling branded consumer products. Depending on their product range and traditional export markets, companies experienced varying effects. Friesland Dairy Foods in the Netherlands and MD Foods (now Arla Foods) in Denmark were vulnerable to the changing import and export regulations. Both firms exported most of their cheese production to the world market, and they were particularly affected by lower subsidies. MD Foods experienced 23 per cent lower export subsidies for mozzarella, 27 per cent for feta, and 58 per cent declines for cream cheese.[21] Lower export subsidies were doubly disadvantageous for Friesland: exports of its cheese and milk powder to the world market became less competitive, and purchases of milk powder by Asian subsidiaries became more expensive as world market prices rose.

MD Foods reacted to reduced export subsidies with a strategic reorientation toward the EU market.[22] Not only exports but international production shifted toward Europe, particularly Germany. Other elements of MD Foods' reorientation included a shift from bulk commodities with low value-added to branded products with higher value-added. However, the latter shift required more investments in marketing, for which MD Foods, as a cooperative, had difficulty obtaining funds. In 1989 it established MD Foods International (MDI) for all its international production activities. As a limited liability company, MDI attracted institutional investors, who owned 49 per cent of its shares. MDI's biggest venture had been the acquisition of several UK dairy companies, making it the third largest dairy company in the United Kingdom (in 1998). By 2003, the British subsidiary of Arla Foods, the successor of MD Foods, merged with the UK Express Dairies to make Arla Foods the largest supplier of dairy products. Obtaining branded food positions in the UK retail sector was one of the key motives for MD Foods/Arla Foods to expand in the UK dairy industry.

Changes in agricultural policies have been a motive for changes in the strategies of the European dairy companies, but they cannot provide a sufficient explanation. Dairy cooperatives also have become more international to benefit from economies of scope in product innovation and marketing. For example, the Dutch dairy cooperative Campina always had a strong position in the consumer market for dairy desserts in the Netherlands, but it has further developed along this path by acquiring dairy companies in Germany and Belgium. In 1994, Campina took over the German dairy company Südmilch, which possessed strong consumer brands (such as *Landliebe*) in its regional and national dairy market. In 1997 Campina formed a joint venture with Milchwerke Köln/Wuppertal, which also was a major producer of dairy desserts. The Milchwerke Köln/Wuppertal cooperative merged with Campina in 2001.

One of the key questions, of course, is whether internationalisation leads to better performance. Measuring performance is always a tricky issue in cooperatives, because financial performance indicators may not be the best criteria for a firm whose main purpose is to serve its member-users. However, Ebneth and Theuvsen[23] find that the most international dairy cooperatives (i.e., Arla Foods, Campina, and Friesland) are also the most successful in terms of financial performance. At the same time, dairy cooperatives in Germany are mainly oriented toward their domestic market. Their cost-leadership strategy and focus on standardised low-cost and low-price mass market products (milk, milk powder, butter) generally leads to lower performance. This approach results in a weak market position and limited financial resources for establishing international business activities. Ebneth and Theuvsen argue that despite the structural changes taking place in the German dairy market, more than 100 dairies are still engaged in ruinous predatory competition. The concentration ratio within the dairy industry is substantially lower in Germany than in other countries.

As Proposition 5 states, market orientation should lead to firm growth through mergers and acquisitions, both domestically and internationally. Although few quantitative studies investigate this relationship, several case studies and anecdotal information suggests that market orientation (e.g., expressed through brand building) correlates with firm growth and internationalisation.

NEW FINANCIAL/OWNERSHIP STRUCTURES

In a cooperative, the capital relationship between members and the cooperative firm is subordinated to the transactional relationship; farmers do not consider themselves investors in the cooperative. In practice, farmers generally receive a fixed return on their capital inlays. This financial relationship is one of the key issues in the restructuring process. Cooperatives seeking international expansion, developing new products, and building consumer brands require substantial risk capital. For the board of directors, this demand implies finding an answer to two questions. First, how can we encourage our members to put additional risk capital in the cooperative? Second, when members cannot provide enough equity capital, how can we invite outsiders to invest in the cooperative without losing member control over the cooperative?

Several authors present overviews of different organisational models for cooperatives that aim to attract additional equity capital. Nilsson distinguishes between five cooperative models (Table 7.2):[24] Model 1 is the traditional cooperative (only members invest in the cooperative and all equity capital is collective). The other models are called entrepreneurial models, because Nilsson emphasises that only when the co-op seeks to become (more) entrepreneurial does it feel the need to restructure its financial basis (i.e., entrepreneurial can be interpreted as market oriented). The main distinction among these five models pertains to ownership rights, which can be collective or individual and can be held only by members or by members and external investors. Model 2, the participation share cooperative, has partly individualised its equity capital; members may also become individual shareholders in the cooperative (and these shares can be sold

Table 7.2   Five types of financial structures

	Type of equity capital		Providers of equity capital	
	Collective	**Individual**	**Members**	**Non-members**
1. Traditional cooperative	X		X	
2. Participation share cooperative	X	X	X	
3. Co-operative with subsidiaries	X	X	X	X
4. Proportional tradable share cooperative (or new generation cooperative)		X	X	
5. PLC cooperative		X	X	X

Source: Based on Nilsson.[24]

or redeemed by an internal share market). Model 3 is the cooperative with a subsidiary in which outside investors participate, such as is the case for MD Foods International. Equity capital is partly collective, partly individual, and both members and outsiders own shares in the subsidiary. Model 4 is the cooperative with proportional tradable shares (in the United States, called a new-generation cooperative). Only members are shareholders, and all equity capital is individually owned. Finally, Model 5 is the PLC cooperative, which means the cooperative is listed on a stock exchange, and outsiders may own and trade the common stock.

Chaddad and Cook[25] also use an ownership perspective in their typology of cooperative organisational models and define ownership as the combination of residual control rights and residual claim rights.[26] The authors distinguish seven discrete ownership models, ranging from the traditional cooperative at one extreme to the investor-owned firm as the other. In between are five models that discriminatively combine the ownership structure of traditional cooperatives (member/users have full residual control and claim rights) with those of investor-owned firms (investors hold all ownership rights). The six alternative cooperative models take the following labels: traditional, proportional investment, member-investor, new generation, capital-seeking entities, and investor-share cooperatives. Chaddad and Cook emphasise that the five non-traditional models can be used by cooperatives to ameliorate perceived financial constraints (while retaining a cooperative structure).

The cooperative with a subsidiary in which outside investors can participate (Nilsson's model 3) has been quite popular in several European countries. The next step in this process of inviting outside investors occurs when the subsidiary becomes listed on the stock market. The latter model has been named the Coop-Plc model[27] or the Irish Model.[28] While most of the activities and assets are transferred from the cooperative to the Plc-firm, the cooperative remains in place as a shareholder organisation, representing its member interests. Although the cooperative initially is the majority shareholder, there is an option of changing into minority shareholder when the public listed subsidiary is issuing new shares.[29] Examples of this Irish model of cooperative ownership structure are:

- the dairy cooperative *Kerry* (which became stock listed in 1986);
- the supply cooperative *IAWS* (1988);
- the dairy cooperatives *Avonmore* and *Waterford* (1989; they merged and became *Glanbia* in 1997); and
- the dairy cooperative *Golden Vale* (1989; acquired by *Kerry* in 2001).

A variant to the Irish Model is the Finnish Model.[30] Three agricultural cooperatives from Finland brought their subsidiaries to the stock market but retained a controlling stake in these companies through the introduction of a separate class of shares, exclusive to members, that carry stronger voting rights. Two of these cooperatives are minority shareholders in terms of number of shares, but they still retain the majority of the votes:

- *Metsäliitto* has 38 per cent of the shares of its subsidiary M-real (listed in June 1987) but retained 60 per cent of the voting rights;
- *LSO Cooperative* has 35 per cent of the shares in HKScan (listed in February 1997) but 73 per cent of the voting rights;
- The three cooperatives that founded the *Atria Group* maintain a majority position both in number of shares (58 per cent) and voting rights (92 per cent).

Kyriakopoulos and colleagues performed one of the few empirical studies on the relationship between market orientation and financial/ownership structure.[31] They compare two models: a traditional cooperative model and an entrepreneurial cooperative model (which could be models 2–5 from Table 7.2). Entrepreneurial culture has been defined as a combination of four attributes: risk-taking attitude of the management, innovative leadership, a commitment of the staff to innovation and development, and growth-oriented strategies.[32] They find that the entrepreneurial firm culture has a systematic influence on cooperatives' market orientation and performance. Thus, the more entrepreneurial the cooperative firm, the more market oriented it is, and the better its performance.

Proposition 4 states that increasing market orientation should lead to new financial and ownership structures, particularly because market orientation requires introducing and maintaining consumer brands, which often requires an expansion of the risk capital. Kyriakopoulos and colleagues show that cooperatives with an entrepreneurial ownership model are more market oriented.[33]

RESTRUCTURING COOPERATIVE CORPORATE GOVERNANCE

As the cooperative firm becomes larger, more international and more market oriented, the relationship between the board of directors and professional management changes. The so-called control problem becomes more serious as the size and complexity of the cooperative increases.[34] Although the formal structure of the cooperative still gives the board of directors the final authority, in large, market-oriented cooperatives, the managers have an information advantage that makes it almost impossible for the board to make independent decisions. Although the board holds formal authority, the managers have real authority.

Changes in the size and strategy of the cooperative may have several repercussions for managing and controlling the cooperative firm. Few empirical studies on this issue have been conducted; I therefore offer conjectures about the following developments when cooperatives embark on more market-oriented strategies. First, the extra marketing knowledge required usually does not reside with the directors (traditionally selected from the membership), so professional marketing management must be hired. Second, even without a strong focus on marketing, a large and diversified cooperative requires professional management expertise that board members do not necessarily possess. Third, as the cooperative grows, internationalises and diversifies, the interests of the members in all of these activities may diverge. As a result, strategic decision making may become more laborious and/or members become less committed. Fourth, and probably most important, when companies become large and diversified, they require top managers of a kind that is only rarely available; moreover, these managers demand autonomy from direct interference in company activities by the board of directors. Giving managers higher discretion and more autonomy reflects traditional organisational theory's prediction that decision rights should be allocated to those actors who have the knowledge to make the decisions. Such autonomy is an important element of the intrinsic rewards that top managers prefer as incentives for their effort. In summary, market-oriented cooperatives likely hire top managers with marketing experience and who require the board of directors to act as a supervising body instead of as a directing body.

One of the examples of restructuring corporate governance in Dutch cooperatives is the introduction of the so-called corporation model (Figure 7.1). This model is similar to the Irish model. However, cooperatives choose the corporation model not to acquire addition equity capital (as in the Irish case) but rather to give the management of the cooperative firm more autonomy (and thereby attract qualified managers). In the corporation model, the cooperative firm takes the legal form of a Plc or Ltd company, and all the activities and assets of the cooperative are placed in this Plc or Ltd company. The firm is legally separated from the association, though the association remains the 100 per cent shareholder of the firm.[†] One of the main characteristics of the corporation model, in terms of corporate governance, is that the board of directors of the association becomes the supervisory board of the cooperative firm; there is no longer a separate supervisory committee at the level of the association. It is a simple model in terms of corporate governance, as there is only one supervisory body, which takes into account both member interests and firm interests. Several large Dutch agricultural cooperatives have obtained this legal structure in the past 15 years, including the two largest dairy cooperatives Friesland Foods and Campina and the largest marketing cooperative for fruits and vegetables, The Greenery.

Proposition 1 states that more market orientation leads to more marketing expertise in the governing bodies of the cooperative, and P2 posits that market orientation shifts control rights from the members (represented by the board of directors) to the professional managers. Proposition 2 seems to receive support from the developments among Dutch cooperatives.

[†]   This legal structure makes it much easier for outside investors to invest in the cooperative firm.

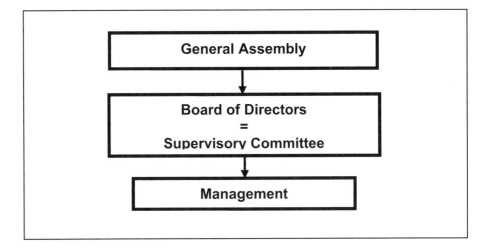

Figure 7.1   The corporation model of cooperative corporate governance

ESTABLISHMENT OF NEW PRODUCER ORGANISATIONS

Changing market conditions post not only threats but also opportunities. Although some opportunities can be exploited within the framework of existing cooperatives, others lead to new establishments. New farmer-owned cooperatives or producer organisations have taken up the marketing of organic products and regional specialties. For example, marketing for high-quality agricultural products (including organics and regional specialties) is often carried out by an organisation owned collectively by the producers.[35]

In the traditional fruit and vegetables industry, many new organisations have emerged, some induced by new European legislation. In the Netherlands, many new marketing cooperatives appeared in the fresh produce industry in the 1990s. Bijman and Hendrikse find that 75 new cooperatives were established in the years 1995–2000.[36] Dutch growers of fruits and vegetables established new marketing cooperatives because they were dissatisfied with existing (auction) cooperatives, which seemed more concerned with keeping the majority of members satisfied than with new market opportunities. The objective of most of those new marketing cooperatives has been to trade directly with retail customers and build a reputation with large retailers or even consumers (e.g., through a brand name). Many of the new cooperatives focus on the top end of the market, selling high-quality, customised (mainly packaging), or exclusive products.

In other EU countries, many new producer organisations represent responses to new EU regulations. On 26 October 1996, the European Council adopted Regulation (EC) no. 2200/96 regarding the common organisation of the market in fruit and vegetables.[37] This common market order meant a major deviation from earlier policies: from the defensive instrument of intervention payments to the offensive instrument of promoting marketing activities and strengthening producer market positions. The main goals of the new Regulation were as follows: first, lower intervention payments that had created structural excess supply by reducing community compensation for product withdrawals. Second, alleviate the negative effects of greater liberalisation of international trade and the accession of new member states by strengthening producers' positions in the more competitive and open market. Third, strengthen the position of relatively small producers compared with large purchasers by encouraging them to establish producer organisations. One of the eligibility requirements for EU financial support demands that producer organisations develop an explicit marketing plan. According to Duponcel, these new producer organisations have led to improved product quality and product innovation.[38]

The new producer organisations marketing fruits and vegetables are not necessarily formal cooperatives (i.e., falling within the domain of national laws on cooperatives). Many organisations have legal forms that do not fall under cooperative legislation, but they are producer-owned organisations with democratic decision making (as required by the EU regulation). Table 7.3 describes the role of producer organisations in the marketing of fruits and vegetables in Europe. Comparing these figure with Table 7.3 reveals not only that more numbers are available but that in many countries – notably Belgium and the Netherlands – the market share of collective producer organisations is greater than the share of the formal cooperatives.

Although these new producer organisations (which include many cooperatives) are not a direct proof of greater market orientation, they certainly suggest indirect proof. The introduction of new fruit and vegetable varieties mainly moves through these new organisations. Many new producer organisations in the fresh produce industry focus their strategies and operations on close collaboration with retail customers, strict quality control, and introducing new consumer products. This finding could be seen as preliminary support for P3, which states that market orientation should lead to greater innovativeness.

Table 7.3 Market share (in %) of producer organisations in marketing fruit and vegetables, 2003

	Fruit and Vegetables
Austria	35
Belgium	85
Denmark	30
Finland	12
France	45
Germany	30
Ireland	75
Netherlands	85
Sweden	45
UK	50

Source: Based on Duponcel.[38]

Conclusions

In response to changes in market conditions and public policies, as well as the rise of new technologies, cooperatives have taken up new tasks, particularly relating to improving and assuring quality, enhancing logistic efficiency, strengthening responsiveness, and increasing innovation activities. The traditional objectives of cooperatives focussed on cost reductions and strengthening market positions; the new objectives target value creation at different stages of the value chain.[39] In the traditional cooperative, attention focussed on horizontal coordination and collective action, whereas in new and restructured cooperatives, the emphasis is on vertical collaboration and coordination. In other words, both supply chain management (cost reduction tool) and innovation (value creation tool) have become relatively more important.

Changing market conditions also have led to restructuring processes among agricultural cooperatives, such as internationalisation, more vertical coordination in supply chains, greater investments in innovation and product development, new financial and ownership structures, new corporate governance structures, and, finally, the establishment of many new cooperatives.

This development toward more market orientation and the resulting restructuring into larger, more international and more corporate structures also has influenced the relationship between members and their cooperative. When the need for more equity capital has been a reason to invite outside investors, it has often been accompanied by a loss of control of the members over their cooperative. The developments of the Irish dairy cooperatives show that once cooperatives list their subsidiary on the stock market, the conflict between users and outside shareholders becomes very real, even when these shareholders have only a minority of votes.

Another effect of more market orientation is the increasing heterogeneity in membership, not only because the co-op is larger and more international but also because as a preferred supplier to retail companies, the co-op must include more and varied products in its product portfolio. Increasing membership heterogeneity is a serious issue for collective action organisations such as marketing cooperatives, because both the efficiency of the decision-making process and the mechanisms for obtaining equity capital depend on member commitment.

Still, the cooperative as an organisational form seems flexible enough to accommodate changes in the economic and political environment. Given the steady market share of cooperatives in the marketing of farm products and the establishment of many new cooperatives in the past 10–15 years, doubts about the future of the agricultural cooperative seem, as yet, unfounded.

References

1. Stee, A.P.J.M.M.van der (1989), 'Om schone zakelijkheid: perspectieven voor de agrarische sector in Nederland'. Rapport van de Adviescommissie Perspectieven voor de Agrarische Sector in Nederland aan het Landbouwschap. Den Haag: Landbouwschap.
2. Dijk, G. van and Mackel, C. (1991), 'Dutch agriculture seeking for market leader strategies', *European Review of Agricultural Economics*, vol. 18, pp. 345–64.

3. Harte, L. and O'Connell, J.J. (2007), 'European dairy cooperative strategies: Horizontal integration versus diversity', in Karantininis, K. and Nilsson, J. (eds), *Vertical Markets and Cooperative Hierarchies. The Role of Cooperatives in the Agri-Food Industry*. Springer, Dordrecht.

4. Draheim, G. (1955), *Die Genossenschaft als Unternehmungstyp*. Göttingen.

5. Dijk, G. van (1997), 'Implementing the sixth reason for co-operation: New generation cooperatives in agribusiness', in Nilsson, J. and Dijk, G. van (eds), *Strategies and Structures in the Agro-food Industries*. Van Gorcum, Assen; Dijk, G. van (1999), 'Evolution of business structure and entrepreneurship of cooperatives in the horti- and agribusiness', *Finnish Journal of Business Economics*, vol. 48, no. 4, pp. 471–83.

6. Barton, D.G. (1989), 'What is a cooperative?', in Cobia, D.W. (ed.) *Cooperatives in Agriculture*. Prentice Hall, Englewood Cliffs, NJ.

7. Ton, G., Bijman, J., and Oorthuizen, J. (eds) (2007), *Producer Organisations and Chain Development: Facilitating Trajectories of Change in Developing Countries*. Wageningen Academic Publishers, Wageningen.

8. The World Bank (2007), *World Development Report 2008: Agriculture for Development*. The World Bank, Washington, DC.

9. Bijman, J. (2000), 'Cooperatives', in Douw, L. and Post, J. (eds), *Growing Strong. The Development of the Dutch Agricultural Sector; Background and Prospects*. Agricultural Economics Research Institute (LEI), The Hague.

10. Bekkum, O.F. van and Dijk, G. van (eds) (1997), *Agricultural Cooperatives in the European Union; Trends and Issues on the Eve of the 21st Century*. Van Gorcum, Assen.

11. Juliá-Igual, J.F. and Meliá, E. (2007), 'Social Economy and the Cooperative Movement in Europe: Input to a New Vision of Agriculture and Rural Development in the Europe of the 25'. CIRIEC Working Paper 2007/06.

12. Bekkum, O.van (2007), 'European top-100 of agrifood cooperatives', *Coöperatie*, no. 593, pp. 22–23. (Dutch)

13. Bekkum, O.F.van (2001), *Cooperative Models and Farm Policy Reform. Exploring Patterns in Structure-Strategy Matches of Dairy Cooperatives in Protected vs. Liberated Markets*. Van Gorcum, Assen.

14. Bijman, J. and Hendrikse, G (2003), 'Cooperatives in chains: institutional restructuring in the Dutch fruit and vegetables industry', *Journal on Chains and Network Science*, vol. 3, no. 2, pp. 95–107.

15. Narver, J.C. and Slater, S.F. (1990), "The effect of a market orientation on business profitability", *Journal of Marketing*, vol. 20, no. 4, pp. 20–35.

16. Beverland, M.B. and Lindgreen, A. (2007), 'Implementing market orientation in industrial firms: A multiple case study', *Industrial Marketing Management*, vol. 36, no. 4, pp. 430–42.

17. Bekkum (2001), op. cit.

18. Bijman, J. (1998) 'Internationalisation of European dairy companies: Strategies and restrictions', in Ziggers, G.W., Trienekens, J.H., and Zuurbier, P.J.P. (eds) 'Proceedings of the Third International Conference on Chain Management in Agribusiness and the Food Industry' (Ede, 28–29 May 1998), Wageningen Agricultural University, Management Studies Group, Wageningen, pp. 769–79.

19. FrieslandCampina (2008), 'Explanatory Notes to the Proposed Merger, April 2008'. Available at www.campina.com.

20. Madsen, O.O. and Nilsson, J. (2007), 'Issues in cross-border mergers between agricultural cooperatives', *Journal of Cooperative Studies*, vol. 40, no. 3, pp. 27–38.

21. MD Food Annual Report, 1995/1996

22. Boon, A. (1997), 'Internationalisation of Denmark's food industry: strategic challenges for cooperatives and implications for their members', in Loader, R.J., Henson, S.J. and Traill, W.B. (eds) 'Globalization of the Food Industry: Policy Implications'; Proceedings, Reading, UK, 18–19 September, 1997,University of Reading, Department of Agricultural and Food Economics, Centre for Food Economics Research, pp. 417–31.

23. Ebneth, O. and Theuvsen, L. (2005), 'Internationalization and Corporate Success; Empirical Evidence from the European Dairy Sector', paper presented at the EAAE XIth International Congress, Copenhagen.

24. Nilsson, J. (1999), 'Cooperative organisational models as reflections of the business environments', *Finnish Journal of Business Economics*, vol. 4, pp. 449–70.

25. Chaddad, F.R. and Cook, M.L. (2004), 'Understanding new cooperative models: An ownership-control rights typology', *Review of Agricultural Economics*, vol. 26, no. 3, pp. 348–60.

26. Hansmann, H. (1988), 'Ownership of the firm', *Journal of Law, Economics and Organization*, vol. 4, no. 2, pp. 267–304.

27. Harte, L.N. (1997), 'Creeping privatisation of Irish cooperatives: A transaction cost explanation', in Nilsson, J. and Dijk, G.van (eds), *Strategies and Structures in the Agro-food Industries*. Van Gorcum, Assen; Nilsson, op. cit.

28. Chaddad and Cook, op. cit.

29. Bekkum, O.F. van and Bijman, J. (2007), 'Innovations in cooperative ownership: Converted and hybrid listed cooperatives', in Rajagopalan, S. (ed.), *Cooperatives in 21st Century. The Road Ahead*, Ifcai University Press, Hyderabad, India, pp. 34–56.

30. Chaddad and Cook, op. cit.

31. Kyriakopoulos, K., Meulenberg, M., and Nilsson, J. (2004), 'The impact of cooperative structure and firm culture on market orientation and performance,' *Agribusiness*, vol. 20, no. 4, pp. 379–96.

32. Moorman, C. (1995), 'Organizational market information processes: Cultural antecedents and new product outcomes', *Journal of Marketing Research*, vol. 32, no. 3, pp. 318–35.

33. Kyriakopoulos et al., op. cit.

34. Cook, M.L. (1995), 'The future of U.S.agricultural cooperatives: A neo-institutional approach', *American Journal of Agricultural Economics*, vol. 77 (December), pp. 1153–59.

35. Raynaud, E., Sauvee, L., and Valceschini, E. (2005), 'Alignment between quality enforcement devices and governance structures in the agro-food vertical chains', *Journal of Management and Governance*, vol. 9, pp. 47–77. Verhaegen, I. and Huylenbroeck, G. van (2002), *Hybrid Governance Structures for Quality Farm Products. A Transaction Cost Perspective*. Shaker Verlag, Aachen.

36. Bijman and Hendrikse, op. cit.

37. Commission of the European Communities (2001), Report from the Commission to the Council on the State of Implementation of Regulation (EC) No 2200/96 on the Common Organisation of the Market in Fruit and Vegetables. Brussels. (COM(2001) 36 final).

38. Duponcel, M. (2006), 'Role and Importance of Producer Organisations in the Fruit and Vegetable Sector in the EU', presentation at the CAL-MED second workshop, Washington, DC. Available at http://ec.europa.eu/agriculture/capreform/fruitveg/slides_en.htm.

39. Cook, M.L. and Plunkett, B. (2006), 'Collective entrepreneurship: An emerging phenomenon in producer-owned organizations', *Journal of Agricultural and Applied Economics*, vol. 38, no. 2, pp. 421–28.

8 Can Cooperatives Build and Sustain Brands?

BY MICHAEL B. BEVERLAND*

Keywords

cooperative structures; brand building; brand success and failure; New Zealand

Abstract

This chapter proposes that cooperative structure is central to sustainable brand positioning. Drawing on cases of New Zealand-based cooperatives, I examine whether cooperatives can develop global brands. Although all the brands studied enjoyed some success, their long-term sustainability was more elusive. Only those cooperatives with transferable ownership structures were able to sustain their brand value, because they could undertake long-term strategies supportive of their customers.

I will identify why agricultural cooperatives need to build brands; examine the branding strategies of New Zealand cooperatives; identify success factors driving long-term brand sustainability; and examine brand supportive structures necessary for long-term success.

The marketing of much New Zealand farm produce traditionally has been left in the hands of farmer/grower cooperatives. Increasingly, concerns have been raised about the effectiveness of such structures. Marketers have identified the need for cooperatives to move from a farmer-centric to a market-centric approach. However, questions have been raised about the viability of traditional cooperative arrangements to support a market-oriented strategy. This chapter therefore examines the ability of traditional and new generation cooperatives to develop and support market-based assets, including brands and long-term relationships with channel buyers, to develop a sustainable position for their members and increase returns. The findings suggest that traditional cooperatives may be able to develop innovative marketing programmes but still struggle to support them over the long term, due to problems in ownership structures. The new generation cooperatives attain more sustained long-term success, because members can capture the

*   Professor Michael B. Beverland, School of Economics, Finance and Marketing, RMIT University, GPO Box 2476V, Melbourne, Victoria 3001, Australia. E-mail: Michael.beverland@rmit.edu.au. Telephone: + 61 3 9925 1475.

equity of intangible assets, such as brand value, thus ensuring that they undertake actions (e.g., channel support) consistent with building a sustainable, long-term position.

Why Do Cooperatives Need Brands?

Brands are a ubiquitous feature of modern markets. Intangible assets such as brands provide firms with strong returns, awareness among both consumers and trade buyers such as retailers, and assets that are difficult to imitate.[1] Brands therefore represent the opposite of a commodity, that is, products that have little differentiation in the eyes of the marketplace and whose value is determined solely by the forces of supply and demand (commodity suppliers are typically price takers[2]). Agri-businesses have often been slow to develop brands,[3] preferring instead to seek government protections, improve efficiency or reduce buyer power through collective supply and marketing arrangements, such as cooperatives or producer boards. For many cooperatives, the recognition of the need to invest in marketing and break the commodity cycle has been belated.[4] External forces, such as changes in consumer demand, retail and the competitive landscape, have also driven firms to take a more market-oriented approach, which includes the use of brands.[5] Yet research on repositioning commodities as brands remains scarce and, to date, researchers have been silent about the effectiveness of agricultural cooperatives in developing market-oriented brand programmes.

Such research is important in light of debates about the effectiveness of traditional cooperative arrangements to deliver sustainable returns.[6] Traditional cooperative structures are limited in their effectiveness because of 'vaguely defined property rights [that] create losses in efficiency because the decision maker no longer bears the full impact of his or her choices.'[7] Because brand leadership requires a supportive firm structure or orientation,[8] an examination of the ability of traditional and new-generation cooperatives to develop and support brand position is timely. This article examines this question using five New Zealand case studies.

CASES OF AGRICULTURAL COOPERATIVE BRANDS

In New Zealand, as in many countries, the export marketing of agricultural products is often left up to cooperatives. These cooperatives have been empowered by the legislature to conduct collective marketing on behalf of all agricultural producers within their industry and often have single-desk selling provisions that give them the sole right to market and sell agricultural produce. Export-grouping schemes set up by governments try to obtain benefits from coordinated actions among members.[9] These benefits include shared information for mutual advantage and pooled resources to gain economies of scale and scope. The groups focus on the creation of collaborative advantage to achieve a competitive advantage, and relationships play vital roles in ensuring the long-term success of these schemes.

For example, the Danish Pork industry's success derived from its coordinated approach to production, processing and marketing, built on an understanding of the requirements of different markets and a dedication to quality to provide reliable and consistent supply, tailored to the needs of individual markets.[10] The umbrella organisation Danske Slagterier was partly responsible for this strategy and encouraged close cooperation across all stages of pork production and the value chain. A clear export goal is necessary to gain export success for these groups, but the interests of members must be balanced with those of the group, and gains are likely to be unequal amongst group members.[11]

Despite some successes, these representational structures have come under increasing criticism.[12] New Zealand agricultural producers (and indeed, the majority of agricultural producers) have been slow to adopt a market orientation that involves the development of strong brands for their products to differentiate their offers in the marketplace. Traditionally, producers of fresh produce have believed that their responsibility for the product end when their produce leaves the farm/orchard gate.[13] Despite qualitative differences between the produce of one country and another, competition generally occurs on the basis of price, with consumers being provided few cues as to why the produce from one country may be better than that of the next. Instead,

> 'Emphasis has been placed on large volume, throughput to a large number of independent buyers who purchased unbranded, undifferentiated produce on an ad hoc, transactional basis from the supplier who offered the best price and quality at the time.'[14]

Recent efforts have focused on increasing product quality as a means of market differentiation. However, it is no longer sufficient for firms to produce a technically superior product.[15] Future success requires the adoption of strategic planning models by agri-business,[16] including the development of augmented products that offer strong brands, close relationships throughout the supply chain, a market orientation, and a unique selling proposition.[17]

LIMITS TO THE COOPERATIVE MODEL

Because of these criticisms, many cooperatives have sought to develop new strategies. Despite some successes, problems remain. Researchers interested in cooperatives have begun to question whether traditional cooperative structures are effective in building sustainable returns, achieving a market orientation, and building and sustaining brands.[18] Others have argued that their compulsory nature ties all members of an industry into one set of strategies, which does not provide an environment for innovation.[19] Economic schools such as transaction cost economics hold that ownership structure affects the ability of firms to position themselves in the market and capture value from their activities.[20] Agricultural cooperatives in theory are 'owned' by the industry or farmers they represent. However, in practice, these boards lack the commercial structure to capture the value of their marketing activities.[21] The political nature of each board often leads to a short-term focus, as farmers demand increased prices or returns, which may come at the expense of long-term value and strong supply chain relationships.[22] This trend has potential negative implications for brands, which represent a promise to consumers and other relevant stakeholders, which means they must be at once relevant and consistent.[23] Yet many cooperatives engage in brand marketing to increase prices at auction, an activity that often undermines the brand promise of stability (particularly for business buyers that require reputable suppliers that ensure stability of price and supply[24]).

This chapter examines five case studies of New Zealand agricultural cooperatives that have attempted to create sustainable competitive advantages through branding. First, I present the details of the cases and the case methodology. Second, I outline the findings. Third, I identify the contributions of this chapter to theory, practice, and policy.

Method

Because this chapter seeks to explore the ability of different cooperative structures to implement brand marketing programmes (i.e., interaction between process and structure), I adopt a qualitative research design to gain a holistic perspective of each case's approach and capture the potentially rich and meaningful characteristics of each brand marketing programme.[25] The sampling technique was purposive; cases were selected on the basis of the likelihood that they would provide useful findings.[26] With this focus, I sampled both traditional and new generation cooperatives that have engaged in brand-building exercises. Traditional cooperatives are characterised by poorly defined property rights,[27] as the case of the New Zealand Game Industry board reveals. Also, I analysed attempts at branding undertaken by Fonterra (or the New Zealand Dairy Board [NZDB], as it was known) and Merino NZ (or the NZ Wool Board), prior to their change into new generation cooperatives. In contrast, new generation cooperatives possess some of the following characteristics: transferable equity shares, appreciable equity shares, defined membership, legally binding contract delivery or uniform grower agreements, and minimum up-front equity investments.[28] The cases of Zespri, Merino NZ, Sealord, and Fonterra represent this type. In total, these five examined cases should be sufficient to attain reliability.[29]

The research design followed the multiple case study approach recommended by Eisenhardt.[30] Each organisation was contacted to set up interviews with key personnel. In all, 16 interviews took place (6 at the Game Industry Board, 5 at Zespri, 2 each at Merino NZ and Fonterra, and 1 at Sealord). Weitz and Jap recommend the use of multiple informants.[31] The number of interviews for each case depended on the complexity of the case, the number of people involved in each strategy, the availability of secondary information, and the size of each organisation. The case details are outlined in Tables 8.1 and 8.2.

Following the interviews, secondary data were sought from each organisation, including a range of company reports, performance information, marketing material and market research studies. For the Merino NZ and NZDB/Fonterra cases, McKinsey and Company[36] provided reviews of their activities and future structure. As part of the Zespri case, a student of the author conducted an initial brand awareness and consumer behaviour study to assist with the introduction of new kiwifruit brands into the New Zealand market. Each industry also formed part of a project conducted by Crocombe et al.[37] An independent search was conducted through the popular business and general press and television to gather further information about each case. All cases have high profiles in New Zealand, and this search resulted in significant new information that confirmed, and in some cases challenged, the views of the participants. This information was reviewed and integrated into a full industry case summary.

To enhance validity and reliability, a standard set of questions was used for each interview.[38] The topics for discussion centred on seven key categories: history and development of the strategy, market environment, market entry, performance, brand management, future aims and challenges, and the content of the strategy. This interview protocol only formed a guide for each interview, and new issues emerged in each case

Table 8.1   Data sources

Organisation	Number of Interviews	Interviewees	Secondary Information
NZGIB (Venison)	6	Concept designer Venison company owner and Cervena franchisee Current Chairperson and Marketing Manager Two former Chairs	Company documents Historical case studies of deer industry Web site information Newspaper and business periodical reports
Zespri (Kiwifruit)	5	Manager turners and growers (wholesaler of fresh produce) Marketing manager, Australasia Market analyst Brand manager Business development manager	Business periodicals Company reports and marketing material Consumer survey of brand awareness among consumers[32] Focus group of consumers of fruit Historical case studies of the kiwifruit industry
Merino NZ (Wool fibre)	2	Chairperson Marketing manager	Newspaper and business periodicals Company documents Historical case studies of wool industry[33]
Fonterra (FMCG)	2	International marketing manager; brand manager	Newspaper articles Historical case studies on the dairy industry[34] Trade and general business press[35] Annual reports
Sealord (Seafood)	1	International marketing manager	Television documentary on Sealord Company information Newspapers articles Business press articles

Table 8.2   Structure and role of each cooperative

Case	NZGIB	Fonterra	Merino NZ	Zespri	Sealord
Established by legislation	Yes	Yes	Yes	Yes	Yes
Single desk	No	Yes	No	Yes	N/A
Funding	Levy	Shareholders	Levy	Shareholders	Shareholders
Membership compulsory	Yes	Yes	Yes	Yes	N/A
Represent	Whole industry– production through to exporters	Dairy companies and farmers	Merino wool growers	KIWIFRUIT GROWERS	Shareholders (private company and Maori tribes whose shareholding was purchased by the government)
Role	To develop and grow the NZ deer industry	To manage and represent NZ dairy industry in export markets and maintain 15 per cent ROA	To develop a unique brand identity for NZ Merino growers in export markets	To increase returns to growers	To increase returns to shareholders and manage fisheries in a sustainable manner

that required further investigation. A draft of the case was sent back to each interviewee for comment. In each case, the interviewees gave extensive feedback, though much of it consisted of correcting dates, answering questions posed by the author, or commenting on interpretations. In each case, the informants would answer some of the challenges posed by the secondary information. Following this step, each case was analysed using the dual process of within- and across-case analysis recommended by Eisenhardt.[39] The author first coded each case, then discussed the codes with two experienced qualitative researchers. Finally, a series of themes were identified and explored across all cases; these formed the basis of the findings and discussion.

Findings

The presentation of the findings follows the two key themes. First, I identify the brand strategies of each case and explore the reasons for success or lack thereof. Second, building on this first section, I explore the interaction between cooperative structure and brand programme outcomes. As a means of categorising the brand programmes, I draw on Aaker and Joachimsthaler's[40] framework of global brand leadership, which proposes that ongoing brand leadership involves four interactive components: brand identity (firm reputation) and position (brand identity and value proposition communicated to the market); brand architecture (formal structure of the firm's brands); brand-building programmes (supportive marketing programmes); and organisational structure and processes.[41] This information is contained in Tables 8.3 and 8.4.

BRAND STRATEGIES AND OUTCOMES

All the sampled cases moved from commodity selling to brand marketing to increase the returns to their members and build a sustainable form of competitive advantage. Each case developed brand identities and supportive programmes targeted at either business customers (e.g., other manufacturers, resellers) or end-consumers. For example, Merino NZ developed an ingredient brand programme targeted at lead business-customers, then supported the programme with a pull-based (consumers pull the product through the channel) public-relations programme directed at consumers. The other cases also sought to build reputations with key business customers but focused most of their efforts on end-consumers. Fonterra developed a range of fast-moving consumer goods brands for the global market; NZGIB targeted restaurant customers; Sealord and Zespri targeted supermarket customers with one brand (and several line extensions). With the exception of Fonterra, all of the brands were developed from scratch as part of a major repositioning effort. Fonterra already had several strong brand lines before changing to a new generation cooperative structure; the other brands were developed either by traditional cooperatives (NZGIB) or as part of a move to new generation structures (the other four cases).

Tables 8.3 and 8.4 identify the position of each brand and their supportive marketing programmes. In relation to each case, I examined initial customer feedback to the programmes and gained assessments from the relevant marketing managers about their effectiveness. Trade customers uniformly indicated that these brand strategies set the standard for cooperative marketing in their respective categories and were enthusiastic about the programmes and their supportive materials. For example, Merino NZ quickly

Table 8.3   Global brand leadership identity/position and architecture for all cases

Global brand leadership component	Merino NZ	NZGIB	Fonterra	Sealord	Zespri
Brand identity/ position	Identity: Pure, innovative, highest quality fibre from one animal and country of origin. Position: Strictly up-market positioning, high price, scarcity, supported by certainty of supply and customised programmes for major buyers.	Identity: Grade-A meat subject to strict quality controls and sourced from one country of origin. Position: Year-round, healthy, versatile product at a high price. Positioning also includes adaptability and preparedness to work with channel members and users.	Identity: Largest and cheapest supplier of a complete range of dairy products globally. Position: Aims to be the global leader in the supply of milk component products and solutions. Desire to work with leading pharmaceutical companies in close partnership.	Identity: Supplier of high-quality seafood ingredients sourced from one location and sustainable management practices of fish-stocks. Position: Aims to be global leader of high-quality, environmentally friendly seafood. Positioning also includes working directly with key customers to build strong co-brands.	Identity: Preeminent supplier of specialty dairy produce for corporate customers. Position: Develops specialist dairy products, sources others globally, and targets top-end of any market. Works directly with key customers to build strong co-brands and with other like-minded suppliers to provide a full offering for customers.
Brand architecture	One umbrella brand to endorse a number of co-brands.	Two brands, each targeted at different channels. Both brands supported by the same quality programmes. Each brand endorses co-brands.	One umbrella brand endorses multiple functional area brands.	One corporate ingredient brand.	One corporate brand.

gained the support of the lead user Loro Piana (a top maker of wool cloth) in Italy, developing a jointly branded product (Zealander). Other secondary reports also identified that this brand was setting the standard for operations in the fine wool market. In one case, a lead German buyer indicated he would not source wool from other buyers unless they copied important quality and performance standards developed by Merino NZ (key components of brand equity[42]).

For Fonterra, branding excellence and leadership resulted in its appointment as category captain by many large retailers in the 'yellow fats' category. Category captains manage the entire category of products in store on behalf of retailers and gain a strategic advantage over competitors because they effectively manage the position of the competing brands as well as their own. Even the case with the least long-term success managed to increase brand awareness among targeted buyers to 69 per cent and usage rates to

Table 8.4 **Global brand leadership programmes and supportive structures and processes for all cases**

Global brand leadership component	Merino NZ	NZGIB	Fonterra	Sealord	Zespri
Brand-building programmes	Accessing multiple media; leveraging network relationships to build awareness; co-branding; public relations; leveraging intangible brand assets. Achieving brilliance by accessing low-cost, high-impact promotional activities and constant increases in product quality; leveraging network resources in new ways. Measuring results and brand performance by developing causal feedback about separate brand activities and customers. Information filtered back to suppliers.	Accessing multiple media; leveraging network relationships to build awareness; co-branding; public relations. Achieving brilliance by accessing low–cost, high-impact promotional activities and constant increases in product quality. Disconnect due to reliance on a single, global measure of brand impact. Information filtered back to suppliers, though information provided is of limited use to enhance customer value.	Accessing multiple media: leveraging network relationships; co-branded alliances; public relations; targeted marketing programmes for lead customers; global integrated marketing communication (IMC) programme. Achieving brilliance through co-branded activities and IMC activities globally. Message adapted to local contexts. Measuring results and brand performance with standard financial and marketing metrics. Information filtered back to suppliers to guide improvements.	Accessing multiple media: leveraging network relationships; co-branded alliances; public relations; targeted marketing programmes for lead users; IMC. Achieving brilliance through co-branded promotions and IMC activities. Message adapted to local context. Works with stakeholders to identify further promotional opportunities. Measuring results and brand performance with standard financial and marketing metrics.	Accessing multiple media: leveraging network relationships; co-branded alliances with customers and suppliers; public relations; targeted marketing support. Achieving brilliance through mass-market campaigns, co-branded promotions, and sponsorship. Message adapted to local context. Measuring results and brand performance with standard financial and marketing metrics.
Organisational structure and processes	Responsibility for brand strategy: cascading ownership for the brand. Management processes: adapt to individual customers and network partners across nations; constant product and marketing innovation.	Responsibility for brand strategy: top leadership support but disconnect with suppliers. Management processes: adaptability to individual customers and network partners across nations; constant marketing innovation.	Responsibility for brand strategy: top leadership support, centralised brand team works with regional areas to develop brand programmes. Management processes: adaptability to key customers and markets; constant marketing innovation.	Responsibility for brand strategy: one global brand manager works directly with regional partners. Management processes: adaptability to key customers and markets; constant marketing innovation; lobbying for sustainable fishing practices.	Responsibility for brand strategy: one global brand management team works directly with lead users, and other alliance partners. Management processes: adaptability to key customers and markets; constant marketing and product innovation; public and trade education on benefits of fruit.

36 per cent within two years of launch. At the time, these increases placed the brand second behind the mass-market brand Angus Beef in the United States.[43] Sealord and Zespri also experienced strong support in the retail context (top shelf space, in-store promotion opportunities) and increased their margins.

Several factors help explain this initial success. First, many of the brands were first-movers and thus set the standard for the category. Second, research by each of the brand teams indicated latent desire among consumers and business customers for more marketing investment, including the development of brands. In the case of business buyers, increased investment in innovation, marketing communications, brands and push programmes (channel promotional support) provided an important point of difference from their competitors and help solved problems regarding certainty of supply and price. For consumers, changes in lifestyles, health concerns, positive country-of-origin images and environmental concerns presented each brand manager with the chance to build a brand identity around these attributes. Third, in the cases of Zespri, NZGIB, Sealord, and Fonterra, product leadership and brand awareness had been established with business buyers to some degree and, in some cases, with end-consumers. Thus, the brand programmes built on existing equity in the marketplace.

The final explanation relates to the integrated nature of the brand programmes. Aaker and Joachimsthaler propose that brand leadership can be achieved only through a comprehensive and consistent brand marketing programme.[44] Consistency results from integrated marketing communications that ensure all deliberate brand messages, regardless of the source or medium, effectively provide a 'one voice, one look' approach.[45] As part of this strategy, brand value can be built through effectively combined push and pull programmes (as opposed to one or the other), because supportive brand programmes at the channel level ensure ongoing uptake and support, while pull programmes with end-consumers help generate awareness and demand, thus reinforcing the commitment of channel buyers to the brand.[46] As Tables 8.3 and 8.4 identify, this combined effort was the approach adopted by all of the brands studied.

For example, even though Merino NZ only developed an ingredient brand, and thus invested the majority of its marketing resources into push programmes, it also recognised the value of making end-consumers aware of the brand story, investing in public relations (e.g., giving free gifts of high-quality cloth to high-profile world leaders) to build consumer awareness of the fibre's positive attributes and developing joint promotional material with key brands (e.g., Smedley, Just Jeans, Icebreaker) to drive demand, which ensured steady orders from channels. The NZGIB adopted a similar strategy, building demand within channels through innovation and customised products and product support while also building brand awareness through sponsorships, public relations, and a high-profile chef competition (Cervena Plates).

The other cases invested more heavily in pull strategies because their products were lower involvement, fast-moving consumables delivered to the consumer in finished form. Nevertheless, the significant investments in packaging, promotions, in-store trials and promotions, and jointly branded material (including exclusive branded lines for retailers) represented significant investments in push strategies and, again, established these brands as leaders in their categories. As such, the initial success of these brands can be explained partially by their successful execution of a comprehensive marketing strategy. Therefore, the cases studies followed recommended best practices.[47]

Although each brand was successfully launched, not all managed to sustain this success. For example, the NZGIB's Cervena brand did not sustain its initial success. Although the other brands faced key challenges, such as the appreciation of the New Zealand dollar, increased competition, increased retailer power, challenging seasons and various levels of imitation, their performance in terms of market share, margin, and sustained returns to growers continued and outperformed those of other commodity producers. In some cases, these natural, market, and macroeconomic variables seemingly had no effect on brand value (particularly for Zespri). Although the NZGIB faced these same forces, its difficulties cannot be attributed to them (given that the other cases managed them effectively), nor can they be attributed to poor management or changes in product demand and preferences among consumers. Central to ongoing brand reinforcement in these markets was a strong reputation for consistency of delivery, supply and price.

Through an analysis of secondary sources, such as trade articles and newspapers, and direct discussions with buyers, it became clear that success across all the cases related to careful relationship management between the cooperatives and their customers. Such activities posed several challenges for industries such as wool and venison that sell using auction pricing. With these arrangements, members judge the success of brand programmes (and any marketing activity) by the increased auction prices received. However, this form of pricing and feedback undermines the brand position and cooperative reputation because customers require certainty of supply and price (and these brands often positioned themselves as 'solution providers' to business customers). The business customers studied faced significant price pressure themselves (e.g., retailers, restaurateurs) or had little ability to pass on extra costs up the channel (e.g., buyers of raw wool). Merino NZ solved this problem by removing sales from auctions and assisting sellers to form five-year, fixed price and supply contracts with buyers (a key characteristic of new generation cooperatives[48]). Such activities often represented the first-time channel members and sellers had interacted. The other cases dealt with these concerns by offering fixed prices to retailers and passing on the benefits of growth and efficiencies to their members in terms of increased share prices and asset values.

Concerns about supply and price variation particularly affected the NZGIB because its strategy targeted up-market restaurants. Although these restaurants charge high menu prices, they also have high costs, and margins are usually made on liquor, not food. As well, because the NZGIB wanted to dominate the game listings on menus and position venison as a healthy red meat alternative, any uncertainty of supply would undermine this strategy, because restaurants do not change their standard menus regularly. Despite initial successes in the late 1990s, by 2004 (and today), prices per kilo of venison had fallen from a high of NZ$10 in 2001 to NZ$3.75, and reports in 2005 saw the price fall further to levels that many in the industry believed were unsustainable (i.e., the economic farm surplus for deer in 2001 was NZ$1,000 per hectare, whereas by 2004 it was NZ$26). The buyers identified price and supply uncertainty as key problems still to be solved by the cooperative. The positive views of product performance held by these buyers indicated, however, latent equity for the brand and demonstrated that potential demand remained high (industry insiders believed that NZ$4 per kilo was a sustainable price for buyers and sellers).

Such declines also prompted some cooperative members to criticise the activities of the marketing team and the Cervena brand programme (accounting for 10 per cent of all venison sales), resulting in decreased funding for the programme. The programme then re-launched under a user-pays principle (entitled 'Who Benefits, Who Pays'), with the US office shut down in favour of Web-based support. The marketing team argued that the brand strategy had lifted prices of New Zealand venison across the board (the marketing strategy involved a strong country-of-origin brand programme[49]), regardless of quality, though they could not substantiate this claim because they did not have a separate price schedule for branded and non-branded products. As a result, in 2006, the outlook for the industry and brand remained uncertain. The next section explores why this cooperative struggled to reinforce its brand position over time and saw declining returns and performance.

INTERACTION BETWEEN STRUCTURE AND BRAND-RELATED OUTCOMES

Traditional cooperative arrangements may be problematic because they involve vaguely defined property rights and other related problems (e.g., free riders[50]). In examining whether such structures undermine brand positions and the process of branding per se, it is instructive to examine the responses of the NZGIB to the fall in price and compare them with the activities of the other new generation cooperatives. Since the price per kilo started to fall for New Zealand venison in 2001, the NZGIB has responded with increased marketing efforts and, more important, calls for farmers to increase herd sizes to smooth out problems in demand. At the height of the commodity price cycle, deer farmers started to sell hinds (female deer) as well as stags (males) for meat processing. The short-term impact of this move was to flood the market with product, and the medium-term impact was to restrict supply due to declines in on-farm breeding. Deer farmers also responded to low prices for venison by diversifying into velvet production (deer antler powder for traditional Asian medicines). Harvesting velvet does not involve the death of the animal, and thus, supply of venison is restricted. In response, the NZGIB urged farmers to solve the supply and demand problem by growing herd numbers by a fixed ratio of 10 per cent year-on-year. This call was also motivated by a desire to stabilise prices at mutually sustainable levels (for buyers and sellers alike). To date, such calls have gone unheeded.

In contrast with other cooperative members, venison farmers can only capture the value of marketing activity through increased prices at auction (which also affect farm value). In contrast with the other new generation cooperatives, the lack of tradable shares among NZGIB members means that brand equity cannot be captured by appreciating assets, and therefore members act in an economically rational manner and attempt to increase auction prices using strategies that undermine the position of the Cervena brand and the espoused statements of the NZGIB, which attempts to build a strong reputation as a 'solutions provider.' This strategy is a common brand position for business-to-business firms.[51] Central to this brand position is adapting the offer (defined as the product, service, logistics, and advice[52]) to each customer to solve important strategic problems faced by these buyers. As identified previously, the NZGIB's key buyers require certainty of supply, quality, and price (certainty is a core part of reputation, a key component of a business marketing offer).[53] For the NZGIB to maintain its brand equity (and build reputational capital), it must reinforce this position with a supportive marketing programme and capabilities.[54] Any disconnect between espoused promises and actual delivery will result in customer dissatisfaction, complaint behaviour and relationship exit.[55] In the case of this traditional cooperative, the lack of defined property rights, fixed terms contracts and obligations, and tradable shares undermines the brand promise, because members have no sustainable way of capturing the value of increasing brand equity and instead demand higher prices at auction – the very thing that customers do not want.

In contrast, the other cooperatives did not suffer this fate. For example, prior to the emergence of Fonterra (new generation cooperative), the New Zealand Dairy Board faced problems associated with poorly defined property rights. A common complaint was that the Dairy Board was required to take all milk available to it, regardless of quality, logistical costs and market demand. Farmers could not exit the marketplace at a price that reflected the value of their farm.[56] In response, the new generation cooperative structure, developed as part of the move from the Dairy Board to Fonterra, provided farmers with tradable shares that enabled them to exit the industry and/or sell to more efficient members. This strategy then enabled Fonterra to invest more strategically in products with high margins and growth and enabled members to benefit directly from branding activities in the form of share price increases. The other new generation cooperatives enjoyed similar successes, providing incentives to move away from a reliance on short-term auction price increases (which undermined brand positions in the long term) and toward more sustainable structures, such as fixed-term contracts and closer relationships with key buyers. In several cases, these closer arrangements led to increased mutual investments in the relationships, thus building in switching costs (protecting either party against short-term changes in competitors' prices or input costs) and allowing for collaboration. In all the new generation cooperatives studied, this collaboration resulted in new product development (e.g., jointly developed Merino cloth, blends such as denim wool) and provided access to new markets or segments.

Finally, such arrangements allowed local farmers to benefit from licensing arrangements. For example, Zespri, in a bid to ensure market leadership by developing a year-round supply base, licensed the use of its new varieties (gold and red) to European and South American growers. In the past, this move would have been seen as commercial suicide, but under current arrangements, licensed foreign growers can support the promise of the brand, expand market coverage, and be required to sell at prices that do not undermine the Zespri brand position. Such licensing fees are then reflected in returns to growers. Fonterra had similar success in acquiring foreign milk cooperatives and licensing their brands to gain access to greater supplies of fresh products, thus ensuring market coverage.

Conclusions

Recognising that these findings are exploratory, this article still contributes in several ways. Primarily, it explores the ability of different cooperative structures to support an important intangible market asset, namely, brands. Given calls for cooperatives to move from a production to a marketing orientation, this contribution is particularly timely.[57] I show that cooperatives can develop innovative branding programmes that sustain long-term customer relationships, deliver increased returns to members, and provide a strong point of difference in the market. Thus, cooperatives can break out of the commodity price cycle. This finding supports those who advocate greater strategic diversity among agricultural cooperatives, including moving up the value chain,[58] though the long-term viability of such a strategy depends on cooperative structure. In particular, I find that long-term brand positioning requires a supportive governance structure that includes some of the characteristics of new generation cooperatives. For example, though not using tradable shares, Merino NZ required fixed-term contracts and exclusive arrangements; thus, each member with customer relationships was able to capture brand value directly through ongoing returns, increases in farm prices, and new market opportunities. Such arrangements were effective in the case of Merino NZ because of its small size. In contrast, larger cooperatives (Sealord, Fonterra, and Zespri) used more characteristics of new generation cooperatives (fixed-term contracts, tradable shares, appreciable equity shares, defined membership and minimum up-front investments) to support their brand and general business strategy. The relationship between cooperative size, degree of new generation cooperative structures, and ongoing brand and market success requires further empirical validation.

Given the findings, I also address two debates in relation to developing a market orientation and supporting brands over time. Edwards and Schultz[59] propose that new competitive conditions in agricultural value chains necessitate a move toward market orientation. These findings suggest that traditional cooperatives are able to develop a market orientation but not support it in the long term because of their inability to maintain ongoing member commitments to key marketing metrics, such as investments in marketing programmes, brands, and customer relationships. The findings also suggest that new generation cooperatives are more effective at sustaining a market orientation. These findings require further longitudinal, empirical support. In relation to supporting brands, I find a similar relationship, in support of Aaker and Joachimsthaler's[60] proposed framework for brand leadership. The relationship between different cooperative structures and ongoing brand equity also requires longitudinal empirical support.

Finally, this study supports prior calls from researchers for changes in cooperative arrangements. A role remains for cooperative structures within global markets. The sustained success of four of the cases and the latent equity attached to the NZGIB's brand provide evidence that cooperative structures with well-defined property rights can not only break commodity price cycles but also become market leaders in terms of both market share and rate of return. As well, these structures can provide the basis for reconfiguring agricultural markets away from antagonistic buyer–seller relationships and toward more cooperative and mutually beneficial relationships. All the cooperatives studied were instrumental in reconfiguring traditional networks characterised by arm's-length relationships into competitive networks consisting of mutually binding obligations and switching costs. The result (even in the case of NZGIB in the short term) was to increase the survivability of the entire network. The relationship between cooperative structures and the ability to maintain downstream relationships characterised by high trust and commitment is another area deserving of further empirical research. In particular, research could identify whether different cooperative structures have more sustained success over the life cycle of the relationship.

The findings also have practical and policy implications. With regard to practice, cooperative leaders and members should push for new generation structures to enhance their long-term chances of survival and market success (whether they are niche players like Merino NZ or global giants like Fonterra). In particular, members should consider seriously the ability of traditional cooperatives to support brands and a more market-oriented approach. Given the inability of traditional cooperatives to sustain a brand's promise, members not desiring more clearly defined property rights perhaps should focus on greater efficiencies and price competitiveness rather than risk a buyer backlash arising from the failure to sustain the expectations raised by branding programmes. With regard to policy, policymakers should encourage more government-empowered cooperatives to move toward new generation structures, and industry bodies should review existing arrangements if they wish to capture greater value from their activities.

References

1. Anderson, J.C. and Narus, J.A. (2004), *Business Market Management: Understanding, Creating, and Delivering Value*, 2nd edition, Pearson, Sydney; Keller, KL; (2003), *Strategic Brand Management: Building, Measuring, and Managing Brand Equity*, 2nd Edition, Prentice Hall, Sydney.
2. Crocombe, G.T., Enright, M.J., and Porter, M.E. (1991), *Upgrading New Zealand's Competitive Advantage*, Oxford University Press, Auckland.
3. Beverland, M.B. (2005a), 'Creating value for channel partners: The Cervena case', *Journal of Business and Industrial Marketing*, vol. 20, No. 3, pp. 127–35.
4. Beverland (2005a), op. cit.; Edwards, M.R. and Shultz, C.J. (2005), 'Reframing agribusiness: Moving from farm to market centric', *Journal of Agribusiness*, vol. 23, No. 1, pp. 57–73.
5. Edwards and Schultz, op. cit.
6. Cook, M.L. and Iliopoulos, C. (1999), 'Beginning to inform the theory of the cooperative firm: Emergence of the new generation cooperative', *The Finnish Journal of Business Economics*, vol. 4, pp. 525–35; Van Bekkum, O.-F. (2001), *Cooperative Models and Farm Policy Reform*, Van Gorcum, the Netherlands.
7. Cook and Iliopoulos, op. cit., p. 528
8. Aaker, D.A. and Joachimsthaler, E. (2000), *Brand Leadership*, Free Press, New York; Jaffee, S. and Masakure, O. (2005), 'Strategic use of private standards to enhance international competitiveness: Vegetable exports from Kenya and elsewhere', *Food Policy*, vol. 30, no. 3, pp. 316–33; Keller, op. cit.; Urde, M. (1999), 'Brand orientation: A mindset into strategic resources', *Journal of Marketing Management*, vol. 15, no. 1–3, pp. 117–33; Yakimova, R. and Beverland, M.B. (2005), 'The brand-supportive firm: An exploration of organizational drivers of brand updating', *Journal of Brand Management*, vol. 12, no. 6, pp. 445–60.
9. Wilkinson, I., Young, L.C., Welch, D., and Welch, L. (1998), 'Dancing to success: Export groups as dance parties and the implications for network development', *Journal of Business and Industrial Marketing*, vol. 13, no. 6, pp. 492–510.
10. Hobbs, J.E., Kerr, W.A., and Klein, K.K. (1998), 'Creating international competitiveness through supply chain management: Danish pork', *Supply Chain Management: An International Journal*, vol. 3, no. 2, pp. 68–78.
11. Wilkinson et al., op. cit.
12. Crocombe et al., op. cit.; van Bekkum, op. cit.
13. Beverland (2005a), op. cit.; Crocombe et al., op. cit.; Edwards and Schultz, op. cit.
14. White, H.M.F. (2000), 'Buyer-supplier relationships in the UK fresh produce industry', *British Food Journal*, vol. 102, no. 1, pp. 6–17.
15. Lichtenthal, J.D. and Long, M.M. (1998), 'Case study: Service support and capital goods – dissolving the resistance to obtaining product acceptance in new business markets', *Journal of Business and Industrial Marketing*, vol. 13, no. 4/5, pp. 356–69.
16. Miles, M.P., White, J.B., and Munilla, L.S. (1997), 'Strategic planning and agribusiness: An exploratory study of the adoption of strategic planning techniques by cooperatives', *British Food Journal*, vol. 99, no. 11, pp. 401–408.

17. Anderson and Narus, op. cit.; Jaffee and Masakure, op. cit.; Webster, F.E. (2000), 'Understanding the relationships among brands, consumers and resellers', *Journal of the Academy of Marketing Science*, vol. 28, no. 1, pp. 17–23.

18. Beverland (2005a), op. cit.; Cook and Iliopoulos, op. cit.; Crocombe et al., op. cit.; Edwards and Shultz, op. cit.; van Bekkum, op. cit.

19. Crocombe et al., op. cit.

20. Hunt, S.D. (2000), *A General Theory of Competition*, Sage Publications, Thousand Oaks, CA.

21. Cook and Iliopoulos, op. cit.

22. Gifford, D., Hall, L., and Ryan, W. (1998), *Chains of Success: Case Studies on International and Australian Food Businesses Cooperating to Compete in the Global Market*, Commonwealth of Australia, Canberra.

23. Keller, op. cit.

24. Beverland, M.B. (2005b), 'Repositioning New Zealand venison: From commodity to brand', *Australasian Marketing Journal*, vol. 13, no. 1, pp. 62–7.

25. Lewin, J.E. and Johnston, W.J. (1997), 'Relationship marketing theory in practice: A case study', Journal of Business Research, vol. 39, no. 1, pp. 23–31; Yin, R.K. (1994), *Case Study Research: Design and Methods*, Sage Publications, Thousand Oaks, CA.

26. Yin, op. cit.

27. Cook and Iliopoulos, op. cit.

28. Ibid., p. 529.

29. Yin, op. cit.

30. Eisenhardt, K.M. (1989), 'Building theories from case study research', *Academy of Management Review*, vol. 14, no. 4, pp. 532–50.

31. Weitz, B.A. and Jap, S.D. (1995), 'Relationship marketing and distribution channels', *Journal of the Academy of Marketing Science*, vol. 23, no. 4, pp. 305–320.

32. Beverland, M.B. (2001), 'Creating value through brands. The Zespri kiwi fruit case', *British Food Journal*, vol. 103, no. 6, pp. 383–99.

33. McKinsey and Company (2000a), *Report to New Zealand Woolgrowers on Improving Profitability*, McKinsey and Company, Auckland.

34. McKinsey and Company (2000b), *Report to New Zealand Dairy Board on the Future Industry Structure*, McKinsey and Company, Auckland.

35. Oram, R. (2000), 'The board, the farmer, his cow and her udder', *Unlimited*, November, pp. 40–8; Webb, B. (1995), New Zealand Dairy Board. Ernst and Young, Auckland.

36. McKinsey and Company (2000a, 2000b), op. cit.

37. Crocombe et al., op. cit.

38. Johnston et al., op. cit.; Yin, op. cit.

39. Eisenhardt, op. cit.

40. Aaker and Joachimsthaler, op. cit.

41. Aaker and Joachimsthaler, op. cit.; Keller, op. cit.

42. Keller, op. cit.

43. Beverland (2005b), op. cit.

44. Aaker and Joachimsthaler, op. cit.

45. Keller, op. cit.

46. Webster, op. cit.; Beverland (2005a), op. cit.

47. Aaker and Joachimsthaler, op. cit.; Keller, op. cit.

48. Cook and Iliopoulos, op. cit.

49. Beverland, M.*B.* and Lindgreen, L. (2002), 'Using country of origin in strategy: The importance of context and strategic action', *Journal of Brand Management*, vol. 10, no. 2, pp. 147–67.

50. Cook and Iliopoulos, op. cit.

51. Ford, D. and Associates (2002), *The Business Marketing Course: Managing in Complex Networks*, John Wiley and Sons, Brisbane.

52. Ibid.

53. Weitz and Jap, op. cit.

54. Aaker and Joachimsthaler, op. cit.

55. Oliver, R.L. (1996), *Satisfaction A Behavioral Perspective on Consumer Behavior*, Irwin/McGraw-Hill, Boston, MA.

56. Oram, op. cit.

57. Edwards and Shultz, op. cit.

58. Crocombe et al., op. cit.; Gifford et al., op. cit.; McKinsey and Company (2000a, 2000b), op. cit.

59. Edwards and Shultz, op. cit.

60. Aaker and Joachimsthaler, op. cit.

CHAPTER 9

Role of Market Orientation in Improving Business Performance: Empirical Evidence from Indian Seafood Processing Firms

BY SMITHA NAIR*

Keywords

market orientation, Indian seafood firms, business performance

Abstract

This chapter proposes to determine whether the market orientation concept, which has been successful in increasing the business performance of firms in developed nations, can bring about a similar consequence in an Indian setting, especially in the case of the Indian seafood processing firms. Empirical evidence proves that the adoption of market orientation principles can help Indian seafood firms enhance their sustainable business performance.

Introduction

The global fisheries sector, and especially the seafood trade sector, is undergoing tremendous changes in response to pressures due to diminishing supply, increasing demand, environmental changes and regulations, and geopolitical events. Demographic trends and preferences are also changing as a result of growing complexities in the lifestyles of people everywhere. Today's fast-paced life, coupled with a general lack of time for cooking compared with the situation in previous years, has resulted in consumers

* Dr Smitha Nair, Prathibha, Ushus Avenue, Pallippuram Post, Palakkad - 678006, Kerala, India. E-mail: smithanair99@yahoo.co.in. Telephone: + 91 491 254 1747.

who demand more convenience in their home cooking than ever before. Therefore, demand for value-added seafood products has been increasing. Recent studies indicate that seafood consumers are very specific in their choices and search for more simple preparations: street foods; fresh, healthy, and functional foods; and diverse flavours.[1] Developing countries contribute 78.5 per cent of world fish production and 57.2 per cent of the world's fish exports.

As the third-largest producer of fish and the second-largest producer of inland fish in the world, India is blessed with vast and varied aquatic resources. The country is rich in marine life, with a potential for approximately 3.9 million tonnes within an Exclusive Economic Zone of 2.025 million sq km. The potential for fish production from inland sources has been estimated at 4.5 million tons.[2] Along India's extensive, 8,118 km coastline, there are as many as 356 seafood-processing units, all of which are completely export-oriented.[3]

Marine-product exporting from India started in 1953. Until the early 1960s, the marine products exported consisted solely of dried items, such as fish, shrimp, shark fins and fish maws to neighbouring Asian countries like Sri Lanka, Burma and Singapore. As prospects for increased returns rose, more entrepreneurs entered the industry. During the 1960s, exporters focused on modernising their plants and adopting canning and freezing technologies in a bid to capture newer and larger markets, which promised higher returns. The new markets, world players in the global fish trade, included Japan, the United States, Europe and Australia. After 1966, following the devaluation of Indian currency, exports increased significantly, and the United States took over as the leading export market for the marine products. Since then, the industry has flourished steadily, and the country has been widely acknowledged as a strong contender in the global marine-products export trade. In the latter part of the 1970s, Japanese and European markets started to replace the US market as top importers. Japan retained its top slot until 2001, when the United States regained its position for two years before falling behind the European Union, which continues to lead.[4]

Growth in the seafood processing industry also has been phenomenal; India crossed the landmark figure of US$1 billion in 1994. Its overall export figures for 2005–06 registered an all-time high level of US$1.6 billion.[5] Approximately 84 per cent of the nation's marine exports consist of frozen shrimp (28 per cent), finfish (36 per cent), cuttlefish (10 per cent), and squid (10 per cent), and in terms of value, frozen shrimp dominated the export scenario, contributing 59 per cent. These figures reflect the industry's excessive dependence on shrimp to prop up its export sales. The European Union (29.46 per cent) accounts for the largest share of India's export of marine products, followed by the United States (22.63 per cent) and Japan (15.96 per cent), in terms of value. In terms of quantity, the EU and China both logged 27 per cent each, thus accounting for more than half of all marine exports.

Although the general trend of seafood exports has increased, the Indian seafood processing industry is not performing as well as it should. Approximately 75 per cent of Indian seafood products are sold in block form, and value-added products account for just 5 per cent of exports. These products are shipped to countries that reprocess and repackage them under different brand names and sell them at high profits to end-customers, usually institutional buyers such as supermarkets and restaurants. Thus, Indian products are faceless in the global seafood trade for the most part. Few products retain their brand names and identity. India thus represents a massive supplier of raw material rather than a processor or marketer. Indian exporters, on their own, lack the ability to enter into collaborations with supply chains in the respective export countries or ensure that their products reach final consumers. Among other problems, the perception of Indian seafood products in the world market is dismal. The Marine Products Export Development Authority (MPEDA), the flagship central government agency dealing with the development of marine product exports, has conducted market surveys in several markets, including the United States. Its US importer survey revealed that several importing firms are unaware of Indian products.[6] Those that are aware are unimpressed with the quality, workmanship, consistency and market promotion, as well as the reliability of shipments and delivery, the lack of cooked and ready-to-eat products, the poor packaging, poor brand image, lack of information about the products and the poor business ethics. On the whole, importers cite inconsistent quality and workmanship, poor visibility of Indian products and lack of proper marketing and promotion as major reasons for choosing not to import from India. Thus, a wide gap exists between the perceptions of Indian exporters and importers. Several importers, however, appear more enthusiastic about Indian products, mainly due to their value, variety, availability, quality standards, processing facilities, improvements in farm-raised products, and excellent potential labour force.

Huge quantities of seafood products from emerging markets like Indonesia, Thailand, Vietnam and Bangladesh, produced at much cheaper costs, also are flooding the global seafood market. Importers can choose from a variety of products that meet their quality, quantity, pricing and food safety requirements. The degree of competition and consumer demand for differentiated products has forced the traditionally product-oriented Indian seafood processing industry to shift focus away from production efficiency, high volume, efficiencies of scale and inventory control and toward value-added products.[7] This shift underscores the need for Indian seafood processing firms to adopt a market-oriented approach to trade. Being market-oriented requires being customer-centric, such that customer satisfaction is assured and the high value of customer investments is guaranteed.

This chapter assesses the role of market orientation in improving business performance among Indian seafood processing firms. Therefore, it attempts to answer a key question: can the market orientation framework, which has been successful in increasing business performance for firms in developed countries, exert a similar influence in an Indian setting?

Market Orientation Framework and Business Performance

The concept of market orientation receives significant academic support. Market orientation can be traced back to its roots in the marketing concept, which holds that the customer provides the reason for the existence of the business and that customer satisfaction is the key to improved business performance and competitive advantages. The objective of market orientation thus is to ensure customer satisfaction by determining their needs and wants and then meeting them.

Peter Drucker was among of the early proponents of the marketing concept and emphasised its role in stating that marketing pertains more to the customer's opinion about the business and less to the firm's specialised functions.[8] Other studies from the 1950s describe the marketing concept variously as a customer- and profit-oriented business philosophy or a corporate-wide mental makeup that integrated and coordinated all marketing functions to increase profits.[9]

The 1990s was marked by a flurry of academic research on market orientation, its meaning, antecedents, consequences, moderating influences, operationalisation, and characteristics, the behaviour of market-oriented firms, and the factors that bear on a firm's orientation. Since then, market orientation has been variously defined as a specific set of firm behaviours, a firm culture and a system. Various academicians posit that a market-oriented firm enjoys improved business performance.[10] Kohli and Jaworski famously define market orientation as the organisation-wide generation of market intelligence, dissemination of that intelligence across departments, and organisation-wide responsiveness to it.[11] Narver and Slater also propose that market-oriented firms, ceteris paribus, will improve their market performance.[12]

Awareness of the relevance of market orientation in food production also is steadily growing. Grunert and colleagues highlight the importance of implementing market orientation in food production, including agriculture and fisheries, and the subsequent processing links in the food value chain.[13] Trondsen examines market orientation in the seafood industry and observes that firms producing low-value products focus on cost factors in production, to obtain competitive advantage, whereas firms that produce high-value products require a much higher degree of market orientation.[14] In another study, Grunert and colleagues suggest extending market orientation from a company-specific level to a value chain level,[15] because the food market consists of several value chains, and the combined effort of all these chain members gives the final customer the strongest perception of the product as a whole.

CONCEPTUAL FRAMEWORK

The proposition that market orientation improves business performance has been borne out by various studies.[16] This chapter follows Kohli and Jaworski's view of market orientation,[17] one of the most widely accepted measures of market orientation.[18] Unlike Narver and Slater,[19] they emphasise customers rather than competitors. Therefore, many previous studies adopt their MARKOR scale.[20]

The market orientation construct

At the core of the framework is the market orientation construct, which consists of three components: intelligence generation, intelligence dissemination and intelligence responsiveness, as shown in Figure 9.1.[21]

Market orientation requires an intelligence system that the firm uses regularly and systematically to collect information about changing consumer preferences, competitors, government regulations, technology and other environmental forces. Houston stresses the need to gather information about consumers' future needs, in addition to their present needs.[22] According to Webster,[23] intelligence generation is not the exclusive function of the marketing department; it is a collective effort, such that the whole organisation is responsible for the collection of information.

After the intelligence is gathered, the need to disseminate it to all departments in the firm is paramount. Changing trends in consumer preferences, for example, must be translated into product innovations by the R&D division, which then demands a change

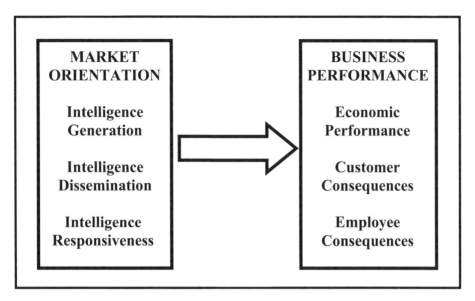

Figure 9.1   Market orientation framework

in the production system, changes in the marketing system, allocation of resources by the finance wing, purchases of raw material, reallocation of distribution channels, and so on. Information dissemination is also necessary so that all the departments in a firm can remain abreast of all new developments in the principal markets the firm serves. This dissemination ensures quicker transitions and faster responses to market changes.

Responsiveness entails the translation of the firm's reaction to the market information collected and internally circulated throughout the firm. The process of information generation and dissemination remains incomplete without a response to that information.[24] In today's changing markets, firms need to stay on their toes to respond immediately to the market signs they receive, because their survival may depend on how fast they are able to respond to customers' demands.

The fourth component of the market orientation model is consequences or business performance. Extant literature shows that business performance improves with the adoption of market-oriented principles. The business performance component consists of three components: economic performance, customer consequences and employee consequences. Economic business performance includes five subjective financial indicators: return on investment (ROI), sales growth relative to competitors, overall performance, overall performance relative to competitors, and overall performance relative to expectations.

Research Hypotheses

H_1: The market orientation of Indian seafood processing firms significantly determines their overall business performance.

This main hypothesis suggests three sub-hypotheses, pertaining to the effect of market orientation on economic performance, customer consequences and employee consequences.

H_2: The market orientation of Indian seafood processing firms significantly determines their economic performance.

Economic business performance consists of five subjective measures: increase in performance compared with the previous year, performance relative to major competitors, ROI relative to all competitors, sales relative to all competitors in the previous year, and business performance relative to the anticipated outcomes.

H_3: The market orientation of Indian seafood processing firms significantly determines their level of customer consequences.

In line with extant literature,[25] the implementation of market orientation in Indian seafood firms should lead to increased customer satisfaction and customer loyalty. The measures of customer consequences therefore include customer satisfaction and repeat customers.

H_4: The market orientation of Indian seafood processing firms significantly determines their level of employee consequences.

Employee consequences include employee commitment, employee job satisfaction and job security, improvement in the equity of the company, and improvements in the training function.

Research Methodology

As its research design, this chapter employs a cross-sectional approach. Consistent with market orientation literature,[26] the applied causal research method involves both hypothesis testing of the relationships and their quantification. The data collection methods focused on both primary and secondary data. Primary data came from a survey of various seafood exporting firms in India. The survey questionnaire was designed to fit with SPSS and focused on the marketing practices of the firms. The chief executives of the firms either completed the questionnaires themselves or sent them to persons who were knowledgeable about the export marketing operations, who were in charge of the daily marketing operations, or who formulated the marketing policies and strategies. The respondents thus included chief executives, managing directors, general managers and marketing managers. The questionnaire included questions about the firms' revenue and competitive advantages, the market orientation scale,[27] respondent details, performance indicators and marketing problems. When possible, the constructs adopted previously developed instruments and multiple indicator items to strengthen their validity. The responses to these items used five-point Likert scales (1 = strongly disagree and 5 = strongly agree). The mean scores of all questions provided the composite score for each variable. This study employed both mail survey and personal interview methods to gather these data. Moreover, the research method employed quantitative data analysis. The survey began in May 2005 and was completed within six months.

SAMPLING FRAME

The sampling frame consists of 356 processing plants listed in the 2004 version of the Indian Seafood Exporters' Directory, published by MPEDA. The questionnaire was mailed to the CEOs of all 356 processing firms, and 120 firms returned the questionnaires. Out of these, 12 questionnaires were incomplete and had to be discarded, for a final sample size of 108 (response rate = 30.34 per cent). The multivariate analyses of the data obtained relies on SPSS (Version 13.0). Stepwise regression analysis represented the major data analysis method.

Results and Discussion

Assessments of the psychometric properties of the market orientation scales can ensure the suitability of the research instrument and confirm whether the data obtained is appropriate and relevant. Measures should be both reliable and valid.[28] This chapter uses Cronbach's alphas to estimate construct reliability. All the measures come from highly reliable borrowed scales, but because these scales measure the constructs in a new setting, the cut-off value is fixed at 0.60. All constructs achieve reliabilities near the acceptable range, as follows: intelligence generation 0.633, intelligence dissemination 0.832, responsiveness 0.716, and business performance 0.876. The low value of the intelligence generation measure may reflect the smaller number of items and reverse-coded nature of some items. Overall, the data collected for this study appear reliable.

The tests of validity of a research instrument help assure that the different constructs are sufficiently well-defined. To assess the convergent validity, a factor analysis checks whether the items that constitute each construct load cleanly on that factor. Hair and colleagues advocate that factor loadings of 0.5 or higher can be considered significant.[29] The factor analysis results indicates that the three factors do not load cleanly on intelligence generation, intelligence dissemination and responsiveness; rather, it reveals a six-component structure with eigenvalues greater than 1 that accounts for 71 per cent of the variation. But the results are not clear.[30] Therefore, a second factor analysis extracts three factors, which resulted in a factor solution of three factors, though the items for each factor still did not load cleanly, and the variance accounted for was 54 per cent. Pulendran and colleagues report only 48 per cent variance in their study;[31] therefore, this study supports their recommendation that future studies explore scale validation and address the substantive and application issues related to the psychometric deficiencies of the MARKOR scale. The scree plot, which indicates the optimal number of factors or components to be retained in the solution, reveals that three main factors contribute to market orientation. Because the internal consistency of the scale is high, the average of the three components of market orientation (generation, dissemination, and responsiveness) provides the independent variable for the empirical analysis, in line with previous studies.[32]

COMPETITIVE ADVANTAGES IN EXPORTING

The respondents of the surveyed firms listed what they considered their major competitive advantages in exporting; the results appear in Table 9.1. On a scale of 1 to 5, the mean scores ranged from 3.55 to 4.62. Relationships with customers, with a mean score of 4.62, represented the highest-ranking competitive advantage, followed closely by product quality (mean = 4.61) and product delivery (mean = 4.23). The competitive advantages of production capacity and marketing capacity both earned mean scores of 4.06. The lowest mean score relates to product uniqueness/differentiation and technology. The low score

Table 9.1   Competitive advantages in exporting

Competitive advantages	N	Mean	Std. deviation
Relationship with customers	108	4.62	0.506
Product quality	108	4.61	0.490
Delivery	108	4.23	0.678
Production capacity	108	4.06	0.915
Marketing capacity	108	4.06	0.708
Cost factor	108	3.71	1.094
After-sales services	108	3.59	1.168
Product uniqueness/differentiation	108	3.57	1.129
Technology	108	3.55	0.813

on product differentiation indicates that these exporters do not concentrate on achieving product uniqueness or product differentiation. The Indian seafood exporting industry as a whole continues to produce mainly block frozen products and provides minimum value addition, unless required to add more by customers.

MARKETING PROBLEMS

The survey also asked about the major marketing problems faced by the firms (Table 9.2). Frequent changes in price trends (mean = 3.85, SD = 1.084), competition from international firms (mean = 3.52, SD = 1.501), competition from domestic firms (mean = 3.33, SD = 1.421), pricing problems (mean = 3.12, SD = 1.309), quality problems (mean = 2.71, SD = 1.381), and frequent changes in consumer trends (mean = 2.64, SD = 1.315) represent the major marketing issues encountered by the Indian seafood exporting companies. Lack of market presence (mean = 2.30, SD = 1.518) and inadaptability of the production system to meet market changes (mean = 2.22, SD = 1.270) also are concerns.

CONSEQUENCES OF MARKET ORIENTATION

The results of the stepwise regression analysis pertaining to the market orientation–business performance relationship appear in Table 9.3. Market orientation accounts for 23 per cent of the variation in overall business performance; that is, there is a significant positive relationship between market orientation and overall business performance for Indian seafood processing firms. This finding confirms the original hypothesis that business performance in Indian seafood firms will increase if firms adopt a market orientation (ß = 0.488, p = 0.000). This result also is consistent with existing literature that reports market orientation relates significantly and positively to business performance.[33]

The adoption of market orientation thus results in improved business performance, customer consequences and employee consequences. The empirical evidence also demonstrates the universal applicability of the MARKOR scale in different settings.

Table 9.2   Ranking of marketing problems

Ranking of marketing problems	N	Mean	S.D
Frequent changes in price trends	108	3.85	1.084
Competition from international firms	108	3.52	1.501
Competition from domestic firms	108	3.33	1.421
Pricing problems	108	3.12	1.309
Quality problems	108	2.71	1.381
Frequent changes in consumer trends	108	2.64	1.315
Lack of market presence	108	2.30	1.518
Production system inability to meet changes	108	2.22	1.270
Packaging and transportation problems	108	2.10	1.199
Promotional problems	108	2.01	1.106
No unique attribute/undifferentiated products	108	1.96	1.260
Inadequate market knowledge	108	1.82	0.936
Poor brand image	108	1.47	0.891
Distribution problems	108	1.36	0.755

Table 9.3   Summary of stepwise regression analysis: consequences of market orientation

Independent variable	Dependent variables/ consequences	Standardised coefficient beta	T value	Significance	R^2	Adjusted R^2
	Overall business performance	0.488	5.752	0.000	0.238	0.231
	Economic performance	0.315	3.420	0.001	0.099	0.091
Market Orientation	Non-economic performance	0.545	6.688	0.000	0.297	0.290
	Customer consequences	0.452	5.220	0.000	0.204	0.197
	Employee consequences	0.453	5.230	0.000	0.205	0.198

Improved business performance

Academic literature has established that firms enjoy improved business performance if they implement the principles of market orientation by testing this claim in different settings in different countries.[34]

Market orientation accounts for 23 per cent of the variance in the business performance variable in this study. That is, market orientation improves the business performance of Indian seafood processing firms, consistent with prior research.[35] Economic business performance includes five subjective financial indicators: ROI,[36] sales growth relative to competitors,[37] overall performance,[38] overall performance relative to competitors, and overall performance related to expectations. The results show that market orientation accounts for 9.1 per cent of the variance in the economic business performance variable. This low percentage may indicate that efforts to develop a market orientation will take some time to yield economic rewards.[39] The lagged effect of market orientation on business performance might not be captured in cross-sectional research, as employed herein.[40] Non-economic performance is a function of two subjective measures, customer consequences and employee consequences, and the results show that market orientation accounts for 29 per cent of the variance in this performance measure. The effect of market orientation thus appears more pronounced for the non-economic factors of business performance than for the economic factors.[41]

Subramanian and Gopalakrishna examine the role of market orientation on business performance in an Indian setting and also find that market orientation plays a positive role in fostering growth in overall revenue, return on capital, success of new products and services, ability to retain customers, and success in controlling expenses.[42]

Customer consequences

Another consequence of adopting market orientation is the benefit accrued from customers in the form of customer satisfaction, repeat buying and retention. Because market orientation emphasises customers, the consequences of customer satisfaction, retention and repeat sales are very important. The adjusted R^2 value is 19.7 per cent, which means that market orientation accounts for 19.7 per cent of the variance in consumer consequences. Thus, if Indian seafood processing firms implement a market orientation, the customer consequences will improve significantly. In the long run, the firm should attempt to maximise this consequence, because customer focus is fast becoming imperative for business survival and augmented profitability. Siguaw and colleagues opine that, in addition to enhancing profitability, market orientation can reduce the costs associated with customer and employee defection,[43] in that acquiring new customers often is considerably more expensive than building customer loyalty among the firm's current customers.[44] Matsuno and Mentzer also highlight the importance of measuring customer consequences, suggesting that studies should investigate the relationship between market orientation and non-economic performance to obtain a more holistic view of the effects of market orientation.[45]

Employee consequences

The employee consequences include employee commitment, employee job satisfaction and job security, equity of the company (e.g., wages, promotions, fringe benefits), and improvements to the employee training function. Market orientation accounts for 19.8 per cent of the variance. Thus, market orientation helps foster psychological and social benefits for the employees, which in turn should result in increased commitment to the firm.[46]

Managerial Implications and Conclusions

Empirical evidence reveals that market-oriented behaviour increases overall business performance ($\beta = 0.488$; $R^2 = 0.23$) and helps seafood processing firms develop sustainable competitive advantages, though the effect on non-economic measures ($\beta = 0.545$; $R^2 = 0.29$) is greater than that on economic measures ($\beta = 0.315$; $R^2 = 0.09$). This result is comparable to values obtained in similar studies, including Jaworski and Kohli ($R^2 = 0.18$ Sample I, $R^2 = 0.25$ Sample II), Pulendran and colleagues ($R^2 = 0.32$ and $R^2 = 0.36$ in two separate studies), Bhuian ($\beta = 0.41$, $p = 0.001$), Subramanian and Gopalakrishna in an Indian setting ($R^2 = 0.33$), and Kirca and colleagues ($r = 0.46$).[47] The positive and significant relationship between market orientation and business performance thus appears to hold in the case of Indian seafood processing firms.

The next question thus becomes how to implement market orientation in seafood firms. The preceding sections offer views into what constitutes market-oriented behaviour. This section in turn offers insights that seafood industry managers can use when they implement a market orientation process.

Indian seafood managers should look for ways and means to improve their image, increase their marketing skills, improve product quality, and develop sustainable competitive advantages to be assured of a steady clientele, repeat sales and increasing profits over the long term. To develop these advantages, firms must adopt a dual strategy of market orientation and the resource-based view offered by strategic management specialists. Market orientation can help firms attain an outside-in view, while the resource-based view offers them an inside-out view. Some Indian seafood firms have well-developed resource-based competencies and produce world-famous brands. The key to their business success thus lies in adopting market-oriented competencies to complement their existing resource-based competencies and thus developing a sustainable competitive advantage over other international firms.

The first task of the top management should be to define the vision and mission statements of the new market-oriented organisation. If such statements already exist, they should be reviewed to ensure they reflect the organisation's new priorities. If not, they should be redefined to state the desired future of the firm clearly. Grant and Spender recognise knowledge as a significant competitive asset that a firm possesses,[48] which also provides the wherewithal to develop into a sustainable competitive advantage. Market orientation stresses the importance of intelligence or knowledge management. Therefore, one of the first prerequisites for market orientation is that the firms should develop and institutionalise an intelligence-based system that collects both internal and external information and collates the results to produce regularly updated information that top management can peruse to enhance the efficacy of decision making. An intelligence-based system also should help pinpoint customer-related processes that need improvement, which can lead to their rectification and the dissemination of this knowledge throughout the organisation.

Another important prerequisite for implementing a market orientation involves guaranteeing the total support of top management. Top management should be not only enthusiastic about the project but also willing to initiate all market-oriented activities, and managers should institutionalise market-oriented behaviour. They must instil in employees recognition of the necessity and value of customer orientations, because a customer focus lies at the centre of any market-oriented activity.

The next step is the process of rigorous assessments of the current level of performance, using either the level they wish to attain or the best performers in the industry as a benchmark.[49] Common performance-indicators include customer satisfaction, customer loyalty, financial performance, sales growth, market growth and market share. Another assessment also should consider the present behaviour system within the firm.

Top management should sent personnel to visit customers at regular intervals, to acquaint themselves with the changing culture and background settings, in addition to the buyers' business practices, product evaluation measures and expectations of products. This step should offset problems arising from changes in consumer trends, as widely faced by Indian seafood processors. Using this information, the firm can adopt a comprehensive marketing plan that incorporates consumer views; in turn, the firms should be able to deliver exactly what the customers want. Delivery also should be on schedule and with the correct product specifications, as well as any other labelling required by the importer. The packaging should be trendy, attractive and strong enough to withstand wear and tear. Indian processors should adopt consumer-oriented packaging. The quality of the product should be competitive with that of products from other markets. Pricing also should be competitive; the firms simultaneously should try to obtain cost advantages to attract new markets that will result in higher profits. Product diversification is another strategy by which the firms can exert stronger competitive pressures.

Instead of concentrating on block frozen products, Indian processors might switch to cooking-lines and value-added items, thus taking advantage of the financial assistance schemes proposed by the MPEDA. Indian seafood exporters should strive to improve their global competitiveness, otherwise they will lose on lucrative customers and deals. Drawing consumer attention to products other than the usual shrimp and cuttlefish might help create awareness about the diversity of Indian seafood. Continuous contacts with the importers, followed by regular feedback, as well as enquiries about the importer's views on product quality should be made, such that the importers believe the processors are genuinely concerned with and committed to them. Processors also should work in conjunction with MPEDA to develop a unique seafood logo for all seafood products, which could raise them to a branded status. Regular attendance at trade fairs, the display of products in attractive packing, and close collaboration with MPEDA for tie-ins in supermarket chains would go a long way toward creating customised packs instead of faceless bulk products. This step also would help command premium pricing, because the products go directly to final customers and eliminate middlemen who demand some of the revenue. This approach can further address problems that arise owing to difficulties with pricing, competition, lack of differentiation and market presence.

Employees should become properly acquainted with the marketing strategy of the firm and encouraged to incorporate a market-oriented perspective in all their behaviour and activities. Top management should regularly provide feedback on employee performance and institute a performance appraisal system that offers timely and fair rewards.[50]

Customer focus is at the heart of any market-oriented activity,[51] so employees can never forget to consider customers' interests as their first priority.[52] Market-oriented activities will also increase employee satisfaction and productivity. Continuous intelligence generation, dissemination, and responsiveness is primary. The intelligence generated should include customer information, competitor information, trade-related news, product developments, details about the dynamic environment, regulations, and so forth.

The adoption of market orientation principles can ensure improved business performance for Indian seafood processing firms. Given the present situation of the industry, the adoption of market orientation principles can introduce a professional outlook to their marketing practices and thereby achieve what the industry sorely needs at the moment: improved sustainable business performance and a competitive edge over other international competitors in the global seafood market.

References

1. Möller, A. (2007), 'Seafood processing-local sources, global markets', Workshop on Opportunities and Challenges of Fisheries Globalization, jointly hosted by the OECD Committee for Fisheries and FAO Fisheries and Aquaculture Department, 16–17 April, p. 36.

2. Anonymous (2004), *MPEDA – An Overview*, MPEDA, Cochin.

3. Anonymous (2007), MPEDA statistics, unpublished data, MPEDA, Cochin.

4. Anonymous (2004), op. cit.

5. Anonymous (2006), MPEDA statistics, MPEDA, Cochin. Available at http://www.mpeda.com/ (accessed April 8, 2007).

6. Johnson, H.M. and Associates, (2000), *Phase I Report, United States Seafood Market Research and Marketing Plan Development*, September, Marine Products Export Development Authority, Cochin.

7. Grunert, K.G., Jeppesen, L.F., Jespersen, K.R., Sonne, A.M., Hansen, K., and Trondsen, T. (2004), 'Four cases on market orientation of value chains in agribusiness and fisheries', *MAPP working paper*, no. 83.

8. Drucker, P.F. (1954), *The Practice of Management*, New York: Harper and Row.

9. McKitterick, J.B. (1957), 'What is the marketing management concept?', in F.M.Bass (ed.), *The Frontiers of Management Thought and Science*, Chicago: American Marketing Association, pp. 71–92; Felton, A.P. (1959), 'Making the marketing concept work', *Harvard Business Review*, (July/August), pp. 55–65.

10. Kohli, A.K., and Jaworski, B.J. (1990), 'Marketing orientation: The construct, research propositions and managerial implications', *Journal of Marketing*, vol. 54, no. 1, pp. 1–18; Narver, J.C. and Slater, S.F. (1990), 'The effect of a market orientation on business profitability', *Journal of Marketing*, vol. 54, no. 4, pp. 20–35; Ruekert, R.W. (1992), 'Developing a market orientation: An organizational strategy perspective', *Journal of Research in Marketing*, vol. 9, pp. 225–45; Jaworski, B.J. and Kohli, A.K. (1993), 'Market orientation: Antecedents and consequences', *Journal of Marketing*, vol. 62 (July), pp. 53–70; Deshpande, R., Farley, J.U., and Webster, F.E. (1993), 'Corporate culture, customer orientation, and innovativeness in Japanese firms', *Journal of Marketing*, vol. 57, no. 1, pp. 23–7.

11. Kohli and Jaworski, op. cit.

12. Narver and Slater, op. cit.

13. Grunert, K.G., Baadsgaard, A., Larsen, H.H., and Madsen, T.K (1996), *Market Orientation in Food and Agriculture*, New York, Kluwer Academic Publishers.

14. Trondsen, T. (2001), 'Fisheries management and market-oriented value adding (MOVA)', *Marine Resource Economics*, vol. 16, pp. 17–37.

15. Grunert, K.G., Jeppesen, L.F., Risom, K., Sonne, A.M., Hansen, K., and Trondsen, T. (2002), 'Market orientation at industry and value chain levels: Concepts, determinants and consequences', *Journal of Customer Behaviour*, vol. 1, no. 2, pp. 167–94.

16. Deshpandé, R. and Webster Jr., F.E. (1989), 'Organizational culture and marketing: Defining the research agenda', *Journal of Marketing*, vol. 53 (January), pp. 3–15; Deng, S. and Dart, J. (1994), 'Measuring market orientation: A multi-factor, multi-item approach', *Journal of Marketing Management*, vol. 10, pp. 725–42; Greenley, G.E. (1995), 'Market orientation and company performance: Empirical evidence from UK companies', *British Journal of Management*, vol. 35, pp. 105–116; Atuahene-Gima, K. (1995), 'An exploratory analysis of the impact of market orientation on new product performance: A contingency approach', *Journal of Product Innovation Management*, vol. 12, pp. 275–93; Pelham, A.M. and Wilson, D.T. (1996), 'A longitudinal study of the impact of market structure, firm structure, strategy, and market orientation culture on dimensions of small-firm performance', *Journal of Academy of Marketing Science*, vol. 24, pp. 27–43; Farrell, M.A. and Oczkowski, E. (1997), 'An analysis of MKTOR and MARKOR measures of market orientation: An Australian perspective', *Marketing Bulletin*, vol. 8, pp. 30–40; Becker, J., Homburg, C. (1999), 'Market-oriented management: A systems-based perspective', *Journal of Market-Focused Management*, vol. 4, no. 1, pp. 17–41; Cadogan, J.W., Diamantopoulos, A., and Siguaw, J.A. (1998), 'Export market-oriented behaviors, their antecedents, performance consequences and the moderating effect of the export environment: Evidence from the UK and US', in P. Anderson, (ed.), *Marketing Research and Practice, Track 2 International Marketing, Proceedings of the 27th EMAC Conference*, Stockholm, pp. 449–52; Aggarwal, N. and Singh, R. (2004), 'Market orientation in Indian organizations: An empirical study', *Marketing Intelligence and Planning*, vol. 22, no. 7, pp. 700–715; Kara, A., Spillan, J.E., and DeShields, O.W. (2005), 'The effect of a market orientation on business performance: A study of small-sized service retailers using MARKOR scale', *Journal of Small Business Management*, vol. 43, no. 2, pp. 105–118; Renko, M., Carsrud, A., Brännback, M., and Jalkanen, J. (2005), 'Building market orientation in biotechnology SMEs: Balancing scientific advances', *International Journal of Biotechnology*, vol. 7, no. 4, pp. 250–68; Paul, D.E (2006), 'Market orientation and performance: A meta-analysis and cross-national comparisons', *Journal of Management Studies*, vol. 43, no. 5, pp. 1089–1107; Kohil and Jaworski, op. cit.; Narver and Slater, op. cit.

17. Kohli and Jaworski, op. cit.; Jaworski and Kohli, op. cit.

18. Felton, op. cit.

19. Narver and Slater, op. cit.

20. Hooley, G., Lynch, J.E., and Shepherd, J. (1990), 'The marketing concept–putting the theory into practice', *European Journal of Marketing*, vol. 24, no. 9, pp. 7–23; Hart, S. and Diamantopoulos, A. (1993), 'Linking market orientation and company performance: preliminary work on Kohli and Jaworski's framework', *Journal of Strategic Marketing*, vol. 1, no. 2, pp. 93–122; Cadogan, J.W. and Diamantopoulos, A. (1995), 'Narver and Slater, Kohli and Jaworski and the market orientation construct: Integration and internationalization', *Journal of Strategic Marketing*, vol. 3, pp. 41–60; Maltz, E. and Kohli, A.K. (1996), 'Market intelligence dissemination across functional boundaries', *Journal of Marketing Research*, vol. 33, no. 1, pp. 47–61; Selnes, F., Jaworski, B.J., and Kohli, A.K. (1996), 'Market orientation in United States and Scandinavian companies: A cross-cultural study', *Marketing Science Institute Report*, Marketing Science Institute, Cambridge, MA, pp. 97–107; Avlonitis, G. and Gounaris, S. (1997), 'Marketing orientation and company performance', *Industrial Marketing Management*, vol. 26, pp. 385–402; Baker, W.E. and

Sinkula, J.M. (1999), 'The synergistic effect of market orientation and learning orientation on organizational performance', *Journal of the Academy of Marketing Science*, vol. 27 (Fall); Pulendran, S., Speed, R., and Widing, R.E. (2000), 'Antecedents and consequences of market orientation in Australia', *Australian Journal of Management*, vol. 25, no. 2, pp. 119–43; Cadogan et al., op. cit.; Paul, op. cit.

21. Kohli and Jaworski, op. cit.

22. Houston, F.S. (1986), 'The marketing concept: What it is and what it is not', *Journal of Marketing*, vol. 50, no. 2, pp. 81–7.

23. Webster, F.E. (1988), 'The rediscovery of the marketing concept', *Business Horizons*, vol. 31, no. 3, pp. 29–39.

24. Kohli and Jaworski, op. cit.

25. Jaworski and Kohli, op. cit.

26. Kohli and Jaworski, op. cit.; Narver and Slater, op. cit.; Greenley, op. cit.; Pulendran et al. (2000), op. cit.; Kumar, K., Subramaniam, R., and Yauger, C. (1998), 'Examining the market orientation: A context-specific study', *Journal of Management*, vol. 24, no. 2, pp. 201–233; Aaker, D.A, Kumar, V., and Day, G.S. (1998), 'Sample size and statistical theory', in *Marketing Research*, 6th edn, John Wiley and Sons, New York, pp. 405–37.

27. Jaworski and Kohli, op. cit.

28. Nunnally, C.J., Jr. (1970), *Introduction to Psychological Measurement*, McGraw-Hill, New York; Parameswaran, R., Greenberg, B.A., Bellenger, D.N., and Robertson, D.H. (1979), 'Measuring reliability: A comparison of alternative techniques', *Journal of Marketing Research*, vol. 16, no. 1, pp. 18–25; Churchill, G.A., Jr. (1979), 'A paradigm for developing better measures of marketing constructs', *Journal of Marketing Research*, vol. 16, no. 1, pp. 64–73.

29. Hair, J.F., Anderson, R.E., Tatham, R.L., and Black, W.C. (1998), *Multivariate Data Analysis*, 5th edn, Prentice Hall, Upper Saddle River, NJ.

30. Pulendran et al. (2000), op. cit.

31. Ibid.

32. Ibid.

33. Mehta, S.C. and Joag, S.G. (1981), 'Marketing orientation in Indian industry', *IIMA Working Paper*, no. 374, IIMA, India; Shoham, A. and Rose, G.M. (2001), 'Marketing orientation: A replication and extension', *Journal of Global Marketing*, vol. 14, no. 4, pp. 2–25; Pulendran et al. (2000), op. cit.; Narver and Slater, op. cit.; Jaworski and Kohli, op. cit.; Selnes et al., op. cit.

34. Diamantopoulos, A. and Hart, S. (1993), 'Linking market orientation and company performance: Preliminary evidence on Kohli and Jaworski's framework', *Journal of Strategic Marketing*, vol. 1, no. 2, pp. 93–121; Diamantopoulos, A. and Cadogan, J.W. (1996), 'Internationalizing the market orientation construct: An in-depth interview approach', *Journal of Strategic Marketing*, vol. 4, no. 1, pp. 23–52; Oczkowski, E. and Farrell, M.A. (1998), 'Discriminating between measurement scales using nonnested tests and two-stage least squares: The case of market orientation', *International Journal of Research in Marketing*, vol. 15, no. 4, pp. 349–67; Kohli and Jaworski, op. cit.; Narver and Slater, op. cit.; Ruekert, op. cit.; Deng and Dart, op. cit.; Greenley, op. cit.; Becker and Homburg, op. cit; Pulendran et al. (2000), op. cit.

35. Narver and Slater, op. cit.; Jaworski and Kohli, op. cit.; Selnes et al., op. cit.; Pulendran et al. (2000), op. cit.; Mehta and Joag, op. cit.; Shoham and Rose, op. cit.; Kumar et al., op. cit.

36. Pelham, A. (1997), 'Mediating influences on the relationship between orientation and profitability in small industrial firms', *Journal of Marketing Theory and Practice*, vol. 5, no. 3, pp. 1–23; Raju, P.S., Lonial, S.C., and Gupta, Y.P. (1995), 'Market orientation and performance

in the hospital industry', *Journal of Health Care Marketing*, vol. 15, no. 4, pp. 34–41; Atuahene-Gima, op. cit.

37. Ngai, J.C.H. and Ellis, P. (1998), 'Market orientation and business performance: Some evidence from Hong Kong', *International Marketing Review*, vol. 15, no. 2, pp. 119–39; Pelham and Wilson, op. cit.

38. Ruekert, op. cit.

39. Gauzente, C. (2001), 'Why should time be considered in market orientation research?', *Academy of Marketing Science Review*, vol. 1. Available at http://www.amsreview.org/amsrev/forum/gauzente01–01.html.

40. Jaworski and Kohli, op. cit.

41. Ibid.

42. Subramanian, R. and Gopalakrishna, P. (2001), 'The marketing orientation-performance relationship in the context of a developing economy–An empirical analysis', *Journal of Business Research*, vol. 53, pp. 1–13.

43. Siguaw, J.A., Brown, G., and Widing, R. (1994), 'The influence of a market orientation of the firm on sales force behavior and attitudes', *Journal of Marketing Research*, vol. 31, no. 1, pp. 106–116.

44. Kotler, P. (2003), *Marketing Management*, 11th edn, Prentice Hall, Englewood Cliffs, NJ.

45. Matsuno, K. and Mentzer, J. (2000), 'The effects of strategy type on the market orientation-performance relationship', *Journal of Marketing*, vol. 64, no. 4, pp. 1–16.

46. Jaworski and Kohli, op. cit.

47. Jaworski and Kohli, op. cit.; Pulendran et al. (2000), op. cit.; Pulendran, S., Speed, R., and Widing, R.E. (2003), 'Marketing planning, market orientation and business performance', *European Journal of Marketing*, vol. 37, no. 3/4, pp. 476–97; Bhuian, S.N. (1998), 'An empirical examination of market orientation in Saudi Arabian manufacturing companies', *Journal of Business Research*, vol.43, pp. 13–25; Subramanian and Gopalakrishna, op. cit.; Kirca, A.H., Jayachandran, S., and Bearden, W.O. (2005), 'Market orientation: A meta-analytic review and assessment of its antecedents and impact on performance', *Journal of Marketing*, vol. 69, no. 2, pp. 24–41.

48. Grant, R.M. (1996), 'Toward a knowledge-based theory of the firm', *Strategic Management Journal*, vol. 17 (Winter special issue), pp. 109–122; Spender, J.C. (1996), 'Making knowledge the basis of a dynamic theory of the firm', *Strategic Management Journal*, vol. 17 (Winter), pp. 45–62.

49. Van Raaij, E.M., Van Engelen, J.M.L., and Stoelhorst, J.W. (1998), 'Market orientation: Exploring the implementation issue', in B.J.Gray and K.R.Deans (eds), *Marketing Connections–Proceedings of the Australia New Zealand Marketing Academy Conference*, Dunedin, 30 November–2 December, pp. 2741–2753.

50. Ibid.

51. Day, G.S. (1990), *Market Driven Strategy: Processes for Creating Value*, The Free Press, New York.

52. Deshpande et al., op. cit.

PART III

Market Orientation in the Downstream Food Chain

10 *Communication Between Actors of Food Chains: Case Studies of Two Organic Food Chains in Finland*

BY MARJA-RIITTA KOTTILA* AND PÄIVI RÖNNI†

Keywords

supply chain, organic products, supply chain management, efficient consumer response, market orientation, communication, collaboration

Abstract

It is challenging to practise market orientation in supply chains of organic products, which are characterised by small market share, limited supply, and restrictive regulation. Increased communication with consumers may provide a means to meet the needs of consumers and deliver the multifaceted value of the organic products. Thus, we will introduce the problems and challenges of organic supply chains; approach market orientation through the information flow and integration of the key processes from farmers to consumer; present the methods and findings of the case study with two organic food chains operating in Finland; discuss the findings; and provide concluding remarks, including managerial recommendations.

Introduction

Meeting the demands of fragmented consumer segments, such as those favouring organic products, is a great challenge for globalised food chains. The market share of organic

---

*   Ms Marja-Riitta Kottila, University of Helsinki, Faculty of Agriculture and Forestry, Department of Applied Biology, Taivaanvuohentie 9 B 26, FIN-00200 Helsinki. E-mail: marja-riitta.kottila@helsinki.fi. Telephone + 358 40 581 9252.

†   Ms Päivi Rönni, University of Helsinki, Faculty of Agriculture and Forestry, Department of Economics and Management, Tinnonpolku 23, FIN-36640 Iltasmäki. E-mail: paivi.ronni@mtk.fi. Telephone + 358 40 563 4074.

products of the total food market varies between 0–5 per cent in Europe, reaching approximately 1 per cent in Finland.[1] However, consumer surveys suggest that the potential demand for organic food is greater than its market share indicates.[2] The supply does not seem to satisfy demand in terms of price–quality ratio, availability, or diversity of the product assortment.[3] The development of the organic food market is hindered by a few well-identified bottlenecks, such as the low level of organic production, imbalance between production and demand,[4] fragmented marketing structures, as well as inadequate cooperation and communication among the market actors,[5] all of which suggest grounds for turning the focus to the supply chain.

The food chains of organic products (hereafter, the organic chain) are often considered alternative food chains, emphasising the close relationship between food producers and consumers.[6] In addition, the pursuit of ecological sustainability encourages local organic chains.[7] However, most organic food sold in Europe, as well as in Finland, goes through general types of retailers, such as supermarkets.[8] Researchers have identified these supermarkets as crucial sales channels when attempting to increase the organic market, because they can provide both easy access for consumers and organic food at reasonable price.[9] The organic chains are therefore integrated into traditional food chains, which has provided opportunities for suppliers to obtain the benefits of scale but also has entailed challenges. The main obstacles pertain to the ability of these chains, with the increased number and imbalanced power of the actors involved, to meet the needs of the small segment of consumers in tandem with maintaining the multifaceted value of the organic products.

Marsden et al. suggest that in spatially extended food chains, information might replace the close relationship between consumers and producers.[10] In Europe, organic food is regulated by Council Regulation,[11] which sets the minimum requirements for organic food, from raw materials to final products. The legislation decreases the agility of the chain to respond to changes in demand but forces actors to ensure that the information of organic origin is passed on from farmers to consumers. The organic labels have played an important role in helping consumers to identify organic products, thereby expanding their markets.[12] However, the existence of several competing organic and quality labels, as well as claims, confuse consumers[13] and result in rather poor knowledge about organic products.[14] Furthermore, the studies have identified obvious shortcomings: traditional communication often fails to deliver the grounds for the superior quality of organic products,[15] and an overly general form of communication misses the product-specific points and may therefore fail to persuade consumers.[16] In addition, the disconnectedness of the chains and the fragmented nature of communication may hamper the information flow.[17]

In some countries, one actor, such as a supermarket chain, has taken an active role in developing the organic market.[18] This approach has not occurred in Finland, where the small population of about five million is geographically spread out, and the retail sector is highly concentrated. Two leading retail groups manage about 70 per cent of the market. All but one of a total of six retail groups sell organic products, but none targets differentiation by specifically providing organics. This outcome is, for example, manifested by the absence of organic private labels that in many countries are common.[19] Contrary to several other European countries, one national organic label is prevalent in Finland.[20]

Despite conflicting views with respect to collaboration and its suitability in food chains in general[21] and organic chains in particular,[22] we believe that the collaborative approach is extremely critical in organic food chains to identify and respond to the needs of consumers for the following reasons: (1) The needs of the small segment of consumers are difficult to meet with the concise supply of organic products; (2) the majority of organic product suppliers are small companies with limited resources; (3) organic regulation restricts the agility of organic chains to respond to changing demands; and (4) the intangible values of organic products, mainly based on organic production and processing methods, are difficult to assess at the point of sale without accompanying information from the production chain. Findings reveal a lack of market information,[23] poor communication,[24] and a need for collaboration along the organic chains,[25] yet very little research focuses on the issue. Furthermore, most studies of collaboration in food chains only consider dyadic relationships.[26] The objective of this study therefore is to assess how the organic chain, from farmers to retailers, manages to function as a collaborative, market-oriented entirety. We address the following questions:

- What kind of prerequisites do communication practices along the chain provide for market orientation?
- How do consumers perceive communication between themselves and the other chain actors?

Basic Concepts

Market orientation has been viewed as a culture or implementation of a set of activities underpinning the production of superior value for customers.[27] Both perspectives include the collection of information about customers and competitors, cross-functional information sharing and coordination of activities, and responsiveness to activities by competitors and changing market needs.[28] Market orientation and supply chain management are partly overlapping concepts.[29] Supply chain management (SCM) is 'the integration of key business processes from the end user through original suppliers that provide products, services and information that add value for customers and other stakeholders.'[30] The market orientation of SCM manifests itself as end user–driven integrated processes, as well as information gathering and sharing along the chain.[31]

The concept of SCM applied to food chains is known as efficient consumer response (ECR), defined as 'a grocery industry strategy in which distributors, suppliers and brokers jointly commit to work closely together to bring greater value to the grocery customer.'[32] Thus, ECR aims at system-level efficiency by implementing the following core processes and securing the flow of accurate information: (1) efficient replenishment (i.e., maintaining high in-stock levels by the provision of the right product to the right place at the right time in the right quantity); (2) efficient promotion (i.e., harmonising promotion between manufactures and retailers); (3) efficient store assortment (i.e., optimal allocation of products, maximising consumer satisfaction, and ensuring the most efficient use of self space); and (4) efficient product introduction (i.e., developing and introducing new products by involvement of actors at an early stage of the development process.[33]

We have approached market orientation in organic chains through communication, because it reveals both the integration of the processes and the information flow along the chain. Although ECR is mainly practised between retailers and manufacturers,[34] we adapted ECR processes to the whole length of the supply chains, from farmers to consumers. We believe that communication presents an appropriate perspective for evaluating the integration of the processes along the entire food chain, because all interaction involves communication, even if the practical undertakings within the processes differ a great deal throughout the chain according to the role of each actor. In this chapter, we define communication as the sharing of any relevant information among the actors of the food chain.[35]

Data and Methods

We chose a qualitative case study method,[36] because the aim of this study was to understand collaboration and communication practices in a real-life context, that is, in relationships between persons who actually interact. We depict the chains from the organic manufacturers' point of view, focused on their main products, and follow the product chain upstream upwards to farms, as well as downstream downwards through retailers all the way to consumers. We include primary actors exclusively[37] to simplify the actual supply networks. For confidentiality, the case organisations remain anonymous, with the exception of the retailers.

The cases, two organic food chains, meet the requirements for comparability, representativeness and diversity. The chains represent typical Finnish organic supply chains; they include a small manufacturer processing daily basic food: muesli and yoghurt. With the exception of the manufacturer and its suppliers, the proportion of organic food is small throughout the chains, as in Finnish food chains in general (Figure 10.1). The share of organic food in the yoghurt category is approximately 1 per cent and that in the muesli category is around 2 per cent.[38] Both case products are among the leading organic products in their categories. The chains differ with respect to the nature of the product as well as the marketing concept.

In both cases, the manufacturers were owned by the farmers and processed three to five special organic products. In the chain providing organic yoghurt (i.e., the yoghurt chain), the manufacturer (i.e., the subcontractor) had a subcontract for processing yoghurts into the narrow organic segment of a larger dairy company (i.e., the brand owner), which processed the majority of the milk produced in Finland. The brand owner had supply contracts with the farmers through regional milk producers' co-ops that jointly owned the company. Within the group of farmers of the yoghurt case, both shareholders and non-shareholders of the subcontractor were represented. In the chain providing organic muesli (i.e., the muesli chain), the manufacturer sold final products under its own brand to one retailer (S) and produced a private-label product for another retailer (K). The farmers had supply contracts with a milling company, which in turn was the main supplier for the organic manufacturer.

The retailer groups S and K represented the two largest grocery retailers in Finland, both with shares of almost 35 per cent of the total market. Retail group S consisted of regional co-ops that owned the central organization SOK (i.e., the central unit), and in the retail group K, independent retailers co-existed with a publicly listed company called Kesko (i.e., the central unit). The retail groups managed their four food-store chains primarily through their central units. Both case products were listed in the nationwide, centrally managed assortments of both retailers. The case stores were supermarkets, located in the Helsinki metropolitan area.

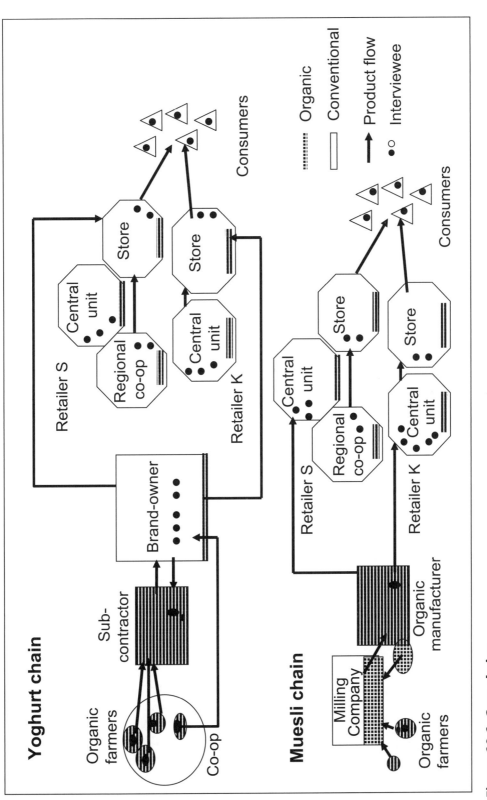

Figure 10.1 Case chains

The interviewed consumers were regular customers of a particular store and those usually responsible for the household's food purchases. They used yoghurt and/or muesli as part of their daily diet and reported using organic products regularly or at least occasionally.

We conducted 28 interviews with actors, from farmers to retailers (altogether, 36 persons), and 5 focus group discussions (altogether, 17 consumers). The interviewees were responsible for the organic products and processes investigated in this study. In the small companies or units, their responsibility covered overall performance, and in the bigger companies, the interviewees were the managers of category, marketing, key account, supply, or environmental issues, or their duties covered marketing research or communication. Consumers were recruited by interviewing 85 customers in the four stores. The interviews then were conducted in the offices of the actors between November 2004 and March 2005, and the group discussions were conducted in the meeting rooms of the focal stores between November 2005 and February 2006.

The semi-structured interviews and the group discussions, lasting one to three hours, were recorded and transcribed. The themes of the interviews captured the interaction in the processes in real-life contexts. We focused on the scopes instead of the activities of the process and therefore use terms such as 'replenishment,' 'assortment,' 'product introduction,' and 'promotion' to refer to the chain-wide processes. After two test interviews, we reduced the themes to address the following topics: the role of the actor in the chain, the position of organic products in the actor's business strategy, the interactions and relationships with other actors during the processes, and the needs and exchange of information within and outside the chain. The group discussions were conducted and devised after the preliminary analysis of the interviews and included the following themes: the use of organic products, interactions with the chain in terms of the four processes, and the needs and flow of information. In addition to interviews, we gathered secondary material such as annual reports and Web pages.

We started the analysis by reading and coding the transcripts to reduce and crystallise the data. The theory and the themes of the interview guided the coding, but we also allowed additional themes to emerge from the data. After coding each interview, we made subsequent memos summarising everything at the actor level. We identified the processes by tabulating, comparing and summarising the interviewers' descriptions of the interactions throughout the processes. Thereafter, we depicted the amount and direction of any communication by summarising the actors' descriptions of the frequency (between chain actors, from farmers to retailers, and feedback from consumers) and the number of communication means used (between the consumers and other chain actors). Finally, we compared the two cases.[39] The memos were sent to the actors (except for consumers) for comments, and the preliminary results were discussed together at a one-day workshop.

Findings

We depict a summary of the findings in Figure 10.2 and then detail the findings by presenting the communication in each of the processes, first from the farmers to the retailers and then between the consumers and any chain actors. We regarded a retail group as one single actor and thus excluded communication among its different members (Figure 10.2). The descriptive quotations were translated from Finnish to English; though they may have lost some of their original qualities in this process, they transmit the key messages.

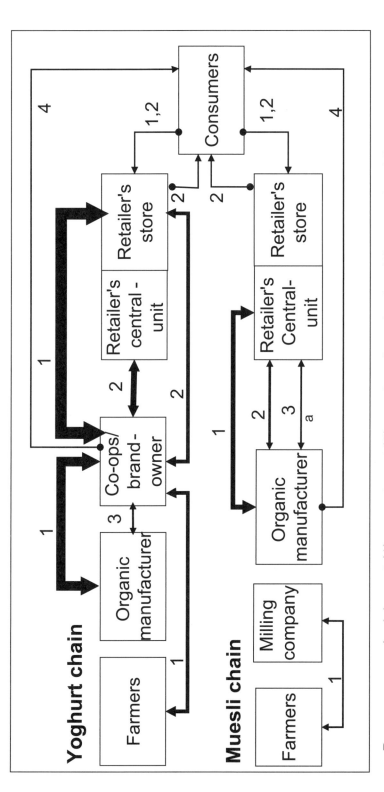

Processes: replenishment (1), assortment (2), product introduction (3) promotion (4).

The width of arrow depicts the frequency and number of communication means.

a) Bidirectional communication in the product introduction process was found only between the organic manufacturer and retailer S.

Figure 10.2 Communication in the key processes between the actors

REPLENISHMENT PROCESS

Farmers–retailers

The replenishment process was the only one in which communication integrated farmers into the chain. However, the communication was much less intensive upstream than downstream. The communication between farmers and their contract partners focused on the terms of the supplying contracts, the deliveries of the production, and feedback about its quality:

> '[E]very spring we have a meeting [with the milling company] ... mainly concerning quality issues, and in addition to that, we farmers try to discuss the fair level of price.' (muesli chain/ farmer).

The communication between retailers and suppliers included placing orders and sharing details of product deliveries. The stores placed the orders daily to the retailers' central unit (muesli chain) or directly to the brand owner (yoghurt chain), according to the anticipated demand and the remaining stock of products. The brand owner forwarded the composite orders to the subcontractor, which then supplied the yoghurts with the related details. Comparing the chains, the stores communicated with different actors because the muesli was delivered to the stores by the retailers' central units, whereas the yoghurt was delivered by the brand owner:

> 'We have the ordering system where we set the desired amounts of the products..., the cash till registers every product sold through it, and when the amount goes under the certain level, the system sends an order to the supplier.' (retailer S, store 1)

The difference in frequency of communication between the cases stemmed from the nature of the product. Whereas cereal was delivered only a few times a year, milk was collected every other day, and hence, relevant information was passed on regularly throughout the year. Likewise, the yoghurt as a fresh product was ordered and delivered daily, which was reflected in the communication.

Consumers

The communication with consumers was incidental and occurred only when consumers came across empty shelf compartments as a result of a disturbed process. These incidents occurred occasionally; owing to the more rapid replacement cycle, they pertained to yoghurt more often than muesli. Few consumers had a habit of asking or informing the staff about missing products:

> 'Well, usually, in case I happen to spot a staff member nearby, I go and ask ...' (X2, group 4)

ASSORTMENT PROCESS

Farmers–retailers

In the assortment process, the communication connected the retailers and the suppliers of the final products. The process mainly related to the evaluation of the demand of the suppliers' products listed and the novelties introduced into the assortment, as well as the terms of the supply contract. In the yoghurt chain, the actors practised much more intense and multi-level communication than in the muesli chain. The yoghurt brand-owner's key account-managers contacted the retailers almost weekly, and four times a year, they presented their proposals for the assortments. However, the organic segment received much less attention, because it represented just a small segment for both parties:

> *'Well, I visit them [retailer's central unit] almost once a week…. How would I say, we have been discussing organic products quite often, not on each 50 visits, but at least 5 times.' (yoghurt chain, brand owner 1)*

> *'[W]e discuss issues concerning the assortment and prices in general, and they usually present their suggestions and views for the assortment and the market three times a year, but I can't recall many recent discussions about organic products.' (yoghurt chain, retailer K3)*

At the store level, the sales promoters visited the largest stores approximately once a month. In the muesli chain, the communication between the manufacturer and the retailers' category manager typically consisted of one meeting and three to five phone calls or e-mails yearly.

Consumers

The retailers communicated their organic assortments to consumers by placing the products on the shelves and emphasising them with green tags on the edge of the shelves, next to the price information. One K store also established a special shelf for introducing the range of available organic products. On average, consumers knew little about the range of organic products in their stores:

> *'I see this product for the first time. I am surprised, because I have been looking for reasonable yoghurt, but haven't found one.' (X1, group 2)*

Consumers perceived that there was no information available about the organic assortment, and the only way to become familiar with it was to scan the whole store. As a consequence, they considered it time consuming to look for organic products and difficult to locate them in the store. Several consumers appreciated the retailers' efforts to ease this search with tags, yet they pointed out problems, such as missing tags or difficulty in noticing or interpreting them correctly. The data indicated that composite assortment information might increase consumers' motivation to buy organic food:

> *'There is no other way to get [information about the assortment of organic products] than going through all shelves searching for the organic products.' (X3, group 4)*

'I am hardly the only one perceiving lack of information [about the assortment], in addition to laziness, a reason for not searching for the organic products.' (X3, group 1)

Only a few consumers expressed their wishes about the assortment to the staff or through feedback boxes. Most consumers doubted if any true impact would be made by giving feedback and felt they needed some proof of having been taken seriously:

'I used to give feedback but gave that up, because it did not help, at least my wishes were not taken into account.' (X1, group 4)

Consumer feedback was collected by both retail groups, but they processed it differently. The stores of retail group S forwarded the feedback to the central units, whereas the stores of group K tried to react to the feedback at the store level. This difference may reflect consumers' perceptions of the effectiveness of giving feedback, which varied according to the customers of the different stores. The retailers' central units noted receiving a lot of consumer feedback, either directly or forwarded from the stores, but found it quite fragmented and laborious to utilise.

PRODUCT INTRODUCTION PROCESS

Farmers–retailers

The product introduction process connected only a few actors at the dyadic level. The brand owner conducted the product development without any systematic information exchange with the other chain actors. Communication toward the subcontractor was restricted to supervising the production process of novelties:

'We haven't at all discussed product development with [the brand owner] and I would be worried if we did. A company like that ought to have interest and competence on their own to practise active product development...'. (yoghurt chain, retailer K3)

On the contrary, in the muesli chain, the process was explicitly discussed between the retailer and the manufacturer. The muesli manufacturer inquired as to the needs of retailer S, whose category manager analysed sales figures to identify and suggest manufacturer potential points to target novel products:

'We have approached retailers and asked what it is that they want. We can't just push products without asking first what they need and/or wish to get.' (muesli chain, manufacturer)

Consumers

The brand owner had a research and development department and conducted extensive consumer research before launching novelties. Within the narrow organic segment, though, products were developed and launched without a systematic process, including consumer surveys.

The consumers trusted the manufacturers' competence in product development and therefore found no need to interfere in the process. They believed, however, that by actively contacting the manufacturer, they could have an influence:

'Yes, I think one could have an impact, if one wants to. Right now I just cannot think of anything that needs my contribution.' (X3, group 4)

The organic manufacturer, the brand owner, and the subcontractor all perceived the amount of consumer feedback as abundant. However, the muesli manufacturer considered the feedback somewhat useless for the development of novelties, mainly because of the unrealistic nature of the suggestions.

PROMOTION PROCESS

Farmers–retailers

We found no systematic interaction among any of the actors, from farmers to retailers, with regard to promotion. The retailers' central unit occasionally discussed the visibility of the products and the need for sales promotion, but no joint actions were taken as a result:

'At this moment we don't promote organic products.' (yoghurt chain, brand owner 3)

'[W]e are tied to the marketing managed by the retail chain, which does not promote organic.' (retailer, store 2)

'I think that successful targeting rather than the amount of [marketing], is the most important as well as the most difficult for us.' (muesli chain, manufacturer)

Consumers

In general, communication toward consumers was rare and random, and even when it occurred, it was often initiated by stakeholders from outside the chain, such as organic associations. The information carried by the package was the only regular and proper means of communication at the consumer end of the chain. In addition to the product specifics, the package displayed the organic label and included a few sentences about organic production methods. Consumers thought the packages communicated the quality of the product as well as its organic origin quite well, but they criticised the lack of distinction from the other products on the shelves. The role of packages in creating value to the products also was questioned:

'[A] consumer with no particular attempt of finding [an organic product]is unlikely to pay attention to this. In that sense, the products are not distinctive.' (X3, group 4)

'I doubt that packages influence buying decisions. A person specifically looking for organic will buy the product regardless of whether there is a picture of a tiger, a dolphin or crop or whatever. As long as the organic label is there, anything goes.' (X2, group 3)

The consumers were quite sceptical of the system of registering purchase details for the purpose of customising the type of communication. Nevertheless, consumers were willing to sign up on special mailing lists to receive additional information about organic products:

'Well, I don't know. My initial reaction is somewhat negative, like somebody is snooping.' (X1, group 2)

Discussion

Most communication appears to take place between two adjacent actors operating in the middle of the chain. The highest level of communication occurred in the replenishment process, involving actors from farming to retailing, which reflects the central role of the efficient replenishment process in ECR.[40] Although the delivery of the products obviously requires intensive communication, the domination of the replenishment process in terms of connecting the actors suggests that the primary focus of the chains is truly the efficient delivery of products. This goal, in addition to the scarce communication with consumers, suggests poor prerequisites for market orientation in relation to collecting information from, and producing collaboratively superior value for, consumers.[41] The findings also reveal that no actor has taken an active role in developing the organic market and strengthening the communication link toward consumers. This outcome may reflect the limited resources of the actors specialising in organic products, but it may just as well reflect the market orientation of the other actors, who focus primarily on the actions of their competitors.[42] Furthermore, the findings may indicate that proper means of communication have not been adapted to this small segment.

The differences between the chains stem mainly from the nature of the products. The product introduction process integrates retailers and suppliers only in the muesli chain and only with the other retailer (S), suggesting that it is the nature of the relationships of the persons interacting rather than the marketing concept that influences integration, especially as viewed through communication.

Our findings also support previous reports: the disconnectedness of food chains creates an obstacle for consumers to gather information about food production.[43] However, the general information provided by third-party actors has increased consumers' knowledge of organic production and the national organic label, creating some kind of a link between consumers and producers.[44] Nevertheless, this general information has not compensated for the scarcity of communication within the chains, which is especially manifested as poor knowledge about the assortment or availability of the organic products and difficulties in finding the products. This conclusion strongly suggests that the promotion of organic food has been too general and not in concrete relation to any particular products.[45] Improvements in the efficiency of the replenishment process among retailers and suppliers obviously has shifted the inefficiency point to the consumers' end of the chain, and therefore, system-level improvements have not been achieved.

Compared with the previous findings that have emphasised the role of varying motivation and poor price–quality ratios, availability, and product assortment diversity,[46] our findings underline the shortage of knowledge about assortments. This result may partly reflect the research design. In our study, consumers represent actors closely related to the food chains, whom we recruited for the group discussions at stores. This approach likely reflects actual buying decisions and therefore can reveal some of the practical obstacles that consumers may meet when shopping.

However, these findings also reveal the challenges involved in communicating with consumers, whose motivation and capacity to receive information varies; their willingness to connect with the chain cannot be taken for granted. Conventional marketing leaves consumers to a passive role of merely serving as objects of activities. Furthermore, food products typically are considered low-involvement goods. Yet consumers appear quite interested in different ethical aspects of food,[47] and such awareness may be able to generate involvement and thus create conditions that favour consumers' integration into the collaborative processes of the chain.[48]

The following limitations should be taken into account when interpreting the outcomes of our study. Consumer respondents were users of organic products, representing a group highly motivated to receive information and look for organic products, which means they do not reflect the views of average consumers. We considered it important to interview real users, because they have experience with the performance of the studied processes. Our findings pertaining to the replenishment process are based on the perceptions of the interviewees, a group that contains no logistics experts. Furthermore, studies conducted in other countries could contribute to a better understanding of the market orientation of organic chains, beyond the characteristics of the Finnish organic market.

Conclusions

Our findings suggest that the scarcity of communication between consumers and the other actors in a supply chain jeopardises consumers' access to organic products. Consumers' poor knowledge about the assortment, as well as the difficulties experienced in finding them, may further undermine consumers' perceptions of the assortment of organic products. These factors are partially responsible for the gap between sales figures and potential demand. To extend the assortment process and include themselves in it, consumers request better placement, assortment information available at stores in print or electronically, in-store product demonstrations, and proper education of the staff. The findings also suggest that if organic perishables cannot be delivered every day, information about the days of delivery and the chance to place orders in advance might be worth considering.

The findings raise a question about regular customers and loyalty card programmes too. Although consumers clearly prefer to do their shopping at one or a couple of stores they know and can easily access, loyalty programmes of these retail groups are managed at the central unit level. Thus, in addition to offering economic benefits for customers that favour the retail group, the programmes might add actions to enhance the relationship of a particular store with its regular customers.

Communication takes place mainly between the manufacturers and the retailers, the central actors in the ECR concept. To ensure the formation and delivery of added value for organic products, consumers and farmers should have effortless access and encouragement to participate in the organic chain processes. The findings suggest a genuine need to develop methods to study collaboration along the whole chain. We consider this goal extremely important in the context of a special segment within a food chain, for which the grounds and origin of added value are, at least partly, based on sustainable farming methods and respect for animal welfare.

References

1. Hamm, U. and Gronefeld, F. (2004), *The European Market for Organic Food: Revised and Updated Analysis*, OMIaRD Publication, vol. 5. School of Management and Business, University of Wales, Aberystwyth, UK. See p. 51; The Organic Farming in the European Union—Facts and Figures (2005), European Commission, <http://ec.europa.eu/agriculture/organic/consumer-confidence/consumer-demand_en>, Accessed November 10, 2008; ACNielsen. (2004), Finfood consumer panel data.

2. Bähr, M., Botschen, M., Laberentz, H., Naspetti, S., Thelen, E., and Zanoli, R. (2004), *The European Consumer and Organic Food*, OMIaRD Publication, vol. 4. School of Management and Business, University of Wales, Aberystwyth, UK; Finfood (2004), *Organic Consumer Barometer 1998–2004*; Wier, M. and Calverley, C. (2002), 'Market potential for organic foods in Europe', *British Food Journal*, vol. 104, no.1, pp. 45–62.

3. Bähr et al., op. cit.; Finfood, op. cit.

4. Hamm and Gronefeld, op. cit., p. 132; Franks, J. (2003), 'Current issues in marketing organic milk in the UK', *British Food Journal*, vol. 105, no. 6/7, pp. 350–62; Baecke, E., Rogiers, G., De Cock, L. and Van Huylenbroeck, G. (2002), 'The supply chain and conversion to organic farming in Belgium or the story of the egg and the chicken', *British Food Journal*, vol. 104, no. 3/4/5, pp. 163–74.

5. Hamm and Gronefeld, op. cit., p. 128.

6. Marsden, T., Banks, J. and Bristov, G. (2000), 'Food supply chain approaches: Exploring their role in rural development', *Sociologia Ruralis*, vol. 4, no. 40, pp. 424–38.

7. Busch, L. and Bain, C. (2004), 'New! Improved? The transformations of the global agrifood system', *Rural Sociology*, vol. 69, no. 3, pp. 321–46; Bähr et al., op. cit.; Guptill, A. and Wilkins, J.L. (2002), 'Buying into the food system: Trends in food retailing in the US and implications for local foods', *Agriculture and Human Values*, vol. 19, no. 1, pp. 39–51.

8. Hamm and Gronefeld, op. cit., p. 53.

9. Hamm and Gronefeld, op. cit., p. 134.

10. Marsden et al., op. cit.

11. Council Regulation (EEC) no. 2092/91, <http://ec.europa.eu/agriculture/organic/eu-policy/legislation>. Accessed November 10, 2008.

12. Hamm, U., Gronefeld, F., and Halpin, D. (2002), *Analysis of the European Market for Organic Food*, School of Management and Business, University of Wales, Aberystwyth, UK; Wier and Calverley, op. cit.

13. McEachern, M.G. and Schroder, M.J. A. (2004), 'Integrating the voice of the consumer within the value chain: a focus on value-based labelling communications in the fresh-meat sector', *Journal of Consumer Marketing*, vol. 21, no. 4, pp. 497–509; Padel, S. and Foster, C. (2005), 'Exploring the gap between attitudes and behaviour: Understanding why consumers buy or do not buy organic food', *British Food Journal*, vol. 107, pp. 606–625.

14. Zanoli, R., Naspetti, S., Vairo, D., Thelen, E., Laberenz, H., and Bähr, M. (2004), 'Potential scope for improved marketing: Considering consumer expectations with regard to organic and regional food', in Schmid, O., Sanders, J., and Midmore, P. (2004), *Organic Marketing Initiatives*

and Rural Development, OMIaRD Publication, vol. 7. School of Management and Business, University of Wales, Aberystwyth, UK, pp. 119–58.

15. Schmid et al., op. cit.; Guptill and Wilkins, op. cit.

16. Peter, P.J., Olson, J.C., and Grunert, K.G. (1999), *Consumer Behaviour and Marketing Strategy*. European Edition, McGraw-Hill, London; Zanoli and Naspetti, op. cit.

17. Duffy, R., Fearne, A. and Healing, V. (2005), 'Reconnection in the UK food chain: Bridging the communication gap between food producers and consumers', *British Food Journal*, vol. 107, no. 1, pp. 17–33.

18. Hamm and Gronefeld, op. cit., p. 135.

19. Jonas, A. and Roosen, J. (2005), 'Private labels for premium products—the example of organic food', *International Journal of Retail and Distribution Management*, vol. 33, pp. 635–53.

20. Hamm and Gronefeld, op. cit.

21. Fearne, A., Hughes, D., and Duffy, R. (2001), 'Concepts of collaboration—supply chain management in a global food industry', in Eastham, J.F., Sharples, L., and Ball, S.D. (eds), *Food and Drink Supply Chain Management—Issues for the Hospitality and Retail Sectors*, Butterworth-Heinemann, Oxford, pp. 55–89; Palmer, A. and Bejou, D. (1994), 'Buyer-seller relationship: A conceptual model and empirical investigation', *Journal of Marketing Management*, vol. 10, pp. 495–512; Peterson, C. (2002), 'The 'learning' supply chain: Pipeline or pipedream?' *American Journal of Agricultural Economics*, vol. 84, pp. 1329–1336; Hingley, M.K. (2005), 'Power imbalance in UK agri-food supply channels: Learning to live with the supermarkets', *Journal of Marketing Management*, vol. 21, pp. 63–88; Lindgreen, A. (2003), 'Trust as a valuable strategic variable in the food industry: Different types of trust and their implementation', *British Food Journal*, vol. 105, pp. 310–27; Glandieres, A. and Sylvander, B. (1999), 'Specific quality and evolution of market coordination forms: The case of environmentally friendly food products', in Galizzi, G. and Venturini, L. (eds), *Vertical Relationships and Coordination in the Food System*, Physica-Verlag, Heidelberg, pp. 547–60; Hingley, M. (2001), 'Relationship management in the supply chain', *International Journal of Logistics Management*, vol. 12, pp. 57–71.

22. Hamm and Gronefeld, op. cit., p. 134; Wycherley, I. (2002), 'Managing relationships in the UK organic food sector', *Journal of Marketing Management*, vol. 18, no. 7/8, pp. 673–92; Kottila, M.-R. and Rönni, P. (2008), 'Collaboration and trust in two organic food chains', *British Food Journal*, vol. 110, no. 4/5, pp. 376–94.

23. Hamm, Gronefeld, and Halpin, op. cit.; Hamm and Gronefeld, op. cit. See p.134.

24. Zanoli et al., op. cit.

25. Hamm and Gronefeld, op. cit., p. 133.

26. Hingley, M. and Lindgreen, A. (2002), 'Marketing of agricultural products: Case findings', *British Food Journal*, vol. 104, no. 10/11, pp. 806–27; Kulp, S.C., Lee, H.L. and Ofek, E. (2004), 'Manufacturer benefits from information integration with retail customers', *Management Science*, vol. 50, no. 4, pp. 431–44; Wycherley, op. cit.

27. Kohli, A.K. and Jaworski, B.J. (1990), 'Market orientation: The construct, research propositions, and managerial implications', *Journal of Marketing*, vol. 54 (April), pp. 1–18; Narver, J.C. and Slater, S.F. (1990), 'The effect of a market orientation on business profitability', *Journal of Marketing*, vol. 54 (October), pp. 20–35; Kennedy, K.N., Goolsby, J.R., and Arnould, E.J. (2003), *Journal of Marketing*, vol. 67 (October), pp. 67–81; Beverland, M.B. and Lindgreen, A. (2007), 'Implementing market orientation in industrial firms: A multiple case study', *Industrial Marketing Management*, vol. 36, pp. 430–42; Heiens, R.A. (2000), 'Market orientation: Toward an integrated framework', *Academy of marketing Science review*, vol. 2000, no. 1, pp. 1–4.

28. Martin, J.H. and Grbac, B. (2003), 'Using supply chain management to leverage a firm's market orientation', *Industrial Marketing Management*, vol. 32, pp. 25–38.

29. Ibid.

30. Lambert, D.M. and Cooper, M.C. (2000), 'Issues in supply chain management', *Industrial Marketing Management*, vol. 29, pp. 65–83.

31. Bechtel, C. and Jayaram, J. (1997), 'Supply chain management: A strategic perspective', *The International Journal of Logistics Management*, vol. 8, no. 1, pp. 15–34; Lambert, D.M., Cooper, M.C. and Pagh, J.D. (1998), 'Supply chain management: Implementation issues and research opportunities', *International Journal of Logistics Management*, vol. 9, no. 2, pp. 1–19; Mentzer, J.T., DeWitt, W., Keebler, J.S., Min, S., Nix, N.W., Smith, C.D., and Zacharia, Z.G. (2001), 'Defining supply chain management', *Journal of Business Logistics*, vol. 22, no. 2, pp. 12–24.

32. Kurt Salomon Associates (KSA) (1993), *Efficient Consumer Response: Enhancing Consumer Value in the Grocery Industry*, Food Marketing Institute, Washington, D.C.

33. Alvarado, U.Y. and Kotzab, H. (2001), 'Supply chain management: The integration of logistics in marketing', *Industrial Marketing Management*, vol. 30, pp. 183–98; Harris, J.K., Swatman, P.M. C., and Kurnia, S. (1999), 'Efficient consumer response (ECR): A survey of the Australian grocery industry', *Supply Chain Management*, vol. 4, pp. 35–42; Hoffman, J.M. and Mehra, S. (2000), 'Efficient consumer response as a supply chain strategy for grocery businesses', *International Journal of Service Industry Management*, vol. 11, no. 4, pp. 365–70.

34. Alvarado and Kotzab, op. cit.; Harris et al., op. cit.; Hoffman and Mehra, op. cit.; Corsten, D. and Kumar, N. (2005), 'Do suppliers benefit from collaborative relationships with large retailers? An empirical investigation of efficient consumer response adoption', *Journal of Marketing*, vol. 69 (July), pp. 80–94.

35. Mohr, J. and Spekman, R. (1994), 'Characteristics of partnership success: Partnership attributes, communication behavior, and conflict resolution techniques', *Strategic Management Journal*, vol.15 (February), pp.135–52; Mohr, J. and Nevin, J.R. (1990), 'Communication strategies in marketing channels: A theoretical perspective', *Journal of Marketing*, vol. 54 (October), pp. 36–51.

36. Yin, R.K. (2003), *Case Study Research. Design and Methods*, 3rd edn, Sage Publications, Thousand Oaks, CA.

37. Lambert et al., op. cit.

38. ACNielsen, op. cit.

39. Miles, M.B. and Huberman, M.A. (1984), *Qualitative Data Analysis. A Sourcebook of New Methods*, Sage Publications, Thousand Oaks, CA; Yin, op. cit.

40. Kennedy et al., op. cit.; Heiens, op. cit.

41. Martin and Grbac, op. cit.; Heiens, op. cit.

42. Harris et al., op. cit.

43. Duffy et al., op. cit.

44. Marsden et al., op. cit.

45. Zanoli and Naspetti, op. cit

46. Padel and Foster, op. cit.; Tarkiainen, A. and Sundqvist, S. (2005), 'Subjective norms, attitudes and intentions of Finnish consumers in buying organic food', *British Food Journal*, vol. 107, pp. 808–822; Bähr et al., op. cit.; Finfood, op cit.

47. Bähr et al., op. cit.

48. Hoffmann, E. (2007), 'Consumer integration in sustainable product development', *Business Strategy and Environment*, vol. 16, pp. 322–38.

11 *Marketing Research and Sensory Analysis: A Reasoned Review and Agenda of Their Contribution to Market Orientation in the Food Industry*

BY ALESSIO CAVICCHI,* MARIA ROSARIA SIMEONE,†
CRISTINA SANTINI,‡ AND LUCIA BAILETTI§

Keywords

qualitative techniques, sensory analysis, new product development (NPD), marketing strategies

Abstract

Innovation development is a fundamental condition for companies to hold their positions in intensified market competition, to stabilise sales, to contribute to a company's growth and to reduce risks through diversification. There are many examples in prior literature of the importance of sensory analysis in terms of designing, testing, launching and rethinking food products. When these tools are conceived of not only as a unique and isolated

* Dr Alessio Cavicchi, Department of Studies on Economic Development, University of Macerata, P.zza Oberdan 3, 62100 Macerata, Italy. E-mail: a.cavicchi@unimc.it. Telephone: + 39 0733 258 3919.

† Dr Maria Rosaria Simeone, Department of Economic and Social Systems Analysis, University of Sannio, Via delle Puglie 82, 82100 Benevento, Italy. E-mail: msimeone@unisannio.it. Telephone: + 39 082 4305714.

‡ Dr Cristina Santini, School of Business Administration, University of Florence, Via delle Pandette 9, 50127 Firenze, Italy. E-mail: cristina.santini@unifi.it. Telephone: + 39 055 43741.

§ Ms Lucia Bailetti, Italian Center of Sensory Analysis, Via Cuoio 19, 62024 Matelica, Italy. E-mail: lbailetti@analisisensoriale.it. Telephone: + 39 0737 84215.

science but also jointly with other fields in marketing research, the interdisciplinary approach increases the potential and efficiency of research and development.

Product innovation can be instrumental for firms that want to respond to changes in the business environment and be market oriented. Thus, after having demonstrated the importance of the 'House of Quality' and 'Buyer Utility Map,' as well as their application to the innovation process, this chapter underlines the usefulness of adopting multiple methods and interdisciplinary approaches, with the support of background information and the use of the latest research in the field of consumer science.

Introduction

It is possible to define a market-based innovation as a product that results from combining skills and resources in such a way that buyers perceive a lasting improvement in the relationship between price and value, relative to competing products.[1] Innovation development is a fundamental condition for companies to hold their positions in intensified market competition, stabilise sales, contribute to a company's growth, and reduce risks through diversification. However, this development process cannot only be 'product oriented'; if it is, it will generate marketing myopia,[2] meaning it has only a partial view of reality.

Instead, product innovation can be instrumental for firms by enabling them to respond to changes in the business environment and become market oriented.[3] Market orientation, in Narver and Slater's[4] view, is the result of three distinct behavioural components: customer orientation, competitor orientation and inter-functional coordination. Even if empirical research has demonstrated low levels of market-oriented new product development (NPD) in the food sector, delivering a superior product and value to end-users and being customer oriented is necessary to achieve higher levels of profitability and sales growth.[5]

Among the factors explored by marketing research (e.g., cognitive perception, affection dimension, sensory sensitivity and their interaction), it is interesting to underline the importance of sensory evaluation for the development of a product (and the process of) innovation.

The human sensory system is a versatile instrument for food companies to measure the sensory aspects of food. Nevertheless, the cooperation between sensory and marketing science has not always been fruitful, because sensory analysis traditionally has been product oriented. Marketing, in contrast, stresses the external validity of test results – or the extent to which test results can be generalised to market behaviour.[6]

Considerable evidence in existing literature highlights the importance of sensory analysis for designing, testing, launching and rethinking food products. When these tools are conceived not only as a unique and isolated science but together with other fields in marketing research, the interdisciplinary approach increases the potential and efficiency of research and development.

Decisions about what food to buy are complex and influenced by many factors in addition to sensory quality. It is broadly agreed that taste and other sensory qualities only partially explain consumer behaviours. It is also important to understand the role of several other factors: convenience, price, production technology, personal health, branding and societal issues. Thus, the complex nature of food choice and consumer behaviour and the influence that non-sensory factors can exert must be considered jointly.[7]

For example, in the market for specialty products, information about food origins is an important variable that indicates consumer preferences. This information generates expectations related to sensory properties and the acceptability of food products; it also influences consumers' evaluations. Thus, considering the data that consumers receive from their own sensory systems determines how they respond to products. It is important to study how using empirical research techniques that include both sensory testing methodologies and psychological and sociological tools might enhance marketing analysis and thus marketing strategy efficiency.

This chapter reviews two main product innovation methods that combine the combined use of sensory analysis and other qualitative research tools. After the construction of a theoretical framework establishing the importance of the 'House of Quality' and the 'Buyer Utility Map,' this chapter notes some empirical applications of these and other tools, summarises some of major marketing and sensory research, and addresses important contemporary sources in the field of consumer science.

FOOD QUALITY PERCEPTION AND CONSUMER'S CHOICE

An innovation refers to any good, service, or idea that is perceived by someone as new. The idea may have a long history, but it is an innovation to the person who sees it as new. In the context of consumer goods, it will mainly depend on whether consumers judge the product as offering more value relative to existing products and prices.[8]

The NPD procedure[9] is very time consuming and costly for food firms. According to estimations by Ernst and Young,[10] more than one-third of products launched every year fail (though due to the nature of confidential data, it is difficult to obtain a clear and definitive proxy of failed launches). To avoid a failed product launch in the market, firms need to conduct effective research of the determinants of consumers' food choice.

Figure 11.1 illustrates a representation of consumers' food choice process. Choice depends on consumer preferences that relate to intrinsic and extrinsic product characteristics. Extrinsic product characteristics, though not physical – such as brands, product origin cues, and quality labels[11] – can influence consumers' perceptions. Intrinsic product characteristics can be evaluated through physical, chemical, microbiological and sensory analyses, though they also may depend on changes to the ingredients. Sensory analysis in food products is naturally integrated with marketing. This discipline not only establishes quality assurance but also performs marketing tests to emphasise people's perceptions of sensory quality rather than effective quality control evaluations.

The perception of food sensory attributes also depends on many factors associated with the sensory sensitivity of consumers, as well as factors that contribute to the identification of sensory attributes and their interpretation.

Personal consumer characteristics may influence the detection and interpretation of sensory attributes. Attitude forms according to the personality of respondents and the importance they grant to the products' intrinsic and extrinsic attributes, depending on the consumer culture and identification with sensory characteristics, as well as other attributes such as price, package, image, and the beliefs and expectations associated with benefits of product usage.[13] This outcome occurs because the real product is the result of properties added to the core product.

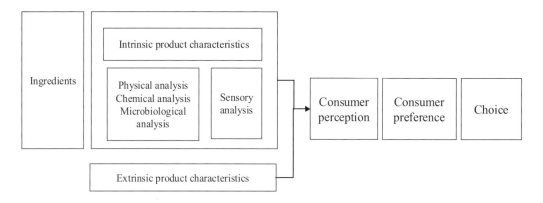

Figure 11.1 Process from primary production to consumption of food products[12]

FOOD QUALITY

Defining food quality is a difficult task; several models and definitions relating to this issue have been presented recently. In 1995 a special issue of the journal *Food Quality and Preference* was dedicated to quality definitions. A book by Monika Schröder[14] also approaches food quality from both technical and consumer satisfaction perspectives. The Total Food Quality Model (Figure 11.2), introduced in 1996 by Grunert and colleagues,[15] is one of the most well-known, studied and applied models of food quality in academic and professional fields. He distinguishes horizontal and vertical dimensions. The first refers to time (quality perceptions before and after the purchase), and the second relates to a sequence of cue detection, quality and purchase motives as a hierarchy of increasingly abstract terms.[16]

From a research point of view, the complexity of quality detection is familiar; considerable empirical evidence shows that consumers, when purchasing a specific food product, express preferences for certain attributes (colour, size, region of origin) that are considered quality cues (signals) of the whole product. Therefore, the concept of perceived quality in food markets is defined as the mediation between a product's characteristics and a consumer's preferences.[18] The quality perception depends on an individual evaluation that relates strictly to the environment and the specific consumption situation and that can be based on incomplete information.[19] The discovery of subjective and objective dimensions and how they interrelate is an important factor in determining the long-term profitability of firms. In this sense, new opportunities arise for those producers willing to take the risk to differentiate their products by adding value for their customers. Adding value is a customer-oriented concept that allows food products to be perceived as having higher quality.[20]

QUALITY ATTRIBUTES

Sensory science is the science of quality perception; by quality, we mean the requirements necessary to satisfy the needs and expectations of consumers.[21] Table 11.1 identifies the key quality sensory and non-sensory attribute subsets for food products.

The first class of requirements in this scheme relates to safety as the absence of food-borne pathogens, pesticide residues and heavy metals, among others. Food safety in a narrow sense is the opposite of food risk, that is, the probability of not contracting a disease as a consequence of consuming a certain food. In the broader sense, food safety

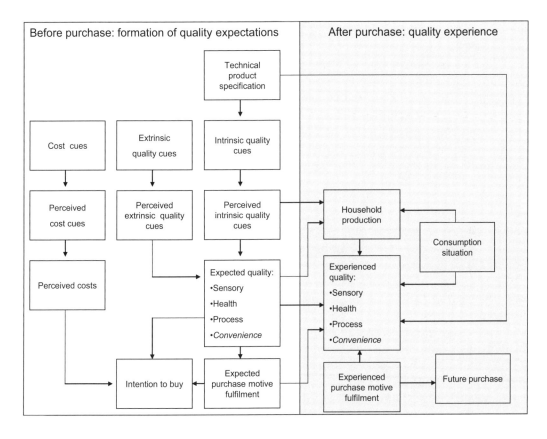

Figure 11.2 Total food quality model[17]

Table 11.1   Quality attribute subsets[24]

Food Safety Attributes/ Food-borne pathogens	Value Attributes	Process Attributes/ Animal welfare
Heavy metals	Purity	Biotechnology
Pesticide residues	Compositional integrity	Environmental impact
Food additives	Size	Pesticide use
Naturally occurring toxins	Appearance	Worker safety
Veterinary residues	Taste	
Nutrition Attributes	Convenience of preparation	
Fat content	Package Attributes	
Calories	Package materials	
Fibre	Labelling	
Sodium	Other information provided	
Vitamins		
Minerals		

is concern for nutritional qualities of food and properties of unfamiliar foods, such as genetically modified food.[22]

Customer attention thus also focuses on nutrition attributes related to conformity standards. More recently in the food industry, innovations increasingly are based on nutritional requirements. In the same product category, it is possible to differentiate two products, such that milk becomes functional, enriched by some properties, and claim that through regular consumption, it helps the consumer attain good health.

Food safety and nutrition attributes, because of the difficulty of forming quality judgements about them, are credence attributes. Consumers often cannot assess quality, even after consumption.[23]

Value attributes, the third macro-category, are search attributes. They facilitate product recognition and can be used to mislead consumers by assigning information to the label and through the aesthetic properties of packaging. These variables are of particular interest for marketing experts. Another important factor is convenience, which refers not only to time efficiency in preparing food but also to the availability of the product in the desired amount. Of particular interest are the trade-offs between convenience properties and other sensory and non-sensory attributes,[25] as well as their interactions. Increasing the level of convenience may, for example, reduce the freshness of the food product and increase its price. Understanding these trade-offs is important for implementing a marketing strategy.

Packaging attributes enable companies to communicate credible information about the quality of the food product. They serve the function of guaranteeing attributes not verifiable otherwise, that is, immaterial features. The label informs consumers about each relevant product's *credence* attributes, which can be ethical, such as those related to the production system, working conditions, environmental standards and other features not strictly related to sensory preferences.

Providing labels and other information can address the market failure that occurs when the consumer is less informed than the seller. This asymmetrical information can lead to decreased quality and the disappearance of the market.[26] Without signals, consumers are uncertain about the quality of products, which creates an incentive for some vendors to sell low-quality products and drive the higher quality products out of the market. The problem with information asymmetry arises when sellers have more information than buyers about the product.

To support higher quality, it is necessary to communicate the uniqueness of products' attributes. Indications of the origin, the tradition of the production process, the use of organic agriculture, animal welfare, defence of the environment and the presence of ethical requirements all play important roles in consumer choices. These process attributes are immaterial, and their existence is guaranteed by certifications.[27]

Non-sensory factors thus can exert important influences over people's relationship with food. Research usually considers the impact of each factor at any given time. This approach conflicts with how people actually make decisions about food.[28]

RECENT MODELS OF FOOD CHOICE DETERMINANTS

Many researchers have tried to elaborate complete models of food choice determinants. One recent attempt, Sobal's Personal Food System Model (probably the most complete attempt to depict the actual context of the food choice process and therefore extremely complex),[29] asserts that the final choice of food depends on three macro-categories. The first, life course, includes relevant prior experiences. The second emerges from this category, namely, influences. The third component is the personal food system, which includes the mental processes by which people translate influences on their food choices into how and what they eat in particular situations. Personal food systems represent the ways that options, trade-offs, and boundaries get constructed in the process of making food choices. In this context, sensory characteristics are a primary determinant of the personal food system, together with convenience, instrumental consequences (health), cost and social influences or management.[30]

Food Product Innovation from a Market-orientation Perspective

Several examples in the food industry show how the development of a new product in the food industry can occur in accordance with consumer needs and wants. This situation describes some products in existing markets, such as muesli or products to treat specific diseases.[31]

Consumer drives for more quality, together with the conventional conviction that product lifespan is shortening,[32] have pushed marketing researchers to look for new tools to implement NPD processes.

The perceived importance of adopting a customer perspective when conceiving of new products has driven many companies to evaluate the potentialities of *sensory analysis*. Big multinational companies have led the way. Without a doubt, sensory analysis has played an important role in quality control for food products, and it has relevant applications to NPD in the food business. Despite its importance, the integration between sensory and marketing practices continues to evolve though. Through the translation of many principles from sensory analysis to NPD, sensory analysis comes closer to marketing. This connection is worth careful examination, because it could increase profits. Examples of sensory analysis applications include the evaluation of a change in input processing (e.g., introduction of a post-harvest treatment for fruits and vegetables), the development of a new functional/probiotic food, or the measurement of preferences for new packaging. According to this perspective, customer feedback obtained by the company through sensory analysis techniques plays a key role when management evaluates marketing decisions; the data gathered generally can be implemented to detect product preferences and clarify the factors that affect product optimisation and consumer acceptance.

Sensory analysis is product-oriented and stresses the relationship between sensory evaluation and product characteristics. Marketing instead emphasises the validity of the results to define the extent to which conclusions explicate market behaviour. Only by integrating the two dimensions does potential for the success of product development result. Unfortunately, their integration in practice is imperfect, because of differences in their basic orientations. To realise this outcome, integration must grant external validity to the sensory analysis. Two tools used to obtain this validity are the 'House of Quality' and the 'Buyer Utility Map.'

THE HOUSE OF QUALITY

The existing framework for linking physical science to the economic approach[33] can improve the physical food product on the basis of consumer demands by taking consumers' quality judgements as a point of departure and relating them to the characteristics of the physical product: Quality function deployment (QFD) is a process, used in the food sector since 1987, that originated in Japan to manage product development.

The QFD tool is designed to help planners focus on the characteristics of a product or service and thereby translate consumers' needs across the supply chain. The QFD approach emphasises the voice of the consumer as the preliminary aim of product development. The approach also is based on identifying the relationship between the sensory aspects of the product, the technical relations and their combination, such that it relates all of the aspects to the voice of the consumer in the product development process. Thus the probability of innovation success increases.

One important QFD tool is the so-called 'House of Quality' (Figure 11.3), also known as the Product Planning Matrix, which translates customer needs into measurable technical attributes with the goal of defining the objective of the product development.[34]

Starting in the mid 1990s, many applications of the QFD have been discussed in literature.[36] Some contributions applies House-of-Quality structures to food product development.[37] The House of Quality consists of several phases. The first is the *voice of the customer*, which indicates product quality requirements. In this step, there are no

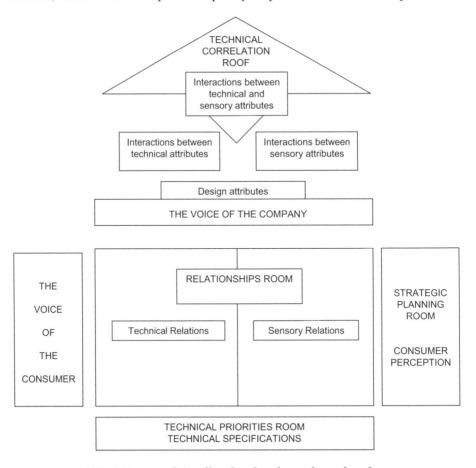

Figure 11.3 Modified House of Quality for food product development: interaction among sensory and technical characteristics[35]

precise or homogeneous statements from consumers. The sources of information include in-depth interviews, focus groups, and other qualitative techniques. This step is very sensitive, because the subsequent steps are based on these results.

After establishing the customer requirements, this information defines the position of the company compared with competitors and in relation to customer perceptions. This process happens in the so-called *strategic planning room*.

By considering the goals of the company, it is possible to identify market opportunities through a comparison of future strategy and customer satisfaction. At this stage, the requirements are listed. Because of the correlation among product characteristics in a product, it is important to specify the degree of interdependence in the *technical correlation room*. This matrix identifies the situations in which the technical requirements support or impede one another in the product design stage.[38]

Changing one product characteristic affects the other attributes. Through marketing research, it is possible to understand the presence of trade-offs and synergies among sensory and non-sensory attributes,[39] as well as interactions, as means for implementing a marketing strategy.

The complex task is to fulfil the aims of the innovation through a checkpoint, namely, the technical competitive assessment of the product's characteristics and the organisational difficulty related with the innovation. These results then are compared with the customer's competitive assessment, which indicates the advantages of a competitive product for fulfilling the customer's needs. This comparison occurs in the *relationship room*. The result arises from a misunderstanding of the previous steps. The last part of the House of Quality is the *technical priorities room*, in which the *technical competitive assessment* is compared with the *customer competitive assessment* to determine the evaluation of the products from both the company's and the customer's point of view. It identifies weaknesses in the other steps. This analysis highlights the target value that represents the performance that the company must accomplish to achieve customer satisfaction.

The application of the House of Quality model to food product development was proposed by Bech.[40] The relationships between sensory attributes, technical attributes and consumer requirements are exhaustive. The modified model for food product development aims to translate consumers' requirements into sensory attributes, measurable by descriptive sensory analysis, to reflect specific food sensory properties.

Figure 11.3 shows the main part of the model. Consumer analysis attempts to understand the importance of the product attributes evaluated by consumers, their knowledge and their behaviour. The design attributes include technical and sensory relations across the quality standards. Their scores become the technical and sensory specifications, with calculations of the improvements. Consumer perceptions derive from their competitive assessments, which indicate any advantage for a particular product because of its ability to fulfil the customer's needs.

Griffin demonstrates how QFD can be implemented to estimate US product development improvements and identifies factors linked to its successful use.[41] The results suggest a variety of improved product development processes, including structuring the decision-making processes across functional groups, building a motivated team, and moving information efficiently from its origin to the ultimate user.

To understand how the House of Quality model helps translate consumer needs for sensory food quality into sensory attributes, measurable by conventional sensory descriptive analysis, an example from existing literature is appropriate. Viaene and Januszewska apply the QFD to chocolate,[42] a complex product whose final quality depends on various intrinsic and extrinsic interactions. Their findings demonstrate that in a more or less conscious way, consumers' judgement is based on intrinsic quality attributes. Promotion and packaging can enhance the expectations, but what ultimately makes the chocolate desirable is its sensory quality. The sensory perception of chocolate in turn depends on four factors: a clear and distinct chocolate aroma, a sweet taste, a smooth texture, and a luxurious melt-in-the-mouth quality. Two articles also review QFD applications in the food innovation process.[43] Tomato ketchup, peas, chocolate cake mix, and sugar-free butter cookies, among others, have been analysed using this technique. Table 11.2 lists QFD strengths and weaknesses, based on available literature.

Table 11.2   Strengths and weaknesses of QFD[44]

Strengths	Weaknesses
Improves communication.	Customer involvement only in the initial phase of the development process. Feedback from customers in the latter stages is not explicitly supported.
Provides a link between consumer wishes and product(ion) characteristics.	Customer wants can be very diverse and variable, resulting in very long lists of 'what' and 'how' that are difficult to capture in a very precise target value.
Matrices permit very complex relationships to be documented and facilitate interpretation.	Food ingredients have a natural variation that may require continuous adaptation based on their specifications.
Helps NPD teams set targets for product characteristics.	Many ingredients interact and affect the way processes should be designed and optimised, giving rise to a very large and complicated relationship matrix.
Helps NPD teams to make trade-off decisions.	It is very time consuming to complete the matrices.
Helps NPD teams gather and structure all relevant information for developing a successful product.	Process-related improvements are more difficult to achieve than product-related ones.
Makes decisions explicit and documents why certain decisions have been made.	Benefits service developers more than product developers.
Allows simultaneous development across functions, and all functions participate from the start.	Improvements more difficult to achieve in projects concerning true innovations. Improvements are increasingly difficult with increasing product complexity.
Enables comparison of a product with competing products on relevant consumer wishes.	More suitable for products that are assemblies of individual components.
Reduces the final production cost because of the high degree of conceptual research.	Can be very hard to establish (and interpret) consumer wishes. With the wrong consumer wishes, the product will not be successful.
Increases the potential market share at the moment of launch because of the consumer segmentation and the consumer analysis.	Very hard to approach the functional properties of food products as detached. A food product cannot be divided into parts (except packaging and content of the packaging).

Table 11.2 *Concluded*

Empowers the NPD team to make decisions.	By putting the emphasis on the 'voice of the consumer,' the interests of the company (policy and profitability) receive less attention.
	Development of QFD usually involves communication of information about skills and resources, future strategies, costs, and current production approaches, so the company may not be willing or able to afford the whole process.
	It demands a new line of thinking and corporate structure.
	Many consumer food product requirements are sensory. Despite extensive research, it is still difficult to measure them or control them, because they depend on multiple variables related to the product, production process, consumer, or surroundings.
	Sensory analysis usually consists of 20 sensory dimensions per product, a large number for consumers to evaluate.

Many opportunities appear to arise for a company that uses the House of Quality model, in terms of cost reduction, wider market share for the launch, and improved communication and information flows across company divisions. However, particular care must be dedicated to the choice of sensory dimension, because large attribute sets are difficult for consumers to evaluate.

THE BUYER UTILITY MAP

Customer orientation can be very helpful for gaining a superior competitive advantage and increasing a company's market share.[45] With the understanding of consumers' needs, products can be reshaped and have a positive effect on a company's competitive strategy.[46] This concept is the foundation for the Buyer Utility Map.

The buying process begins with six activities: purchasing the good, delivering, using the product, using tools required by the product, maintenance, and final disposal of the product. Each activity is associated with a different level of buyer utility. In this map, overall buyer utility consists of six levers: customer productivity, simplicity, convenience, risk, fun and image, environment, and friendliness. By adopting this perspective, companies gain a clearer idea of buyers' utility and can offer different characteristics to satisfy particular customer needs.

The intensity of product diversification depends on the number of levers in each stage of the buying process. By reshaping buyers' utility, companies can achieve 'value innovation.' Focusing on value innovation on the one hand implies that companies should be customer oriented, but on the other hand, it leads to specific competitive advantages. Value innovators can reshape competition and achieve new market opportunities with less intense competition. Value innovation makes competition irrelevant by offering fundamentally new and superior buyer value in existing markets and enabling a quantum leap in buyer value to create a new market.[47]

Although the concept of value innovation has diffused widely, few empirical applications of the Buyer Utility Map appear in existing literature. Sensory analysis can indicate which levers of the Buyer Utility Map should be reshaped. In particular, sensory analysis provides useful information for both packaging optimisation and convenience during the purchasing phase. When using the Buyer Utility Map in food and beverage sectors, sensory analysis can also provide interesting information about the maintenance process.

One example for understanding the interactions between innovative patterns and consumer perceptions comes from the *International Journal of Wine Business Research*. Italian wineries want to launch premium bag-in-box (PBIB) wine into the US market. Figure 11.4 shows possible ways to create new utility propositions. The introduction of PBIB enhances customer productivity. At the earliest stage of the buyer experience – purchasing – the customer saves time and money, because buying boxed wine for a special occasion means spending less time choosing a wine and paying less for the equivalent of four bottles.

Customers save time during purchasing because the limited number of PBIB labels makes the choice easier. Furthermore, when consumers buy four bottles of wine, they must spend time choosing two to four different wines, whereas usually one variety is sufficient when they buy PBIB. After the carton has been opened, the customer can pour a single glass of wine, and the PBIB ensures proper maintenance for a longer time than bottles by reducing the risk of oxidation. Among other utilities, the maintenance of taste and flavour and the reduced risk of a cork smell are sensory characteristics that can be evaluated through sensory analysis; such an objective evaluation helps underscore the innovative potential inherent in this product.

The Six Stage of the Buyer Experience Cycle

	Purchase	Delivery	Use	Supplements	Maintenance	Disposal
Customer Productivity	Time reduction in buying wine; 3 litres PBIB is cheaper than 4 glass premium wine bottles		No wine wasted: you can pull a glass of wine	No need to buy any other supplements (corkscrew)	Wine can be maintained for a long time reducing waste	The container can be recycled
Simplicity	The box can be easily carried in shopping cart	Easy to transport	Although innovative, PBIB contains wine and the customer should not change all his consumption habits		3 litres PBIB can be easily kept in the fridge or in the kitchen	Disposal doesn't require any complex or particular operation
Convenience	In supermarkets PBIB is near bottled wines and purchase is facilitated by some elements contained on the carton (cartons explain that 3l are equivalent to 4 bottles)	Mail delivery is cheaper	Use is easy: just pull the tap		Some PBIB have a nice design and can be kept on the kitchen shelves or in the fridge without any problem	After use cartons can be easily disposed saving space (more easily than with bottles)
Risk	No cork smell	Glass is more fragile than a carton box	Container safety reduces risk for self damaging		Risk of oxidation is reduced	No risk of getting injured with broken glasses
Fun and Image			New idea of premium wine		The box can be nice to be seen in the kitchen	
Environment and Friendliness			Everyone can serve themselves			

The Six Utility Lever (vertical axis label)

Figure 11.4 Buyer's Utility Map for PBIB[48]

Developing a Buyer Utility Map requires focus groups or semi-structured interviews with a tasting session. Both methods can detect the quality attributes perceived by consumers. In the first case, a discussion conducted by a trained moderator among a small group of respondents facilitates conversation about the focal subject in an unstructured, natural manner.[49]

The main variables that emerge from the focus groups provide the starting points for a laddering technique, which detects cognitive and affective linkages between the product and the consumer.

Other Methods Combining Sensory Analysis and Marketing Research Techniques

A huge range of research methods support innovation strategy.[50] Figure 11.5 depicts 10 of these methods according to two dimensions: newness of the product considered and actionability.

The methods located on the left side are suitable for incremental innovations, repositioning and updates of existing products. Their advantage is to understand the current needs of the consumers and optimise existing products. To compete in the market, these tools should follow ordinary production and R&D activities. Those situated on the right side are appropriate for radical innovation and are risky and costly activities because the consumer is not familiar with the use of the product; it does not exist in the market. To give some examples of how these tools might work, the figure cites some research projects related to incremental new products.

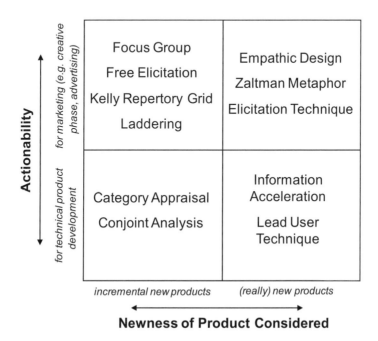

Figure 11.5 Recommended consumer research methods for opportunity identification[51]

Roininen and colleagues[52] established the personal values, meanings and specific benefits that consumers relate to local food products by comparing two different qualitative interview techniques: laddering and word association methods. Word association provides an effective and quick method for gathering information about consumer perceptions of local foods. Laddering interviews, which are time consuming and require laborious analysis, provide an in-depth analysis and important information about the relationship between perceived attributes and the reasons for choices. The link between consumer value, or buying motives, and concrete product attributes may be analysed using laddering.[53] This technique enables means-end chain (MEC) analysis as well, which is a good tool for identifying how product or service selection facilitates a desired end-state, or value, through the meanings that consumers attribute to the product's features. Moreover, MEC can analyse the motivations on which consumers base their purchasing behaviour. Analyses of the MEC require interviews with consumers, which enable the researcher to understand the meanings they assign to attributes and consequences.

For example, semi-skimmed yoghurt contains fewer fats and therefore fewer calories than whole yoghurt. Using laddering techniques, these product attributes may be linked in consumers' minds to the consequence of less cholesterol, and beyond that to increased longevity as a value. The full spectrum of stimuli coming from these products determines consumer perceptions. Poor packaging likely increases the perception of a tasteless yoghurt (i.e., a product with something less). Therefore, consumer product requirements must be defined such that all of them can be incorporated into product development. For these purposes, qualitative and quantitative approaches must be used; a combination of both approaches enables product developers to explore sensory performance and general perceptions. This technique applies to the creation of new products, as well as repositioning or retargeting familiar food products that are approaching the end of their life cycle.[54]

A conjoint analysis also can incorporate tasting sessions.[55] Conjoint analysis measures respondents' judgments of the similarities or differences between different product profiles. This assessment becomes possible by studying the joint effects of the attributes on consumer evaluation.[56] The experimental design is based on four steps: (1) the selection of attributes relevant to the product category through a focus group interview with target consumers, (2) selection of levels of each attribute to be used in the study, (3) presentation of the hypothetical profiles to respondents, and (4) respondent rankings of the stimuli according to an overall criterion, such as preference, acceptability, or likelihood of purchase.[57]

Health messages, packaging, and other information influences initial the purchase of yoghurt, but repeated purchases are driven by the product's sensory properties. Mapping the preference structures that underlie purchase intentions toward foods can be executed most reliably including product tasting and simultaneous modelling of the effects that contribute to the products' other attributes and consumer characteristics.[58]

The methods used depend on the research questions, products, consumer segments and consumption situation. An up-to-date source of such research and experiments, *Food Quality and Preference*, provides interesting examples that underline the necessity of interdisciplinary approaches.

Péneau and colleagues investigate the importance of freshness for consumers and its relationship with sensory and non-sensory properties of apples.[59] Respondents indicated taste, aroma, and freshness as their key attributes for selecting apples. Freshness was important independent of a person's age, gender and apple consumption. Perceptions of freshness depended on taste, crispness and juiciness. Other aspects, such as appearance, storage time, nutritional value and organic status, were less important. Freshness also appeared associated with crispness, juiciness, aroma and preference for apples. Thus, optimal sensory quality contributes to the perception and expectation of the freshness of apples.

Petit and colleagues also investigate the effect of the physical testing environment on preference and consumption of iced coffee by French consumers.[60] A water-based and a milk-based product were tested, and four testing procedures were compared:

- a standard laboratory test in a controlled environment,
- two consumer tests performed outside the laboratory in natural drinking situations,
- a situational laboratory consumer test in an environment modified to remind the participants of the natural consumption condition.

Preference and consumption both depended on the testing situation, the specific location and surroundings, and the consumer's demographics. The two laboratory tests provided very similar results that did not match those from the tests outside the laboratory. This work raises methodological questions about laboratory versus natural consumption situations and highlights the importance of variables that may affect taste even though they are not sensory.

Finally, experimental auctions collect hedonic ratings for the same consumers and the same products in different information conditions. They thus can compare characterisations of the product area origin compared with those of similar products with different origins. For example, Stefani and colleagues conducted experimental sessions to investigate the direct effect of geographical origin on consumer evaluations of and willingness to pay for products.[61] The outcomes show that the more narrowly defined the area of origin, the greater the expected value of the food.

Discussion: Managerial Implications

Many studies note that innovation succeeds when there is a proven added value for the consumer; that is, when innovation processes are structured, there are timely and pointed product definitions, and the collaboration between marketing and R&D gets facilitated.[62]

There are three critical issues for NPD: communication barriers between consumer researchers and natural scientists or product developers; uncertainty about validity; and the subjective and qualitative origin of complementary methods and approaches.[63] Therefore, food-oriented NPD processes must:[64]

- use multi-method and interdisciplinary approaches,
- use tools and techniques that are tailored to food-related research, and
- consider contextual factors to avoid threats related to the validity of food-related research.

It is necessary next to focus on a few aspects from a managerial point of view.

UNDERSTANDING COMPANY RISK

Understanding customers' needs is particularly important in specific stages of a company's lifespan. For example, there is no doubt about the opportunity for exploring a customer's needs when the company is developing a new product.[65] A focus on consumers thus is as a valid tool for avoiding the risk of market failure.

If the philosophy that drives a company's investment in consumer research and sensory analysis pertains to reducing risk, a question arises immediately: *when* does the company take risks? Different kinds of risks relate to each stage of the product life cycle. When a new product enters the market, the risk of market failure is high. Sensory analysis can help create a more successful new product. After the introduction, the company should invest in marketing and sales to improve the availability of the product on the market and promote it. Although the initial risk of market failure has been overcome, the company still faces the threat of ineffective promotion; sensory analysis that monitors customers can help companies achieve their market share targets. When the product has gained wide diffusion, the problem is retaining loyal customers. In this phase, sensory analysis provides information about consumers' needs and expectations. When a product reaches maturity, managers should invest in marketing to increase customer loyalty. The decline phase is the most delicate: The key decision is whether to abandon the product or invest in its rejuvenation. Consumer research in this phase can lead to the development of a value innovation. Thus, the importance of consumer orientation across all phases of product life cycle, not only at the beginning, is clear.

WHY BEING CONSUMER ORIENTED HELPS

Competition in the food industry has increased over time. The profitability of the industry has attracted new entrants, and the presence of a vast number of brands has lowered consumers' switching costs. Without describing in detail the characteristics of competition in the food industry, we note that most companies succeed by choosing a differentiation strategy. A strategy based on a generic concept of differentiation risks failure, whereas differentiating through a focus on consumers helps the company gain a superior competitive advantage.

COOPERATIVE BEHAVIOUR AND ORGANISATIONAL CULTURE

How can a company afford to spend money on consumer research? The issue of R&D investment has always been linked to the size of the company: Bigger companies tend to spend more money in R&D compared with smaller companies. Cooperative behaviour can improve companies' performance,[66] such as when companies share R&D investments and improve the relative flow of information. Cooperating with competitors solves the problem of capital required for R&D in consumer research. In other words, because most small firms cannot afford market research themselves, they should rely on collective bodies to generate this type of information.[67]

What is required for the successful implementation of a combined used of sensory analysis and consumer research thus is a change in the company's organisational culture. Consumer research should involve all the activities of the company, according to the product life cycle.

Conclusions

Adopting a market-oriented perspective helps NPD. Since the earliest research on market orientation, studies have underlined the positive role of market orientation within the firm. The development of a market orientation culture can be facilitated by the implementation of specific tools, such as the House of Quality and the Buyer Utility Map. We also underline how the adoption of a multi-disciplinary approach fosters successful NPD. In particular, we have discussed how sensory analysis can be combined with other marketing research techniques.

The examples in this chapter show how being market oriented is a cultural matter that also implies the implementation of practical instruments. This orientation clearly emerges from the latest research in the food industry, such as work by Frewer and van Trijp,[68] Shepherd and Raats,[69] and MacFie.[70] Shepherd and Raats use a multidisciplinary approach to understand human food choice, as suggested by the multitude of factors that affect food choice. MacFie highlights the importance of implementing practitioner tools in NPD. The scientific journal *Food Quality and Preference* and these three works offer useful training tools for all managers and organisations that want to adopt a multidisciplinary NPD approach.

References

1. Grunert, K.G. and Baadsgaard, A. (1992), 'Market-based process and product innovation in the food sector: A Danish research programme', *MAPP Working Paper*, no. 1, MAPP, Copenhagen.
2. Richard, M., Womack, J., and Allaway, A. (1992), 'An integrated view of Marketing Myopia', *The Journal of Consumer Marketing*, vol. 9, no. 3, pp. 65–71.
3. Ibid.
4. Narver, J. and Slater, S. (1990), 'The effect of a market orientation on business profitability', *Journal of Marketing*, no. 54, pp. 20–35.
5. Bogue, J. and Ritson, C. (2006), 'Integrating consumer information with the new product development process: The development of lighter dairy products', *International Journal of Consumer Studies*, vol. 30, no. 1, pp. 44–54.
6. Van Trijp, H.C. M. and Schifferstein, H.N. J. (1995), 'Sensory analysis in marketing practice: Comparison and integration', *Journal of Sensory Studies*, vol. 10, no. 2, pp. 127–47.
7. Jaeger, S. (2006), 'Non-sensory factors in sensory science research', *Food Quality and Preference*, vol. 17, no. 1–2, pp. 132–44.
8. Grunert, K.G., Harmsen, H., Meulenberg, M.T. G, Kuiper, E., Ottowitz, T., Declerck, F., Traill, B., and Göransson, G. (1997), 'A framework for analysing innovation in the food sector', in Traill, B. and Grunert, K.G. (eds), *Product and Process Innovation in the Food Industry*, Blackie Academic and Professional, London, pp. 1–37.
9. Van Trijp, H.C. M. and Steenkamp, J.B. E.M. (2005), 'Consumer oriented new product development: Principles and practice', in Jongen, W.M. F. and Meulenberg, M.T. G. (eds), *Innovation in Agri-food Systems. Product Quality and Consumer Acceptance*, Wageningen Academic Publishers, Wageningen, pp. 87–119.
10. Ernst and Young and ACNielsen (2000), *New Product Introduction. Successful Innovation/Failure: A Fragile Boundary*. Ernst and Young Global Client Consulting, Paris.

11. Grunert, K.G. (2005), 'Food quality and safety: Consumer perception and demand', *European Review of Agricultural Economics*, vol. 32, no. 3, pp. 369–91.

12. Van Trijp and Schifferstein, op. cit.

13. Stevens, D.A., Dooley, D.A. and Laird, J.D. (1989), 'Explaining individual differences in flavour perception and food acceptability', in Thomson, D.M. H. (ed.), *Food Acceptance*, Elsevier Applied Science, London; Januszewska, R., Viaene, J., and Verbeke, W. (2000), 'Market segmentation for chocolate in Belgium and Poland', *Journal of Euromarketing*, vol. 9, no. 3, pp. 1–25.

14. Schröder, M. (2002), *Food quality and consumer value*. Springer, Berlin.

15. Grunert, K.G., Larsen, H.H., Madsen, T.K., and Baadsgaard, A. (1996), *Market Orientation in Food and Agriculture*. Kluwer, Boston.

16. Grunert, op. cit.

17. Ibid.

18. Steenkamp, J.E. M. (1989), *Product Quality: An Investigation into the Concept and How it is Perceived by Consumers*. Van Gorcum, Assen.

19. Holm, L. and Kildevang, H. (1996), 'Consumers' views on food quality: A qualitative interview study', *Appetite*, vol. 27, no. 1, pp. 1–14.

20. Grunert, op.cit.

21. Peri, C. (2006), 'The universe of food quality', *Food Quality and Preference*, vol. 17, no. 1–2, pp. 3–8.

22. Grunert, op. cit.

23. Hooker, N.H. and Caswell, J.A. (1996), 'Regulatory targets and regimes for food safety: A comparison of North American and European approaches', in Caswell, J.A. (ed.), *Economics of Reducing Health Risk from Food*, Food Marketing Policy Center, Storrs, CT, pp. 3–17.

24. Hooker and Caswell, op. cit.

25. Jaeger, op. cit.

26. Akerlof, G.A. (1970), 'The market for 'lemons': Quality uncertainty and the market mechanism', *The Quarterly Journal of Economics*, vol. 84, no. 3, pp. 488–500.

27. Peri, op. cit.

28. Jaeger, op. cit.

29. Sobal, J., Bisogni, C.A. Devine, C.M. and Jastran, M. (2006), 'A conceptual model of the food choice process over the life course', in Sheperd, R. and Raats, M. (eds), *The Psychology of Food Choice*, Cabi Press, Oxfordshire, U.K., pp. 1–18; Rozin, P. (2007), 'Food choice: an introduction', in Frewer, L. and Van Trip, H.C. M (eds), *Understanding Consumers of Food Product*, Woodhead Pub., Cambridge, pp. 3–29.

30. Rozin, op. cit.

31. Unitech International, *Innovation in the Food Sector*, April 2003; Tanyeri, D. (2006), '10 food innovations', *Restaurant Business*, vol. 105, no. 9, pp. 22–28.

32. Bayus, B.L. (1998), 'An analysis of product lifetimes in a technologically dynamic industry', *Management Science*, vol. 44, no. 6, pp. 763–75.

33. Steenkamp, J.B. E.M. and Van Trijp, H.C. M. (1996), 'Quality guidance: A consumer based approach to food quality improvement using partial least squares', *European Review of Agricultural Economics*, vol. 23, no. 2, pp. 195–215

34. Bech, A.C., Hansen, M. and Wienberg, L. (1997), 'Application of house of quality in translation of consumer needs into sensory attributes measurable by descriptive sensory analysis', *Food Quality and Preference*, vol. 8, no. 5–6, pp. 329–48.

35. Viaene J. and Januszewska, R. (1999), 'Quality function deployment in the chocolate industry', *Food Quality and Preference*, vol. 10, no. 4, pp. 377–85.

36. Charteris, W.P., Kennedy, P.M., Heapes, M., and Reville, W. (1992), 'A new very low fat table spread', *Farm and Food*, no. 2, pp. 18–19; Swackhamer, R. (1995), 'Responding to customer requirements for improved frying system performance', *Food Technology*, vol. 49, no. 4, pp. 151–2.

37. Holmen, E. and Kristensen, P.S. (1996), 'Downstream and upstream extension of the House of Quality', *MAPP Working Paper*, no. 37, MAPP, Copenhagen.

38. Chakraborty, S. and Dey, S. (2007), 'QFD-based expert system for non-traditional machining processes selection', *Expert Systems with Applications*, vol. 32, no. 4, pp. 1208–1217.

39. Jaeger, op. cit.

40. Bech, A.C., Engelund, E., Juhl, H.J., Kristensen, K., and Poulsen, C.S. (1994), 'Optimal design of food products', *MAPP Working Paper*, no. 19, MAPP, Copenhagen.

41. Griffin, A. (1992), 'Evaluating QFD's use in US firms as a process for developing products', *Journal of Product Innovation Management*, vol. 9, no. 3, pp. 171–87.

42. Viaene and Januszewska, op. cit.

43. Benner, M., Linnemann, A.R., Jongen, W.M. F., and Folstar, P. (2003), 'Quality function deployment (QFD) – Can it be used to develop food products?', *Food Quality and Preference*, vol. 14, no. 4, pp. 327–39; Costa, A.I. A., Dekker, M., and Jongen, W.M. F. (2001), 'Quality function deployment in the food industry: A review', *Trends in Food Science and Technology*, vol. 11, no. 9–10, pp. 306–314.

44. Benner et al., op. cit.

45. Narver and Slater, op. cit.

46. Hunt, S.D. and Morgan, R.M. (1995), 'The comparative advantage theory of competition', *Journal of Marketing*, vol. 59, no. 2, pp. 1–15.

47. Kim, C.W., and Mauborgne, R. (1997), 'Value innovation: The strategic logic of high growth', *Harvard Business Review*, pp. 103–112; Kim, C.W. and Mauborgne, R. (1999), 'Strategy, value innovation and the knowledge economy', *Sloan Management Review* (Spring), pp. 41–54; Kim, C.W. and Mauborgne, R. (2001), 'Knowing a winning business idea when you see one', *Harvard Business Review Paperback*, *Harvard Business Review on Innovation*, HBS, Boston.

48. Santini, C., Cavicchi, A., and Rocchi, B. (2007), 'Italian wineries and strategic options: The role of Premium Bag in Box', *International Journal of Wine Business Research*, vol. 19, no. 3, pp. 216–30.

49. Malhotra, N. and Birks, D. (2006), *Marketing Research: An Applied Approach*, European edition. Pearson Education, Harlow, UK.

50. Van Kleef, E., Van Trijp, H.C. M., and Luning, P. (2005), 'Consumer research in the early stages of new product development: A critical review of methods and techniques', *Food Quality and Preference*, vol. 16, no. 3, pp. 181–201.

51. Ibid.

52. Roininen, K., Arvola, A., and Lähteenmäki, L. (2005), 'Exploring consumers perceptions of local food with two different qualitative techniques: Laddering and word association', *Food Quality and Preference*, vol. 17, no. 1–2, pp. 20–30.

53. Schröder, op. cit.

54. Ibid.

55. Haddad, Y., Haddad, J., Olabi, A., Shuayto, N., Haddad, T., and Toufeili, I. (2007), 'Mapping determinants of purchase intent of concentrated yogurt (Labneh) by conjoint analysis', *Food Quality and Preference*, vol. 18, no. 5, pp. 795–802.

56. Hair, J., Anderson, R., Tatham, R., and Black, W. (1995), *Multivariate Data Analysis*, 4th edn, Prentice-Hall Publishers, Englewood Cliffs, NJ.

57. Van Kleef et al., op. cit.

58. Haddad et al., op. cit.

59. Péneau, S., Hoehn, E., Roth, H.-R., Escher, F., and Nuessli, J. (2006), 'Importance and consumer perception of freshness of apples', *Food Quality and Preference*, vol. 17, no. 1–2, pp. 9–19.

60. Petit, C. and Sieffermann, J.M. (2006), 'Testing consumer preferences for iced-coffee: Does the drinking environment have any influence?', *Food Quality and Preference*, vol. 18, no. 1, pp. 161–72.

61. Stefani, G., Romano, D., and Cavicchi, A. (2005), 'Consumer expectations, preference and willingness to pay for specialty foods: Do sensory characteristics tell the whole story?', *Food Quality and Preference*, vol. 17, no. 1–2, pp. 53–62.

62. Bruhn, C.M. (2008), 'Consumer acceptance of food innovations', *Innovation: Management, Policy and Practice*, vol. 10, no. 1, pp. 91–5.

63. Van Kleef et al., op. cit.

64. Jaeger, op. cit.

65. Rochford, L. (1991), 'Generating and screening new product ideas', *Industrial Marketing Management*, vol. 20, no. 4, pp. 287–96.

66. Porter, M.E. (2000), 'Clusters and competition,' in Clark, G.E., Feldman, M.P., and Gertler, M.S. (eds), *Oxford Handbook of Economic Geography*, Oxford University Press, Oxford, pp. 253–74.

67. Verhees, F.J. H.M. (2005), *Market-oriented Product Innovation in Small Firms*. Ph.D.thesis, Wageningen University.

68. Frewer, L. and Van Trip, H.C. M. (2007), *Understanding Consumers of Food Product*. Woodhead Pub., Cambridge, UK.

69. Shepherd, R. and Raats, M. (2006), *The Psychology of Food Choice*. Cabi Press, Oxfordshire, UK.

70. MacFie, H. (2007), *Consumer-led Food Product Development*. Woodhead Pub., Cambridge, UK.

CHAPTER 12

Market Orientation When Customers Seem Content With the Status Quo: Observations From Indian Agri-business and a Case Study

BY S.P. RAJ* AND ATANU ADHIKARI†

Keywords

market orientation, agri-business, India, emerging markets, services, value chain, cold storage, post-harvest management, public policy

Abstract

In the agri-business arena, many products have descended to commodity status. The challenges in rising above commoditisation are accentuated in emerging markets such as India, which face the dual responsibilities of serving customers while enhancing the welfare of growers, who are typically small-scale farmers. Indian consumers, with their rising aspirations and incomes, present additional opportunities and challenges. In this chapter, we bring together these diverse considerations and their interplay and describe recent efforts by several Indian agri-business firms to become more market oriented. We then present an in-depth case study of a cold storage firm that is attempting to break out of the commodity trap. In closing, we discuss public policy initiatives underway in India to facilitate an economy-wide transition to market-driven orientation.

*   Professor S.P. Raj, Distinguished Professor of Marketing, Whitman School of Management, Syracuse University, Syracuse, NY, USA. 13244-2450. Telephone: +1 315-443-3147; and Visiting Professor, International Programs, Cornell University, Ithaca, NY, USA. 14853. E-mail: spraj@syr.edu.

†   Professor Atanu Adhikari, Assistant Professor, ICFAI Business School, 3rd Floor, Astral Heights, Road # 1, Banjara Hills, Hyderabad, 500 034, India. Telephone: +91 9966 5476 33. E-mail: adhatanu@yahoo.com.

Introduction

Market orientation quintessentially calls for detecting customers' articulated needs, divining their latent needs, and successfully delivering value better than the competition. Various requirements of market orientation in existing literature include understanding the market through systematic intelligence gathering, creating competitive differentiation, nurturing competencies, and cultivating an organisational culture that can successfully satisfy market needs.[1]

In detecting and acting on evolving customer needs, timing is a critical factor – too early, and customers are not yet ready to appreciate the differentiator being offered; too late, and it becomes easy for competitors to gain a head start. In essence, the challenge is sensing when a latent customer need 'turns the corner' and becomes a felt or articulated need, and to what extent the firm can help customers realise the need.[2]

Although straightforward in terms of the concept, these maxims present considerable challenges in the agri-business sector, where commoditisation is pervasive[3] and product differentiation difficult. The challenge is magnified in emerging markets where chronic shortages and poverty have nurtured an economic system and business culture with a prolonged emphasis on increasing production and decreasing cost, and where governmental policy may mandate the dual responsibilities of serving customers while also providing better support services to small-scale farmers.

Developing a market orientation in agri-business firms in emerging economies is also hampered by imperfections in the market system as a whole. The chain of players is long – farmers, collators, or aggregators at the farm level, transporters, brokers, traders and retailers – and all may not be ready to accept new ideas or may have self-interest in the status quo. Furthermore, considerable information asymmetry exists in such markets. Ignorance of market information, misinformation, and market imperfections fostered by other channel intermediaries stem from the physical isolation of the farmers in succeeding stages of the supply chain.

Although customers and competitors are the main focus of a market orientation strategy, exogenous but related factors that influence the long chain of entities in the producer–consumer link become important when a firm's actions have significant socio-economic and public policy repercussions along the entire value chain. In emerging markets, a system-wide market orientation takes on added importance, and any single firm may need to work at inculcating a market orientation in both its immediate customers and suppliers.[4] To mould an entire value chain, governmental policy initiatives aim at stimulating and fostering such an orientation may be necessary. Organisational cultures within firms, of course, are important and challenging[5] – even more so when a firm tries to create a market-driven culture in isolation, with few firms that can serve as role models.

These issues collectively present a picture of riddles. Yet this picture also promises enormous potential for firms that can envision a future market situation – one that both is economically lucrative to the players in the value chain and serves to improve economic welfare. In India, visionary market orientation will be achieved by firms that are 'doing well by doing good.'

For individual agri-business firms along the value chain in an emerging economy, two slivers of opportunity offer potential for substantial growth: the rising aspirations of domestic customers, whose increasing incomes make them more aware of global trends, and exploring 'service' offerings as differentiators, rather than only product offerings, which will allow for price-discrimination strategies among those customer segments willing to pay more for better or added service.

In this chapter, we bring together these diverse considerations and their interplay in one Indian firm's efforts in the cold storage industry to become market oriented. The journey along the value chain from the fields to the retail produce bins is arduous in India. Each stage offers challenges but therefore the *potential* to become more market oriented. We preface the discussion by describing recent efforts by several Indian agri-business firms to become more market oriented; we close the chapter with public policy initiatives that are being undertaken to speed this transition.

Background of Agricultural Transformation in India

Agriculture is arguably the oldest organised productive endeavour in the world. In stark contrast to the developed world, India has been slow to adopt new agricultural practices and effectively organise value-added activities system-wide. Value-adding activities in what we broadly refer to as *agri-business* require investments in agricultural inputs, methods of transport, cold chains, processing, retail and so forth. As the government of India's 11th Planning Commission notes, 'Since the green revolution in the sixties, there has been no major technological innovation which could give fresh impetus to agricultural productivity.'[6] To ensure adequate production of good quality produce, it is imperative to adopt new practices without jeopardising the welfare of cultivators and agricultural labour, which involve an estimated 89 million of India's 150 million rural households.[7] The terrain of development in India is rather uneven, but agriculture remains the most powerful lever for improving the quality of life. As a recent World Bank report notes, 'For the poorest people, GDP growth originating in agriculture is about four times more effective in raising incomes of extremely poor people than GDP growth originating outside the sector,' and it deserves greater investments in financial, managerial, and entrepreneurial talent.[8]

Several factors at play in India's emerging economy present challenges and opportunities for adding value for the customer. First, the increasing physical distance between production and consumption and the resulting rise in unmonitored and 'nil' market-oriented entrepreneurial intermediaries not only strain the system by reducing the transparency of information but are, in effect, value-eroding. Second, India's recent economic boom has fuelled a desire for value-added activities such as cold storage, processing and packaging, and enhanced retail experiences. Third, as basic food necessities to satisfy hunger are met for most income levels, inadequacies in meeting nutritional needs gain attention.

INCREASING DISTANCE BETWEEN FARM AND FRIDGE

From about 1951 to the year 2001, the Indian population grew from just over 350 million to more than 1 billion,[9] and the dire need to bridge the gap between agricultural production and population was addressed with single-minded attention. India grew its agricultural output dramatically as a result of the green revolution in the three to four decades after its independence. During this period, the production of staple grains rose from 50 million tonnes to 206 million tonnes; fruit and vegetable production increased from 22 to 113 million tonnes. This remarkable growth occurred without the widespread acquisition and consolidation of small farms by conglomerates. About 233 million people were engaged in agriculture in 2001 on 170,000 hectares of agricultural land. A total of 116,000 operational holdings had an average size of about 1.4 hectares (3.5 acres) with more than 60 per cent of them less than about 1 hectare (2.5 acres). Medium and large holdings averaged about 6 and 17 hectares, respectively, and constituted 5.4 per cent and 1.2 per cent of the total number of holdings.[10] The drive for more effective agricultural practices was tempered by the need to ensure the welfare of small farmers. This demand limited the adoption of many advanced methods that require larger scales of operation. As the population grew exponentially, employment opportunities in manufacturing and service enterprises became concentrated in urban centres and encouraged migration from farming areas to cities. In 1951 the split between rural and urban populations was about 300 million rural and 62 million urban; by 2001 it was close to 750 million rural and 300 million urban.[11] Large segments of the population were steadily but inexorably becoming separated from the proximity of food production with little time for the system to adjust. Earlier in its march to freedom from hunger, India had focused on volume of production; now the challenge was transporting food to the people.

CHALLENGES POSED BY THE WIDENING GAP BETWEEN PRODUCTION AND CONSUMPTION

Transportation

The increased distance between food production and points of consumption has precipitated several challenges. Foremost is the limited availability of infrastructure and transportation required to serve the urban consumer population. Serious obstacles appear in the very 'first mile' of the slow road from the farm to consumer, given the absence of well-organised aggregation centres to hold the marketable surplus grown by small farmers. The situation is aggravated by a weak transportation infrastructure; the lack of paved roadways means great delay in moving produce to markets. Modes of transport, such as trucks, are likely to be of inferior quality compared with those in developed markets. Although India's freight rates are among the lowest in the world – costs range from 0.019 to 0.027 US$ for an average ton-km, compared with 0.025 to 0.050 US$[12] in the United States – a World Bank report notes that 'service quality is poor, with low reliability and transit times nearly double that of developed countries. It is not adequate for higher-value manufacturers or the time-sensitive export trade which comprise a growing share of the Indian economy.' The freight industry may be market driven, but it does not appear to be market driving.

Maintaining quality and availability

The first imperative of a free India was to ensure the availability of food throughout the year and throughout the country. As attaining this goal has become more possible, the quality and variety of agricultural produce become factors for consideration. Quality can be judged by freshness, uniformity, and, perhaps most important, nutritional value and food safety. Eradicating hunger or under-nourishment is a somewhat different goal than eradicating malnutrition. Specific micronutrient deficiencies, such as vitamin A, which is important for eyesight and the immune system,[13] lead to demand for food fortification processes. The road from farm to fridge is fraught with severe value-eroding forces; indeed, almost one-third of fruit and vegetables grown in India are estimated to perish, and more lose nutritive value as they wend their way to market in the absence of processing or cold store facilities. One experimental study that sought to establish the benefits of well-preserved frozen vegetables found that whole green beans initially lost up to 30 per cent of their ascorbic acid (vitamin C) after 1 day, and after 14 days, they retained only 40 per cent at ambient temperatures of about 20 degrees C. (Note that ambient temperatures in India are typically much higher.) Spinach retained only 10 per cent after 3 days. Frozen versions of these vegetables, however, fared much better.[14] In terms of food safety, the growing awareness in India of the potential side effects of agricultural inputs, such as chemical fertilizers, is driving a desire for organic produce, certainly for exports at the present time, and even for domestic consumption at a future date.

As noted in estimates from studies by Rabobank,[15] only 2.2 per cent of fruits and vegetables in India undergo any processing and value addition, with less than 1.3 per cent benefiting from secondary processing, compared with 12 per cent in the Netherlands and as high as 80 per cent in the United States. As economies develop, this study notes, value-added food processing evolves in predictable ways – from staples, to processed mass market foods, to snacks and prepared meals, to convenience foods, and, finally, at the top of the ladder, to diet/functional/organic foods. The study estimates India is entering the stage of snacks and prepared meals.

Availability of storage and processing facilities for post-harvest management of food is severely limited and has become more noticeable as consumers expect more. This rise in expectations is driven by a steadily improving economy, particularly in urban areas. Per capita GDP in India has increased from a little over Rs. 10,000 in 1980 to close to Rs. 27,000 in 2006, in constant prices.[16] Dual-income middle-class households – with their busier lifestyles and contacts with world markets through television, the Internet, and international travel – have further fuelled the demand for convenience and processed food products, a greater variety of food products, and healthy fruits and vegetables throughout the year. Consumers' changing desires demand an agri-business infrastructure that can meet their expectations. For most fruits and vegetables, that means a cold chain able to deliver an acceptable quality of value-added products and ensure off-season availability of produce.

Intermediaries in the supply chain

When growers or small-scale food processors sell directly to consumers who are geographically proximate, they are able to assess demand and other market information readily. However, when this proximity disappears, the task of relaying market information to the producers (i.e., farmers) is left to intermediaries. These intermediaries are subject to a variety of conflicting pressures including, of course self-interest, which result in the producer receiving distorted market information. Although it can be a concern even in developed economies, the communication infrastructure and abundant technologies tend to ameliorate this problem. In India, however, players at each stage of the chain, and most certainly the farmer, are likely cut off from the realities of the end-market and the end-consumer. In addition, as the number of 'hands' touching the produce or relaying the information increases, the quality of both the goods transported and the information transmitted declines.

Brief Overview of Market Orientation and Barriers

A firm that practices the philosophy of market orientation seeks to sense, serve and satisfy customer needs better than the competition. It is as much a philosophy as a practice. Some of the key aspects of market orientation involve systematised attention to acquiring and using market intelligence, investing in product and service developments to create competitive differentiation, and fostering an organisational culture that internalises this philosophy and expresses it in all its actions. Marketing literature has identified high market and technological turbulence as ripe conditions for enjoying the full benefits of market orientation on firm performance.[17] In the case of emerging markets, domestic firms are typically protected by government regulations from the winds of change from the outside – cocooned from potential external threat, which often lulls these firms into preserving the status quo. When governments decide that economic growth is possible only by opening markets to international influences, they set up the conditions for turbulence, often unleashing a tsunami of change for existing businesses. Unfamiliar with sophisticated business practices, such as systematically gathering market intelligence or investing in ideas and technology for new products and processes, these businesses flail about in their efforts to become more market oriented. In developed markets, information gathering, methods of information transmission, and information utilisation in product and process development have been refined over the decades. In emerging markets, the historical lack of demand for such market information has severely limited the information industry and information infrastructure. Occasional entrepreneurial insight and vision coupled with limited trial and error experimentation become substitutes for the systematic development and use of the information infrastructure.

The systematic 'sensing' of market needs, trends, and competitive offerings yields substantial benefits for staying competitive, and the widespread availability of such information to all competitors in the market levels the playing field and raises the overall quality of business practices. Firms with the advantage of unique insights and new product and process development capabilities are bestowed with a competitive edge in 'serving' the market through effective differentiation. Access to technological inputs and the skills to convert ideas into products are essential ingredients for successful differentiation. In emerging countries, even when entrepreneurial vision provides the insight, the absence of technical know-how and skilled labour severely limit the ability to offer innovative solutions to the market.

A market orientation requires an organisational culture that values inter-functional coordination and empowers all levels of employees to focus on customers' needs and develop innovative solutions to satisfy them.[18] Inter-functional coordination can produce superior results in customer satisfaction by developing customer solutions that are holistic in approach.[19] In emerging economies, rampant unemployment and concerns about job security may produce a deference to senior management that inhibits the free flow of ideas upward from the front lines of customer contact – communication essential for a thriving market-orientation culture. This barrier calls for leadership that is visionary in sensing the market, actively encourages such an orientation,[20] and strives to earn the trust of employees to encourage them to be open in sharing their ideas.

SOME EXAMPLES OF VALUE-ADDING INITIATIVES THROUGHOUT THE VALUE-CHAIN

There is considerable potential in India's agri-business sector for adding value at each stage of the chain. To set the backdrop for our case study of the firm Indraprastha Ice and Cold Storage (IPCSL), we share a few value-added initiatives undertaken by other Indian firms – ideas that could be initiated by firms of all sizes, from entrepreneurial ventures to huge companies, and illustrate market-driving behaviour.[21] What makes any of these exemplars interesting is that the 'status quo' is usually profitable, yet these organisations reached out to offer even more value with a vision of the future.

Table 12.1 provides an overview of the various stages in the value chain and examples of value-adding initiatives. Value addition can occur at all stages, from pre-harvest soil management and seed selection to post-harvest transport, storage, processing, and retail sales. A brief description of what each of these initiatives offers in terms of value enhancement to the system follows. This discussion is not intended to be an exhaustive enumeration of value-adding possibilities but merely seeks to highlight some of the potential.

KanBiosys (http://www.kanbiosys.com) is a small, R&D-driven biotech company that is revisiting the development of microbe-based fertilizers first introduced in India in the 1960s but later ignored as chemical-based fertilizers overwhelmed the market. Today, alternative approaches are gaining renewed interest as the issue of overuse of chemicals in agriculture has become prominent, including the causative role it plays in yield fatigue.[22] With its own R&D efforts in formulating microbes for various applications, Kanbiosys has launched specialty safe inputs for conventional and organic farming that target soil health, nutrient, and pest management. Clearly, introducing these kinds of agricultural inputs that are 'certified for use' can add value to market segments interested in organically grown produce, thereby yielding much better prices for the grower and all players in the agri-business sector.

Table 12.1 Examples of recent value-adding initiatives

	Value Added	Growing	Transport	Storage	Processing	Retail
KanBiosys	Organic inputs	XXX				
Golden rice	Fortified seeds	XXX				
ITC eChoupal	Market information	XXX				
IPCSL	Cold chain		XXX	XXX		
VitaRice	Fortification process				XXX	
Spencer's Daily	Ambience, hygiene, quality		XXX			XXX

New technologies offer tremendous potential for adding value as consumers and policymakers in emerging countries begin to appreciate the importance of nutrition and food fortification. A pre-harvest initiative is Golden Rice (http://www.goldenrice.org), which addresses the nutritive value of a key Indian staple by introducing beta-carotene fortification in its seeds, thereby reducing the need for vitamin A micronutrient supplements. VitaRice,[23] for example, is a developing technology that converts broken rice into whole kernels fortified with iron, zinc, vitamin A, and other micronutrients and ensures a low glycemic value. Starting with broken rice that is 30 per cent cheaper than whole rice, the final product offers micronutrients to aid in the struggle to eradicate malnutrition, at the same price as whole grain rice. The value added is less visible but perhaps more important for the health of the consumer and the resulting savings in health care costs.

ITC (http://www.itcportal.com) is a large conglomerate with a portfolio of businesses in various sectors, including agri-business. Interest in assuring consistently high-quality cereal crops for its own processed packaged food products led the company to develop an innovative service that bridges the distance between farmers in rural areas and their markets. Known as eChoupals, these Internet information kiosks or centres attempt to streamline the supply chain. With a presence in about 24,000 villages, commission-based operators or franchisees manage 42,000 kiosks that provide key value to growers, giving farmers online access to real-time market information about prices, demand and advice on farm inputs.

The effect of consumer 'pull' for value-added processed food products and retail experience is best seen in the burgeoning organised retail supermarket industry. The offers of convenience, better hygiene, and an attractive store atmosphere are some of the factors fuelling this growth. One example is Spencer's retail (RPG Group, http://www.rpggroup.com), which now operates approximately 400 stores in 65 cities in a variety of formats – hypermarkets, supermarkets, daily and express – each serving a consumer segment and shopping occasion. Such chains are only the beginning of the potential growth; as noted in a recent study,[24] 97 per cent of Indian food retail still occurs in traditional outlets. With the reduced barriers to foreign direct investment though, growth in this sector is likely to be brisk. From the perspective of value to the customer, many aspects of the supply chain will improve as larger companies become involved in this sector and strive to provide customers with competitive offerings.

These examples exemplify a healthy mix of systematic market research, entrepreneurial 'sensing' of market needs and how they might evolve, and innovations in research and business processes to provide value to consumers. To best appreciate the richness of the intricate web of issues involved, we now discuss in-depth a specific case study of a small and medium enterprise (SME) in the cold-storage business at the post-harvest end of the agri-business chain.

Data Collection

The emergence of interactions between several entities in the value chain can be addressed through a case study approach.[25] The case study approach can help clarify the dynamics present in the economic and cultural aspects of stakeholders.[26] Researchers propose several methods of data collection for case study research,[27] but they generally recommend using two to three data collection procedures, such as archives, interviews, and observations, to obtain a complete picture.

For this study, several members of the company, including the CEO of IPCSL, were interviewed with both researchers present so that one could take notes. A similar interview was conducted with the CEO of a competitor company. In addition, interviews included several customers of IPCSL who serve as commission agents of the growers and make the decision to use the services of IPCSL. We also visited the *mandi*, a fruit and vegetable wholesale market, to observe first-hand how business was conducted and learn about business practices. Written documents pertaining to IPCSL were also obtained from the CEO to round out what we had learned from the interviews. Some of the interviewees were not familiar with English, so the interviews and information gathering was conducted in Hindi, taped, and translated to English by one of the researchers who is familiar with both languages. Beyond the firm, other individuals from government and industry who are knowledgeable about agri-business were consulted to validate our understanding of the overall business context in this industry sector in India.

The interviews, their transcription and translation, and other information were gathered and assimilated during the course of several trips to the company site over several months.

Detailed Case Study: Indraprastha Ice and Cold Storage Limited (IPCSL)[‡]

This company decided to make a bold break from the prevailing business model and sought to change the competitive field by introducing modern cold storage technology to a rather staid service sector, in which lack of infrastructure resulted in considerable wastage. The loss in India has been estimated in the range of Rs. 750 to Rs. 1000 billion annually, varying from 10 per cent for cereals and pulses to 20 per cent for perishables such as fruits and vegetables,[29] with some studies placing estimates even as high as 40 per cent.[30] Lack of good storage prevents the sale of produce off-season at higher prices, perhaps 20–30 per cent higher depending on the grade, and maintaining prices in-season by smoothing out the supply to the market. As such, cold storage facilities offer tremendous potential for preserving value and generating profit. The authors studied this company to develop a case that would illustrate these issues and thus have gained an in-depth understanding and appreciation for the context, challenges, and the firm's efforts to find a way out of the quagmire. Although this one firm's initiatives are specific to its regional market, its practices may be viewed as a prototype of how to lead market-oriented change in the face of myriad obstacles. The firm's immediate customers are middlemen with entrenched business practices and interests; suppliers of fruits, the growers, are unaware of market conditions; and IPCSL is, in effect, a facilitator in preserving value of their produce. To change the business model, the firm needed to look at product and information flows both upstream and downstream along the entire supply chain to understand the motivations and preferences of the various players. The challenge in this context has been to educate the growers who are small farmers to create 'pull' and to 'push' the value

---

[‡]   This section draws from the case study written by the authors and available from the European Case Clearing House.[28]

proposition of the new technology despite its higher price to the firm's customers, who are predominantly middlemen brokers or agents, as well as to find ways to motivate their self-interests. IPCSL is a player caught in the middle of the value chain and has to either optimise the status quo or work toward a vision of the future.

THE ENTREPRENEURIAL VISION

In 1977, Sanjay Aggarwal was given the task of supervising the construction of a new, 2500 MT (Metric Tonnes) cold storage facility in Azadpur by his family, who had been engaged in the cold storage business since 1944. Their business, Indraprastha Ice and Cold Storage, was relocating to the New Sabzi Mandi (a new vegetable wholesale market) that the government of Delhi had just established in Azadpur, and in 1980, their new facility was inaugurated. In 1998 Aggarwal, who was now CEO, realised that Delhi's cold storage service had become commoditised. Price was the main factor that growers' agents, that is, IPCSL's customers, used to select a cold storage service for their produce. Aggarwal, feeling that he had to break away from the pack, gathered his senior staff for a meeting to announce:

> 'We must differentiate ourselves from the rest to avoid a price war. And for that we have to transform the company in the next three years. We have to modernise our cold storage facility with new technology to provide higher value to the customer and charge a premium for the better service. This is the only way we can survive in this environment.'

Sizing up the competition

In Delhi's cold storage market, out of 91 storage facilities, 26 units operate more than 2000 MT of storage capacity. Most of these units are located in the New Sabji Market in Azadpur and in the Lawrence Road Industrial Area, about 15 km away. All these units have multipurpose facilities storing several horticultural and milk products in separate compartments. However, none of the others offered the advanced technology that was planned by IPCSL. According to one of IPCSL's competitors, service was the differentiating factor. Thus, he had introduced 24/7 service at his cold storage to cater to traders who could not keep their produce waiting outside, especially when storing high-value imported fruits like red grapes and kiwi.

Sensing market need

Aggarwal and his staff spoke with several customers in the *mandi* who were the commission agents for the growers, but most of them did not see the need for an improved facility that meant higher costs for them. For example, one man who had started working at a *mandi* several decades ago at the age of 8 years and was now a successful commission agent opined that there were minimal quality differences between produce kept in ordinary cold storage and that kept in storage with modern technology, especially if storage needs were only short-term. What was crucial was the ambient temperature. In winter, when the outside temperature is low, he believed that fruit remains good in most of the cold storages. However, when the outside temperature shoots up, the difference in quality is visible.

MARKET-DRIVING INITIATIVE – A LEAP OF FAITH?

Despite tepid market feedback, IPCSL embarked on a major technological transformation. Aggarwal felt it was the right move for the long term and that customers who balked at the concept now would react differently when the facility became available.[31] He also saw the end-market evolving in a different way – sensing the desires building in the rising middle- and upper-middle classes and the eventual growth of an organised retail sector to satisfy them. He hoped to redefine the cold storage market with a disruptive business model that catered to the emerging higher-income, but under-served, market segment.[32]

IPCSL's immediate customers are primarily commission agents and some wholesalers. The end customers are the Indian consumers who expect higher-quality food both in- and off-season, a market profile that has been changing significantly in the past six years. The increased income of the Indian middle- and upper-middle classes (see Table 12.2) had influenced consumption patterns, with up to 60 per cent of available income being spent on discretionary items like consumer goods, health, and entertainment. These higher-income classes are becoming increasingly health conscious, and those in the urban and semi-urban areas are consuming more fruits daily. The preferred place to purchase fruits and vegetables remains the neighbourhood market, small street shop, or pushcart vendor. In India, almost 90 per cent of fruits are sold through such 'un-organised' retail outlets, for a number of reasons. Consumers prefer the convenience of the pushcart vendor who visits the residential areas daily. Consumers also perceive the fruits available in neighbourhood markets are fresher and of a higher quality than those in retail stores – a perception that seems well-founded, given the lack of quality cold storage facilities. Not only is organised retail not well established, but the prices of fruit in organised retail stores are considerably higher than the prices prevailing in neighbourhood markets or those charged by pushcart vendors.

In season, organised retail chains purchase directly from the markets, while their off-season purchases are made through wholesalers and sub-wholesalers. Wholesalers also supply fruit to hotels and restaurants in Delhi and nearby cities.

Table 12.2 India's emerging middle class[33]

Number of Households (in '000s)	% Growth from 1995–1996 to 2001–2002	Annual Household Income in Rs (in '000s)
20	300	>10,000
40	256	5000–10000
201	219	2000–5000
546	189	1000–2000
1,712	163	500–1000
9034	133	200–500
41,262	43	90–200
1,35,378	3	<90

The value offering

Anticipating a sizable demand for superior cold storage services, IPCSL completed expansion of its cold storage by doubling capacity to 5000 MT by early 2002 and upgraded it in the next couple of years by installing Controlled Atmosphere (CA) and Gas-Controlled (GC) cooling units.[§] Aggarwal invested a total of Rs. 60 million ($1.144 million). Owing to restrictions instituted by the government-sponsored bank, NABARD,[¶] he financed the project with loans from national banks.

Fruit like mangoes, bananas and grapes perish fast. For instance, grapes last about ten days after harvest in ordinary cold storage but could last up to two months in GC storage. Apples can be stored up to two months in ordinary storage, four months in GC storage, and even longer in CA storage (see Figure 12.1a and b). Although most fruits are sold during harvest season, apples are sold throughout the year in India.

Figure 12.1a   Appearance of apples stored in CA/GC storage and ordinary storage – 1 month old apples. Apple of the left GC Storage; apple on the right RA Storage

Source: Photograph by S.P. Raj.

---

§   Approximately 3900 MT of the company's cold storage capacity operated under the gas-controlled technology with equipment imported from the United States. This technology reduces the level of ethylene, a gas produced naturally by fruits and present in the atmosphere that serves as a trigger for ripening most fruits. With this technology, fruits like kiwi and red grapes can be stored for up to two months, whereas they would spoil in ten days in ordinary cold storages. About 100 MT of the company's cold storage capacity operates under the controlled atmosphere technology. This technology is widely used for varied produce and is the most commonly used method of storing apples. It employs a natural process of storage that changes the ratio of the constituent gases in the atmosphere to minimise the respiratory activity of the stored produce.

¶   In 1998, the National Bank for Agriculture and Rural Development (NABARD) introduced a new Capital Investment Subsidy Scheme for the construction, expansion, and/or modification of cold storage units catering to horticultural produce. NABARD-approved projects become eligible for subsidies from the National Horticulture Board, for up to 25 per cent of the project cost to a maximum of Rupee 1 per kilogram (kg) of storage and 5000 MT of total project storage capacity. However, storage facilities built to international quality standards usually cost several times more than ordinary storage.

Figure 12.1b   Appearance of apples stored in CA/GC storage and ordinary storage – 8 month old apples. Apple on the left CA Storage; apple on the right RA Storage

Source: Photograph courtesy David Bishop, International Controlled Atmosphere, Ltd., UK.

Challenges in the system

Transport and cold chain: In most parts of India, growers harvest fruits manually, placing them in padded baskets to avoid damage, and the fruits are then transported from the orchards for grading and packing. During grading, they are placed in different lots based on their size and quality. Packing is mostly done in layers within corrugated paperboard cartons. On average, fruits are handled five to six times before being dispatched to consumer markets for sale. Since most of the fruit produced in India is transported by road, this system has posed severe challenges to the industry. Road infrastructure, especially in the remote but premier fruit-producing areas, is still developing. Alternative transportation, like the railways, though cheap for smaller consignments, is not available for bulk transportation due to the hilly terrain. Kashmir and Himachal, the primary apple growing areas, are quite distant from Delhi (about 900 km and 560 km, respectively), and poor road conditions coupled with insufficient national and state highways result in slow transportation of fruits. The complexity of the system places added costs on growers.

In contrast, a cold chain is an integrated system in which fresh produce is chilled at very low temperatures (0–4° Celsius or 32–40° Fahrenheit for apples) at specific humidity rates (95 per cent RH for apples), in an unbroken link from harvesting through the stages of transport, storage, distribution, and retail sale to the consumer. Since fruits provide a good medium for bacterial growth, their removal from the cold chain at any point enables rapid growth of pathogenic cells at higher temperatures. Thus the cold chain system ensures a cleaner, healthier, and better quality of fruit to the consumer. Very few transportation companies in India maintain refrigerated trucks in their fleet, and the few that do use them for transportation of milk products. Many agents are involved in the movement of produce from grower to the customer, each of whom plays a critical role in maintaining the freshness, nutritional quality, and firmness of produce. Horticulture produce that is transported by refrigerated truck commands a premium in the market, both during the season and in the off-season – as much as 20–30 per cent for apples.

Intermediaries in the distribution channel: Various market intermediaries are involved in transferring produce from the growers to the end consumer. The typical marketing channel for produce like apples is shown Figure 12.2. Delhi's Azadpur market is the main market for apples; more than 80 per cent of the apples sold at Azadpur are dispatched from Himachal Pradesh and Kashmir. Because it is not economically viable for each apple grower to hire transportation separately, fruit produce is aggregated by forwarding agents who transport the produce of several farmers on a trip. These agents send the produce, most often by open trucks, to nodal points like Delhi's Azadpur market, charging approximately 2 per cent for this service.

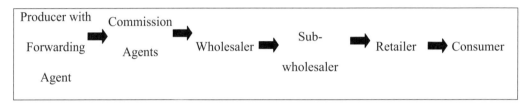

Figure 12.2 Typical marketing channel for produce

Growers designate their preferred commission agents, locally known as *arti*, who receive and unload the produce and auction it to wholesalers. Excess produce is kept in cold storage, and the market is monitored until its sale to wholesalers at off-season rates. This business, with the commission agent as customer, is that for which IPCSL and its competitor firms vie.

The designated commission agents charge the growers 6 per cent of the sales amount. These fees, along with the costs of packing, loading, and transporting, are borne by the growers (see Table 12.3). Wholesalers in this distribution chain buy produce from commission agents and dispatch it to the different states of India for further sales by sub-wholesalers. Sub-wholesalers sell the produce to retailers within their area, who finally market it to end consumers.

Information asymmetry: Very few growers are aware of prices prevailing in different wholesale or retail markets across the country. Few growers even visit Delhi's major market during sale of their produce, and they remain largely unaware of the higher off-season rates at which agents sell their produce, which has been kept in cold storages. Usually only 50 per cent of the fruits are sold the same day, 35–40 per cent is sold within a month, and a small percentage (10–15 per cent) is kept in cold storage to be sold up to six months later.

In Delhi's Azadpur market, for example, the prescribed commission rate is 6 per cent, though in practice, up to 10 per cent is paid to compensate for advances taken by the growers when the growing season begins. The *mandi* functions as a regulated market to safeguard the interests of producers, traders, and consumers by checking unethical and illegal practices like under-weighing, short payment, delayed payments, unauthorised deductions and exploitation by intermediaries.

Table 12.3 Typical expenses borne by Himachal apple grower: farm to Azadpur mandi[34]

	Expense Item	Rupees/Box (20 kg)
1	Cost of grading/packing	12
2	Cost of box	45
3	Transportation cost up to loading point in the orchard	2
4	Forwarding agent's commission	2
5	Transportation cost from farm to main road head	10
6	Loading and Chaukidari or watchman at loading point	2
7	Transportation cost up to Azadpur Market by ordinary truck	20
8	Octoroi or tax by government of India for inter-state transport	2
9	Market fee	2
10	Dalla or unloading agent in ports, mandis, and transport points	2
11	Commission 6% (also charges 4% to wholesaler borne by grower)	60*
12	Gate pass at APMC	5
13	Mandi Charges at NSM, Azadpur @1%	6
14	Misc. Expenditure (Telephone, Postal Dharmada,[a] etc.)	5
	Total Cost	**175**

[a] A levy charged by the market authority on growers to carry out various public charitable activities. It is one kind of double taxation, not authorized by government.

Sales are conducted by *artis* in two ways: closed auction and open auction. The former is effected through handshakes, with the commission agent taking the hand of the bidding wholesaler and signalling the price under a handkerchief, thereby keeping the transaction price confidential. This approach raises the possibility that growers may not know the actual transaction price either. Open auctions are similar to any public auction, with the *arti* calling out the highest bid received and exhorting other buyers to bid higher.

ECONOMIC VALUE OFFERED TO THE CUSTOMER BY IPCSL

Following its modernisation and expansion drive, IPCSL priced its services at Rs. 20 per month per box of Indian fruit and imported apples and Rs 60 per month per box for imported fruit like kiwi and red grapes. This pricing compared unfavourably with the other cold storages at the *mandi*, which were still charging Rs. 10 per box for Indian fruit with no premium for imported fruit. Aggarwal felt his higher prices were necessary as the variable cost of operating IPCSL's upgraded facilities was approximately Rs. 7.50 per box per month – double that of the competition.

A simple back-of-the-envelope calculation approach was used (Table 12.4) to educate the customer about the true value of storing at IPCSL. The savings from reducing fruit wastage far exceeded the higher rates charged by IPCSL and resulted in the customer actually gaining about Rs. 28 per kg of kiwi fruit, for example.

Table 12.4 Illustration of economic value to the customer

Life of kiwi in ordinary cold store is 10 days

At most two-thirds of a container can be sold within 10 days, so about one-third would perish. Price of kiwi is Rs.100 per kg, so Rs. 33 per kg is wasted.

In IPCSL cold storage kiwi can be stored for 2 months and the entire container can be sold.

Ordinary cold storage rate is Re. 0.50/kg/month, whereas IPCSL charges Rs. 3/kg/month.

Hence, over 2 months, IPCSL charges Rs. 6 (Rs. 5 extra over ordinary storage) as rent but saves Rs. 33 in value of fruit saved from spoilage.

So, IPCSL storage adds value of Rs. (33–5) or Rs. 28 per kg of kiwi fruit to the grower.

Although the logic was clear, the new facility did not gain traction in the market. Aggarwal speculated on two reasons for this response. First, the market under-estimated the value addition of IPCSL's upgraded cold storage. Customers continued to be more concerned about short-term costs instead of long-term benefits. The commission agent was more interested in quick turnover of produce and was quite busy and profitable using current business practices. Furthermore, growers were unaware of the potential benefits to them, as they had delegated rights to the commission agent. Second, there was rampant and unsustainable price-cutting in the market. As Aggarwal complained, 'Our competitors have reduced their rates to such a level that it is almost at our cost.' He also suspected that some of the competition was accepting only complete lot orders, and so customers were unable to split their produce into smaller lots for varied storage.

EMERGING FROM THE PARADOX

IPCSL continued to face adversity for a few more months. Customers were reluctant to shift their cold storage to IPCSL because of the high price. Wholesalers of imported fruits and a few big sellers of off-season fruit, however, began to use IPCSL – they recognised its value and had an immediate need for high-quality storage. At this time IPCSL extended a promotional offering to customers at the *mandi*: free storage on an experimental basis so they could tangibly experience the added value. They could store a portion of their produce at IPCSL free of cost and compare the difference after a few weeks with that put in an ordinary storage. Even this offer did not work. Aggarwal, however, was adamant about not reducing the price. He felt he would be acknowledging that the improved quality of service was not worth every extra rupee. Convinced of his strategy, he wanted to stay true to the price.

IPCSL also started storing and selling imported fruits and vegetables directly to wholesalers – like kiwi, grapes, Washington apples, citrus, pears, super sweet corn and green peas. IPCSL's modernised operation could store this produce much longer than ordinary cold storages with little loss of quality. IPCSL also received two large orders for milk and milk products for its deep freezer section, from Amul and Mother Dairy.** Slowly but surely, IPCSL

** Amul is a milk and milk products brand with sales of over Rs. 22 billion and a nationwide marketing network. It is manufactured by Gujarat Cooperative Milk Marketing Federation. Mother Dairy is a Rs. 10 billion company with 1100 retail outlets in and around Delhi.

began to refine and refocus its customer–product mix to align with the new technology and service offerings and began to tiptoe into backward and forward integration.

Subsequently, Aggarwal solidified this evolving strategy and diversified in a planned manner into higher-value fruit and vegetables while focusing on coordinating and integrating the backward and forward supply linkages for fruit. In August 2003 he started a holding company, Dev Bhumi Cold Chain Pvt. Ltd., whose scope extended to establishing a network of collection points in Himachal, with IPCSL continuing under this umbrella as its cold storage facility in Delhi. By the end of 2005, more than 50 per cent of the cold storage capacity was being utilised for Dev Bhumi's fruit. Aggarwal set up a procurement network by coordinating a dedicated farmers' group in Himachal Pradesh (HP). Providing them with pre- and post-harvest support enabled him to gain their loyalty. Dev Bhumi soon began supplying fruit to more than 1000 small street-side retail shops and to two larger retail chains in Delhi. The company also employed six salespeople to sell fruit to retailers.

The new strategy began working for IPCSL as there was a visible difference in the quality of fruits that came out of IPCSL's storage compared with other cold storages. Apples stored in other cold storages used to sell at Rs. 70 per kg, whereas retailers procuring from Dev Bhumi sold their apples at Rs. 88 per kg. A similar difference was observed in the wholesale market as well. A 20 kg box of apples held in other cold storages sold at Rs. 900, compared with Rs. 1100 for a box from IPCSL. As the additional value created by IPCSL became visible to many sellers, they began to change their minds. They slowly started storing their produce with IPCSL to obtain the added value; the market was beginning to value the change.

Perhaps the best endorsement of this initiative to improve the state of affairs in the cold storage business was evident in the actions taken by the commission agent whose original scepticism was noted previously in this section. He is now building a small short-term (few days of storage) cold storage facility in the basement of his *mandi* office with a view to preserving value, even as he tries to get rapid daily turnover in produce. Some of the older storages at the *mandi* are being renovated. One of the earliest stages in the transition to a market orientation is recognising how customer needs can differ by segments and fine-tuning offerings to meet those needs. This stage is most certainly happening at the *mandi*, where a new equilibrium is being established that allows many cold storages to operate and survive at different profitability levels.

LEARNING BY EXPERIMENTATION

IPCSL initially purchased three large refrigerated trucks for transporting apples directly from growers in Himachal Pradesh. These refrigerated trucks were intended to maintain the cold chain from orchard to storage in Delhi, where the fruit could then be safely stored for longer periods. Aggarwal soon realised that for long-term storage to operate at scale, produce is better stored at the point of harvest and then sent on a just-in-time inventory basis to retail points. This method shortens the critical cold chain and provides other efficiencies, like cheaper rural labour, land and operating costs. He therefore developed a plan to set up a large cold storage in Himachal and dispatch fruits only as needed by non-refrigerated trucks to the Delhi cold storage, where products could then be sold well within their shelf-life.

Another insight he developed was that CA technology, essential for long-term storage, is useful for domestic produce only when the integrity of the cold chain is maintained right from the farm, especially important for vegetables and delicate fruit like kiwi. As a result of fine tuning the strategy, a proper alignment of technology and product segments was achieved, with about 20 per cent of India's imported kiwi now being directly procured and sold to wholesalers by Dev Bhumi.

The resistance from the commission agents and learning by experimentation has led Dev Bhumi to refine its strategy continually. It is now completing a very large CA and GC cold storage in Himachal Pradesh to solve the issues of aggregating farm produce at source, storing it for long periods, and supplying customers who may be wholesalers at the Delhi *mandi* or the rapidly expanding retail chains, where high quality has become the expectation.

Throughout this process Aggarwal has gradually built a culture of tenacity and faith in the new mission. He had recruited some new employees who brought with them new skills and knowledge, as well as new cultures. In some instances, they did not work out due to the mismatch, and as a family-run business, Aggarwal began to appreciate even more the importance of preserving certain aspects of the old culture but nudging it to evolve in new directions. Building on tradition but reaching to the future requires a patient and enlightened approach to human resource development, one that prods change without jeopardising the stability of the business.

'Being the change you wish to see' came with enormous financial risk for this small business, but it surely brought about a systemic market orientation. This effort, and the other examples provided in Table 12.1, are but ripples that must be encouraged to spread and coaxed to become waves of a sea change in agri-business in India, with support from government in the form of investment, intervention, and incentives. In the concluding section, we briefly outline some emerging ideas from a public policy outlook that are being planned to facilitate market orientation and provide incentives to take private enterprise efforts to new heights.

Conclusions

At this stage, India requires sophisticated thinking and ideas that can lead to pragmatic implementation. India's stated public policy objectives and plans are reflected well in the 11th Five Year Plan of the Planning Commission of the Government of India.[35] The more detailed thinking on agri-business issues that informed the development of the Plan document is reported by the Working Group on Agricultural Marketing,[36] which also takes stock of what initiatives are already in place and how well they are performing. We draw upon these sources here and note that the reports seek to address many of the issues raised in this chapter. The main issues of relevance for the topic at hand may be categorised as government initiatives and investments on three inter-related fronts: infrastructure, information and incentives.

Broadly, the *infrastructure* initiatives that are recommended by the Working Group span several areas; we briefly highlight a few. Developing farm road infrastructure, establishing cold storages and centres for perishable cargo, and providing farmers access to wholesale market information from across the country in real time via AGMARKNET, which is more comprehensive in scope than ITC's eChoupal, are projects underway.

Institutional infrastructure to develop human resource skills is already present and being aggressively expanded at the district level through Krishi Vigyan Kendras (KVK) units, which train extension workers in agricultural technology, marketing, and business practices; and Agricultural Technology Management Agency (ATMA) offices, which bridge the gap between research and extension services. Parallel initiatives at the state level focus on State Agricultural Universities and Management Extension Training institutes, reorienting their efforts to promote market orientation; at the national level, the National Institute for Extension and the National Institute for Agricultural Marketing have launched new mission-based initiatives. Just as we might speak of inter-functional coordination as an essential ingredient for successful innovation and market orientation, the 11th Planning Commission observes the lack of coordination between KVK and ATMA and emphasises the importance of inter-agency coordination: 'Due to this lack of coordination not only are farming practices in large parts of the country sub-optimal, our plans and programmes are failing to converge technical and development aspects even across Centrally Sponsored Schemes of the Ministry of Agriculture ... especially the extension system, which is the key to bridging the knowledge gap.' Echoing these views, Dr. Sadamate,[37] advisor for agriculture to the Planning Commission, states, 'KVK and ATMA are the frontline organisations that will help implement a greater market orientation; the state level initiatives are required to assure that they are reoriented to promote this perspective.'

In terms of *information*, as the recommendations of the Working Committee note, 'there is a need to develop a comprehensive agricultural marketing information system that can be used to deliver a package of information to assist small farmers and entrepreneurs at the village level.' Based on the government's Situation Assessment Survey of Farmers in 2003, the International Food Policy Research Institute Discussion Paper notes: 'Farmers were asked to identify which, if any, of the sources they had accessed during the past 365 days to obtain information on modern agricultural technology. Nearly 60 percent of the farmers had not accessed any sources.... When asked to make suggestions for improvement, about one-third of the farmers mentioned improvement in the quality and reliability of the information provided.'[38] Noteworthy challenges in this regard are that the information being disseminated must be viewed as reliable, relevant, and user friendly. Rather than overwhelming the user, careful coordination of data generated by numerous agencies and consolidation of the information are essential. Indeed, there is a market for information, and adopting a market-oriented approach to information is critical, because it forces attention toward what information is generated and how it is disseminated, with the objective of being accessible and relevant to users at every step of the agri-business value chain. A cautionary balancing note is to avoid a deluge of data and not fall into the trap so well articulated by Nobel Laureate Herbert Simon[39] in an organisational context: 'a wealth of information creates a poverty of attention.'

Lastly, with the liberalisation of the economy, greater value can come from embracing the efforts of private enterprise alongside public investments. The recommendations of the Working Committee estimate the total investment in agricultural marketing infrastructure during the XI Five Year Plan to be on the order of Rs 643 billion. of which close to 50 per cent is expected to be private sector outlay. In light of the detailed case we have discussed, the largest single item recommended is to increase the number and quality of cold storages, to the tune of Rs. 157 billion, with substantial private participation.

If India is to achieve its 11th Plan target of 4 per cent growth in agriculture, the importance of developing strategies that are guided by a philosophy of market orientation in the agri-business sector cannot be overstated. This goal calls for vibrant private enterprise that is driven by the entrepreneurial spirit and supported and nurtured by thoughtful and implementable governmental initiatives. In an arena as critical to human welfare as agriculture is to an emerging country like India, the agriculture-dependent population is a large asset but also an asset whose productivity must be unlocked by empowering it to become market orientated.

Acknowledgments

The authors acknowledge the valuable public policy insights of Dr. V.V. Sadamate, Adviser (Agriculture), Planning Commission, Government of India; Dr. Sudha Raj, College of Human Ecology, Syracuse University, for help in refining the ideas, especially on nutritional issues; and Dr. Frances Tucker, Whitman School of Management, Syracuse University, for constructive comments on this chapter.

References

1. Day, G. (1994), 'The capabilities of market-driven organizations', *Journal of Marketing*, vol. 58, No 4, pp. 37–51; Kohli, A.K. and Jaworski, B.J. (1990), 'Market orientation: the construct, research propositions and managerial implications', *Journal of Marketing*, vol. 54, no. 3, pp. 1–18; Narver, J.C. and Slater, S.F. (1990), 'The effect of marketing orientation on business profitability', *Journal of Marketing*, vol. 54, no. 5, pp. 20–35.

2. Slater, S.F. and Narver, J.C. (1999), 'Market oriented is more than being customer-led', *Strategic Management Journal*, vol. 20, no. 12, pp. 1165–68.

3. Beverland, M.B. and Lindgreen, A. (2007), 'Implementing market orientation in industrial firms: A multiple case study', *Industrial Marketing Management*, vol. 36, no. 4, pp. 430–46.

4. Hooley et al. (2003), 'Market orientation in the service sector of the transition economies of central Europe', *European Journal of Marketing*, vol. 37, no.1/2, pp. 86–106.

5. Beverland and Lindgreen, op. cit.

6. Planning Commission (2006), 'Towards faster and more inclusive growth: An approach to the 11th five year plan', Government of India, Yojana Bhavan, New Delhi, (December).

7. NSSO (2005), 'Indebtedness of farmer households, NSS59th Round, January-December 2003', Report no. 498(59/33/1), National Sample Survey Organisation, Ministry of Statistics and Programme Implementation, Government of India, (May).

8. World Bank (2007), *World Development Report 2008: Agriculture for Development*, Washington, DC.

9. Census India (2001), Available at http://www.censusindia.gov.in/Census_Data_2001/Census_Data_Online/Population/Total_Population.aspx.

10. Agricultural Census Division (2004), 'Agricultural statistics at a glance', Ministry of Agriculture, New Delhi, (August).

11. Census India, op. cit.

12. Harral, C., Jenkins, I., Terry, J. and Sharp, R. (2003), 'The efficiency of road transport in India: The trucking industry,' WB Background Paper, 2003. Cited in Road Transport Service Efficiency Study, World Bank (2005), Report no. 34220-IN (November).

13. The Micronutrient Initiative (2007), 'India micronutrient national investment plan: 2007–2011.' Available at http://www.micronutrient.org/NewsRoom/Indiapercent20MNper cent20Investment per cent20Plan.pdf.

14. Favell D.J. (1998), 'A comparison of the vitamin C content of fresh and frozen vegetables', *Food Chemistry*, vol. 62, no. 1, pp. 59–64.

15. Rabo India Finance Pvt. Ltd. (2005), 'Vision, strategy and action plan for food processing industries in India, vol. 1', prepared for Ministry of Food Processing Industries Government of India (April); Shrivastava, R. (2006) 'Role of logistics in foods business', Rabo India Finance and Regional Head, Asia, Food and Agribusiness, presentation at ISB, Hyderabad (August). Available at www.isb.edu/simc2006/htmls/pdffiles/rajeshsrivastava_hyderabad.pdf

16. International Monetary Fund (2006), 'World Economic Outlook Database', (April).

17. Slater, S.F. and Narver, J.C. (1994), 'Does competitive environment moderate the market orientation-performance relationship?' *Journal of Marketing*, vol. 58, no. 1, pp. 44–55.

18. Becherer, R.C., Halstead, D. and Hynes, P. (2001), 'Market orientation in SMEs: Effects of internal environment', *Journal of Research in Marketing and Entrepreneurship*, vol. 3, no. 1, pp. 1–17; Gupta, A.K., Raj, S.P., and Wilemon, D.L. (1986), 'A model for studying R&D-marketing interface in the product innovation process', *Journal of Marketing*, vol. 50, no. 2, pp. 7–17

19. Jaworski, B.J. and Kohli, A.K. (1993), 'Market orientation: antecedents and consequences', *Journal of Marketing*, vol. 57, no. 3, pp. 53–70; Slater, S.F. and Narver, J.C. (1995), 'Market orientation and the learning organization', *Journal of Marketing*, vol. 59, no. 3, pp. 63–74.

20. Becherer et al., op. cit.

21. Schindehutte, M., Morris, M.H., and Kocak, A. (2008), 'Understanding market-driving behavior: the role of entrepreneurship', *Journal of Small Business Management*, vol. 46, no. 1, pp. 4–26.

22. Ayala, S. and Prakasa Rao, E.V.S. (2002) 'Perspectives of soil fertility management with a focus on fertilizer use for crop productivity', *Current Science*, vol. 82, no. 7, pp. 797–808.

23. Rizvi, S. and Chen, K.H. (1992), 'Manufacture of a fiber enhanced, nutritionally superior rice product by supercritical extrusion (SCFX)', Docket no. D3724, Cornell University. Available at http://myip.cctec.cornell.edu/index.cfm/ts.details?tk=K4AB1552111611337517.

24. Ernst and Young Report (2007), 'The great Indian retail story', Available at http://www.ey.com/Global/assets.nsf/India/Retail_brochure/$file/Retail per cent20brochure.pdf.

25. Eisenhardt, K.M. (1989), 'Building theories from case study research', *Academy of Management Review*, vol. 14, no. 4, pp. 532–50.

26. Strauss, A (1987), *Qualitative Analysis for Social Scientists*, Cambridge University Press, Cambridge, UK.

27. Miles, M. (1979), 'Qualitative data as an attractive nuisance: The problem of analysis', *Administrative Science Quarterly*, vol. 24, pp. 590–601; Eisenhardt, K. and Bourgeois, L.J. (1988), 'Politics of strategic decision making in high velocity environment: Toward a mid-range theory', *Academy of Management Journal*, vol. 31, pp. 737–70.

28. Raj, S.P. and Adhikari, A. 'Indraprastha Ice and Cold Storage (A),' European Case Clearing House (ECCH) Sl. no. 507–144–1

29. Kachru, R.P. (2007), 'Agro-processing industries in India–growth, status and prospects', in *Status of Farm Mechanization in India*, Director IASRI, New Delhi, Available at http://agricoop.nic.in/Farm20Mech.%20PDF/contents.htm.

30. PricewaterhouseCoopers (2005), 'The rising elephant: Benefits of modern trade to Indian economy.'

31. Narver, J.C., Slater, S.F., and MacLachlan, D.L. (2004), 'Responsive and proactive market orientation and new-product success', *Journal of Product Innovation Management*, vol. 21, no. 5, pp. 334–7.

32. Christensen, C.M. (1997), *The Innovator's Dilemma: When New Technologies Cause Great Firms to Fail*, Harvard Business School Press, Boston, MA.

33. National Council of Applied Economic Research (2003), India Market Demographics Report, NCAER report (May). http://www.ncaer.org/downloads/PPT/TheGreatIndianMarket.pdf, page 8

34. Dhankar, G.H. and Rai, L. (2002), 'Marketing intervention in Shimla district', *Agricultural Marketing*, vol. 45, no. 1, Directorate of Marketing and Inspection, Ministry of Agriculture, Government of India. http://www.agmarknet.nic.in/Journal.htm

35. Planning Commission, op. cit.

36. Acharya, S.S. et al. (2007), 'Report of the working group on agricultural marketing infrastructure and policy required for internal and external trade for the XI Five Year Plan 2007–12', Agriculture Division, Planning Commission, Government of India (January).

37. Sadamate, V.V. (2008), Personal conversation with first author, May 28.

38. Birner, R. and Anderson, J.R. (2007), 'How to make agricultural extension demand-driven? The case of India's agricultural extension policy', IFPRI Discussion Paper 00729, International Food Policy Research Institute (November).

39. Simon, Herbert A. (1971), 'Designing organizations for an information-rich world', in Greenberger, M. (ed.), *Computers, Communication, and the Public Interest*, Johns Hopkins Press, Baltimore, MD.

13 *Breaking the Mould: Characteristics and Consequences of Becoming Market Oriented in Australian Meat Retailing*

BY ANDREA INSCH*

.

Keywords

meat, retailing, market orientation, Australia

Abstract

This chapter discusses the challenges faced by Australian meat retailers in responding to shifting consumer preferences for meat products. Whereas Australia's major retail chains were slow to respond to changing consumer lifestyles, a handful of independent and franchise retailers emerged to break the mould and adopt a market-oriented approach to creating and delivering value-added products. Through a comparative case study analysis, this chapter reveals the characteristics and consequences of adopting a market-oriented approach in this context. The key features and outcomes identified include value-added solutions, consumer understanding, leading the market, flexible responsiveness and whole-of-chain coordination.

*   Dr Andrea Insch, Department of Marketing, School of Business, University of Otago, P.O. Box 56, Dunedin, New Zealand. E-mail: ainsch@business.otago.ac.nz. Telephone: + 64 3 479 4005.

Introduction and Background

A notable trend in meat consumption that warrants attention is the displacement of beef by chicken meat in many Western diets since the mid 1970s. The decline in beef and veal consumption in Australia, previously the mainstay, mirrors international trends as consumers replace their purchases with chicken meat.[1] Consumers lost confidence in beef for several reasons. In particular, it was perceived as inconsistent in quality, expensive in comparison with chicken meat, old-fashioned, boring, and fatty. Many Australian consumers believed that too much meat was not good for them. Beyond these detrimental health connotations, red meat was also perceived as having a limited product range and as inconvenient to purchase from both butchers and supermarkets. Beef was slow to respond to these concerns or the shift to lighter meals and ethnic foods, animal welfare issues and improved food merchandising.[2]

Throughout the 1980s and into the 1990s, meat retailing in Australia could be classified as fitting an outdated mode of marketing most closely resembling a sales orientation. In this competitive landscape, several meat retailers emerged to challenge the traditional orthodoxy and adopt a market-oriented approach to creating and delivering value to consumers. A handful of the 6,600 independent butchers in Australia are breaking the mould to deliver value-added offerings, or meal solutions, to both mainstream and niche market segments and rival the 1,700 supermarket outlets as consumers' preferred places for purchasing meat products.[3]

Objectives and Organisation of the Chapter

The purpose of this chapter is to identify the characteristics and consequences of these meat retailers as they become market oriented. In doing so, this chapter will classify and evaluate the firm- and industry-level determinants and outcomes of the market orientation development process in Australian meat retailing. Insights directly apply to meat retailers in this context and potentially to other markets and agri-food product sectors. To achieve these objectives, the chapter is organised as follows: first, I position the research problem investigated herein against previous research and theoretical perspectives on market orientation in agri-food sectors. Second, I explain and justify the research methodology adopted to address this problem. Third, I outline an overview of the research setting, Australia's meat industry. Fourth, I discuss the key findings from the case study analysis. Fifth and finally, the last section offers conclusions and specific recommendations for managers and researchers.

Literature Review

WHAT DOES IT MEAN TO BE MARKET ORIENTED?

Simply defined, market orientation refers to the activities involved in implementing the marketing concept.[4] The philosophical underpinnings of the market orientation construct – the marketing concept – include an ideal or policy statement,[5] a distinct organisational culture,[6] and a philosophy of business management[7] that comprises three pillars. These pillars focus on (1) customers' current and future needs and the factors affecting them; (2) the ability of the organisation to generate, disseminate, and make use of superior information about customers and competitors;[8] and (3) the integration of organisational efforts to create superior customer value.[9]

With these foundations, market orientation is defined as 'the organisationwide generation of market intelligence pertaining to current and future customer needs, dissemination of intelligence across departments and organisationwide responsiveness to it.'[10] A similar definition by Narver and Slater[11] argues that 'the three hypothesized behavioural components of a market orientation comprehend the activities of market information acquisition and dissemination and the coordinated creation of customer value.' These two widely accepted and empirically applied conceptualisations of market orientation share two common components: continual monitoring of product market conditions, and responses to changing conditions and adaptation of responses to particular market segments.[12] The market intelligence function of an organisation assumes a central role in the processes of acquiring and generating market information, disseminating market information across departments, and coordinating and responding to information to create and distribute value.[13] Sustained investments in marketing information systems and information technologies that facilitate sourcing and dissemination of market intelligence thus are necessary to support and execute a market orientation.[14]

Market-oriented behaviour and activities differ in terms of their responsiveness to changing product market conditions. Proactive actions differ from reactive responses.[15] That is, whereas reactive behaviour responds to signals in the marketplace, proactive responses reflect a continuous search for opportunities and experimentation with responses to changing marketplace conditions.[16] Strong inter-functional cohesiveness and a market focus underpin proactiveness, which is more representative of a market orientation than are reactive responses.[17] Even though customer orientation is a critical component of market orientation, an exclusive focus on customers may be detrimental, because they are 'notoriously lacking in foresight.'[18] The 'tyranny of the served market' may constrain the search for and detection of novel opportunities and threats.[19] A market orientation also affords a broader, long-term focus on creating value to meet customers' expressed and latent needs by learning from competitors and other aspects of the marketing environment.[20]

ARE AGRI-FOOD BUSINESSES MARKET ORIENTED?

Despite the reported business benefits of becoming market oriented, food companies and agri-businesses have been slow to implement this approach.[21] Several reasons have been suggested for their entrenched commodity mindset, which is akin to a production orientation. Among these factors is the intensification of farming practices to maximise output, which has lengthened the distance between producers and consumers. Resource constraints often prevent agri-food producers from retaining ownership of their produce through the entire chain, to trade and end consumers. This inability to engage directly with their buyers is a major barrier to acquiring knowledge and developing an understanding of customer needs and preferences.[22]

Parallel to prior research on the adoption of a market orientation among business-to-consumer firms, a handful of studies investigate the facilitators, barriers, and cultural characteristics of adopting a market orientation in business-to-business firms, such as agri-business and food and beverages companies.[23] In particular, Beverland and Lindgreen's study of the transition of two of New Zealand's agricultural marketing boards to a market-oriented culture demonstrates that the change process is characterised by various stages – unfreezing, movement and refreezing – that place differing emphasis on particular styles of leadership, learning, forms of inter-functional coordination and nature and use of market intelligence.[24]

Whether these features of the market-oriented change process apply in other contexts requires investigation. In addition, the industry-wide conditions that support or inhibit the movement toward a market-oriented culture among agri-businesses must be considered. Finally, the consequences – outcomes for firms and reactions of competitors – of adopting a market orientation need to be explored within the context of food- and agriculture-based industries.

Research Methodology

This chapter uses a case-study approach to identify and categorise the requisite characteristics and consequences of adopting market-oriented behaviours in meat retailing. Specifically, I present an in-depth comparative case study analysis of two meat retailers in the broader context of Australia's meat retailing sector. Primary and secondary sources of information have been gathered and analysed. The primary sources consist of four categories: archival records and documents, corporate communications, media and press reports, and interviews. Interviews conducted with key informants represent each company and members of Australia's beef and chicken meat industries.

Secondary data come from statistical databases: published books, articles, and pamphlets, and unpublished theses. The statistical data come from industry associations, government departments and agencies, international agencies and consultancy reports. Two procedures serve to verify the factual accuracy of the information. First, data from different sources are triangulated to substantiate facts when possible.[25] Second, participants reviewed drafts of the cases to corroborate the reporting and sequencing of facts.[26] The analysis of multiple case studies based on multiple sources of information follows Eisenhardt's advice.[27] This approach supports the development of richer theoretical insights. The case analysis identifies several distinguishing features of each meat retailer as characteristic of a market orientation in this industry. Each feature is discussed in turn, followed by a synthesis of the consequences of this shift.

RESEARCH SETTING – AUSTRALIA'S MEAT RETAIL SECTOR

Most meat products consumed in Australia, similar to other Western markets, are purchased from meat retailers such as major supermarkets, smaller retail chains, independent butchers and specialist delicatessens. Independent butchers, once the main outlet for meat purchases, face strong competition from supermarkets, a convenient one-stop shop for time-limited consumers. The dominance of these large chains has contributed to a rapid decline in the number of independent retail butcher shops.[28] The immediacy of supermarkets' relationships with consumers is apparent in the beef product category: Their share of beef and veal retail sales rose from 23 per cent in 1987–88 to 40 per cent in 1997–98 and to 70 per cent by 2002.[29] Independent butcher shops located in suburban strip malls and major shopping centres collectively account for the majority of the remaining 30 per cent of total retail sales.[30]

A similar trend is observed in white meat products. From the 1970s onward, Australian consumers, particularly housewives, shifted away from local grocers and butchers to supermarket retailers for fresh and frozen chicken meat products. However, this trend stabilised through the adoption of a market-oriented approach by some smaller, independent meat retailers. By 2005, in the case of red meat, Woolworths supermarket chain remained the leading retailer of beef/veal and lamb (30 per cent and 31 per cent share of purchases respectively), with butchers (28 per cent, 29 per cent) in second and Coles supermarket chain in the third position (19 per cent, 20 per cent).[31] In terms of total meat sales, the two major supermarket chains accounted for 51 per cent of national meat sales in the third quarter of 2006 (Woolworths 27 per cent; Coles 24 per cent). Butchers accounted for approximately 31 per cent of sales, other stores accounted for 8 per cent, and independent and other supermarkets accounted for approximately 10 per cent of national meat sales.[32]

To enhance the reliability of supply, both major retailers moved to source greater volumes of grain-finished stock, invested in further processing, and crafted preferred supplier arrangements or vertical partnerships. For example, Woolworths created preferred supplier contracts to rationalise its supply base. It purchases live cattle from a pool of 500 producers, in addition to 27 dedicated or 'valued' producers that have supplied Woolworths for anywhere from 2 to 20 years through forward contracting. These supplies are supplemented with purchases made on-farm, at regional saleyard auctions, and over the hooks. Woolworths also forged preferred processor agreements in each state. Woolworths purchased carcasses from preferred processors through its subsidiaries, which perform further processing. Since July 2001, Woolworths has provided online feedback to producers supplying Cargill's Tamworth plant regarding how well their cattle conform to product specifications. This approach underpins a performance-based payment model, in which returns to producers are based on carcass conformity to Woolworths' criteria.[33]

As the last link in the chain, retailers hold a privileged position in accessing information about consumer meat purchasing, yet they often do not apply this information effectively to deliver added value. Both butchers and supermarkets tend to promote specials on price alone rather than featuring new meal ideas. The positioning of beef as 'lean' since the mid 1980s cemented this attribute in the Australian psyche. Studies of Australian consumer attitudes toward red meat in the mid 1990s confirmed that fat content was the most important factor influencing fresh beef purchases for home consumption.[34] In summary, over time, most major meat retailers in Australia have became complacent about consumer attitudes and preferences, as well as their competitors' marketing strategies. These factors have contributed to the continuing fall in red meat consumption.

Findings

CASE A: FRANCHISE MEAT RETAILER (FMR)

Overview and synopsis

The Franchise Meat Retailer (FMR) is worth examining because of its success in securing a formidable share of the ready-to-cook and heat-and-eat meat product category in Australia's meat retail sector. By 2000, just over a decade after it began, this 'meal solution' franchise helped raise consumer responsiveness in the segment. The founder and managing director of FMR, a Melbourne butcher who migrated to Queensland, opened the first outlet in a Brisbane suburb in 1987. The FMR's growth has continued, reaching almost 190 stores in 2007, which serve 10 million customers each year.

To add value to fresh produce, FMR offers a range of meals for the kitchen-ready market. The range, including kebabs, marinated chicken breast, and seasoned chicken pieces, was developed by FMR's founder, who recognised a trend toward longer working hours, busier lifestyles, and a corresponding increase in the demand for convenient meal solutions. Continuous innovation drives the development of new products, with the proportion of value-added products increasing from 10 per cent to 90 per cent of the product range between 1987 and 2002.[35] The chain sits at the forefront of the home meal replacement (HMR) market in Australia, and it has developed a wide range of pre-assembled foods. These meals involve pre-prepared components that consumers combine at home (e.g., a Salad Kit is a pre-assembled food; Easy Living Beef Lasagne is an HMR).

KEY CHARACTERISTICS OF MARKET-ORIENTED BEHAVIOUR

Corporate philosophy

The company attributes its success and reputation to its commitment to 'fresh, quality products that represent good value for money.'[36] Customer service and convenience are also key features of FMR's corporate philosophy. These components of a customer-focused philosophy are based on the realisation that 'time is precious,' so it aims 'to make planning and cooking meals as easy and convenient as possible.'[37] To achieve this value proposition, FMR built a strong internal infrastructure, as the director explains: 'It is the company culture of teamwork, innovation, service and fun that makes us work hard to ensure we succeed. Our franchisees appreciate the depth of support they receive and the strength of the team which assists them.'[38]

Continuous innovation – 'Always something new on the menu'

Staying ahead of the market and anticipating customers' needs is also a cornerstone of FMR's system, as the director comments: 'We innovate new product lines to suit our customers' lifestyles and they reward our innovation by returning to the counter time and time again.'[39] The chain is at the forefront of the HMR market in Australia. It also developed a wide range of pre-assembled foods. In 1998, FMR joined forces with the Pork Council of Australia to launch a new range of pork meals.[40] It also develops a special Christmas range of pre-cooked products each year, with a special focus on turkey products.

Value adding for profit maximisation

A cornerstone of FMR's continual growth is the application of the concept of added value to all of cuts of meat that constitute a carcass. In particular, FMR focuses on the cheaper cuts first, where the potential lies for the greatest value adding; 'marketing the whole animal is where the profits lie – not in just selling the high value cuts at a premium price.'[41] In addition, FMR applies this approach to both chicken products and red meat cuts, as the director explains:

'We need to add value to the cheaper cuts first. We develop our system with the cheapest cuts and getting the consumer wanting casserole or an Irish stew or a beef stroganoff. As a quality assurance process for the beef industry MSA (Meat Standards Australia) is superb. But somebody still has to deliver the cheaper cuts to the consumer in a way that is memorable to the consumer. And that's a casserole, or a rib on the bone. And the only way to do that is to cook it for them because they don't know how to do it themselves.'[42]

Store system – marketing support and customer responsiveness

The best description of FMR is a meals solution franchise organised as a master franchisee system. In this system, a 'master franchisee,' known as a territory manager, purchases a specific territory and may sub-franchise stores within this territory. A major advantage of this system is the commitment of the master franchisee to assist individual franchisees and ensure that their store operates at optimum efficiency and effectiveness.[43] In addition to this support, FMR staff based at the national head office – retail, marketing, product development, financial, and legal specialists – provide dedicated support to individual franchisees.[44] According to the director, 'owners who follow the system and put in the required effort can expect a return on their investment of between 30 per cent and 50 per cent but they do need to work hard, particularly in the early stages.'[45]

Even though product development and marketing functions are centrally controlled, individual store owners determine their product range from a bank of 150 recipes. In a week, a franchisee likely selects up to one-third of the recipes. Decisions about merchandising, layout, and presentation of the products in the cabinet are also discussed with the franchisee. Store owners are urged to apply the principle of variety so as not to undermine the whole system.[46]

Franchisees offer greater product knowledge than most supermarket staff, because they pre-prepare the products. All value added to the core meat product is performed in-store. This practice contrasts with the organisation of the major retail chains in Australia, in which these tasks are carried out at a central processing plant that produces a retail-ready package and then ships the packages from a central distribution centre to individual outlets. Thus, FMRs have an advantage over the larger retailers, namely, a true just-in-time (JIT) inventory system that is more responsive to changing consumer needs. Major advantages of this system include product freshness and waste minimisation. Wastage is less than 1 per cent in FMRs, compared with 6–8 per cent routinely recorded by supermarkets. This cost saving can amount to as much as $5 million per year.[47] However, through their significant economies of scale, the large retail chains' centralised processing and distribution systems offer substantial costs savings across multiple product categories and the supply networks that they lead. As Hingley argues, the profitability of supermarket retailers must take broader perspective than a single supply chain or product category, to understand the 'broader context of reduced transactional costs and reduced overheads derived from efficiencies of channel consolidation.'[48]

CONSEQUENCES

FMR is the market leader in specialist meat retailing and a growing outlet for value-added chicken meat products in Australia. The growth of this business into a network of 200 franchised stores in its home market and overseas relates to the chain's responsiveness to changing consumer lifestyles. The success of this system is demonstrated not only by the franchisees' growth and expansion but also through peer recognition and industry accolades, including a string of awards.

Even though the chain's red meat sales constitute approximately 35 per cent of company turnover, the focus remains on its core offering – fresh, value-added chicken meat. Despite its success in expanding its range by introducing new species, cuts, and products, FMRs faced a further challenge, as the director explains:

> 'We could not stay like this – we had to look forward to see what consumers want. So for the last four years we've been searching for the next step of what consumers want. So we went into red meat. And we repeated the same formula. We stuffed a leg of lamb, or we rolled it and marinated it. And it went all right but didn't do anything really fantastic. But when we analysed it we didn't do what consumers wanted us to do. We didn't take it to the next level. We had to get down to ten minutes meal preparation time – not 20 minutes. We had to cook it for them.'[49]

Turnover, which was $5 million in 1989, has increased to more than $100 million and continues to grow at a rate of 17 per cent per year. The use of chicken meat has grown from approximately 500,000 fresh chickens in 1989 to currently 9.5 million chickens per year.[50] In 2003, the first FMR store-in-store outlet was opened in an IGA store east of Melbourne. Similar store-in-stores have been opened and received awards for their innovation.[51]

CASE B: INDEPENDENT GOURMET BUTCHER (IGB)

Overview and synopsis

A gourmet butcher located in Melbourne, Victoria, IGB is the brainchild of Mike IGB (names changed for confidentiality). Together with co-founder Susan, IGB opened his first store in 2001.[52] This flagship outlet resulted from years of scientific research, marketing knowledge, and expertise, designed to ensure that customers could trust that the beef they purchased would consistently meet or exceed their expectations. IGB proclaims it has 'the most advanced retailing systems and brand development in Australia' and is the 'new face of meat retailing.'[53] Development of the IGB retail concept and outlet is the culmination of Mike's 25 years of research and experience in the industry, including his lead role developing MSA, an internationally acclaimed beef quality grading system.

This independent meat retailer's aim is to simplify a complex product, production and marketing system to provide a range of added-value products. Fresh beef products sell according to five different cooking methods: grill, BBQ, stir-fry, casserole and roast. In-store butchers add value by preparing beef for cooking, matching each cut to the cooking style for which it is best suited. There are a variety of products sold under IGB's store brand – Marrinya Grills, thinly sliced Shumi, 'Mikez' tender-tasters, Aga cubes for casseroles, flavoured beef for a roast, and 'Farmhouse ground' mince. As quality assured, added-value products, these items attract a price premium from IGB's up-market clientele.

KEY CHARACTERISTICS OF MARKET-ORIENTED BEHAVIOUR

Corporate philosophy

The IGB motto – 'everything that can be done, we've done, to bring you a great beef meal' – reflects a desire to assure customers of the quality of their eating experience. This philosophy underlies the brand, expressed as the 'IGB Promise.' Six elements constitute IGB's brand guarantee: (1) independent grading, (2) scientific testing, (3) sustainable farming methods, (4) a money-back guarantee, (5) sale by cooking method, and (6) whole-of-chain product control. The overarching aim of IGB's system is to reduce complexity and deliver customers a range of value-added beef products. At the same time, Mike believes that beef's inherent complexity gives it greater potential for value-adding. He explains:

> 'Beef has flavour – it has its own flavour. It is not just a carrier. Chicken is just a carrier for flavour. We are trying to add flavour that supplements the natural flavour of the beef. We can develop flavour around roasts, casseroles, curries, and pies. And thinly sliced beef is a whole new world of opportunity.'[54]

Whole of chain innovation – cooking and eating quality

Unlike traditional retail butchers' and major supermarket chains' approaches to selling cuts of meat according to the anatomy of the animal, IGB beef is sold according to cooking method. The focus of IGB's system is not which part of the animal the muscle comes from but the meat's cooking and eating quality. This innovative approach to meat-based meal merchandising assures consumers of the quality of their eating experience, providing they cook the meat as recommended.

One of the keys to IGB's ability to add value is maximising the value that can be obtained from a carcass. This approach requires breaking down the value of the parts of the whole carcass and adding specific value to those parts that will be most valued by consumers. Determinations of what is valued come from information captured about the consumer. The first step involves understanding what consumers want: Do they want a chuck steak, or do they want a good curry? Mike realized that the challenge was to add value in the middle, where there is much consumer uncertainty about buying meat, because most consumers are confident only when buying fillet and mince.[55] Value can be added to both fresh meat products and pre-cooked meat products (e.g., cooked home-style meals). For example, premium trim as a core beef product can be sold as a fresh meat product or a processed and pre-cooked meal depending on the nature and amount of value added.

Adding value through cooking method and meal solutions

Viewing beef in terms of consumers' needs, that is, as a meal, and not as a piece of meat, introduces a fresh perspective to meat retailing. Similar to a chef's approach, this perspective means that each product is prepared according to cooking style rather than specific cuts. Each link in the value chain must adopt a new role, as meal providers rather than as beef producers or butchers. Another way that this retailer adds customer value is through greater personalised service. Trained staff assist customers with ideas and suggestions in making selections. They also offer advice on how best to cook different cuts. This approach removes the 'guesswork' from buying beef, saving consumers time and effort in their purchases and meal planning. As Mike explains, 'people want a meal, not an anatomy lesson.'[57]

Because it is impossible to predict the eating quality of a piece of meat based on the retailer, price, cut, or even inspection of its physical appearance, a proven prediction method was needed that could be easily understood by consumers. This method comes in the form of IGB's tenderness rating system. Each beef product is described and priced according to the real outcome – the cooked result.

Store system – quality control and assurance systems

The impetus for the creation of IGB's system was the lack of incentive to produce high-quality beef when it was lumped indiscriminately with other beef in the butcher's shop.[59] Mike identified the need to gain control over the supply chain and the need to re-establish a direct link between the farmer and the customer. Specifically, the operation maintains control and high standards of care at every stage, to deliver meals to the store through vertical enabling. The backbone of IGB's brand guarantee is the quality assurance system that Mike developed, known as IGB's Pathway. This system is based on the MSA grading assurance scheme and incorporates all the parts that the customer does not see to provide total systematic control and optimise eating quality. The grading system consists of four sensory dimensions, each assigned a weight, that combine into a single palatability or meat quality score (MSA score): tenderness (0.4 weighting), juiciness (0.1 weighting), flavour (0.2 weighting), and overall liking (0.3 weighting). This score then translates into a star rating (3–5 to qualify for MSA standard). A three-star rating means excellent everyday quality, four stars means superb, premium quality, and five stars means melt-in-your-mouth beef. All beef sold through IGB is sourced from either Mike's herd or selected local growers, both small family-based farms and larger corporate herds. In total, the supply base represents about 2,000 cattle.[60]

To be marketed through the IGB outlet, all beef must achieve an eating quality of three stars or higher under the MSA scheme. Furthermore, all beef can be traced back to the specific animal from which it originated, using DNA technology.[61] This trace-back system and close working relationship with abattoir partners allows control over the supply chain and identity preservation beyond the retail level.

Another important facet of the IGB system is optimising inventory management. Whereas most fresh produce deteriorates over time and is best consumed as soon as possible, the quality of the eating experience for some cuts of beef relates to ageing. An inventory information system was designed to enhance the eating quality experience for consumers, while also maintaining the optimum flow of inventory for the store, as Mike describes:

'We have perfect yield data – when we turn on the computer in the store in the morning, all the information is there at our fingertips. Every cut in stock and its eating quality today is there

for us. For example, this cut might not be quite ready – but this bit is at its absolute best and should be sold today. We now have to really use this information within the retail store – as a profit centre'.[62]

CONSEQUENCES

According to IGB, sales growth is steady, and turnover has doubled since it opened in 2001. The standalone outlet recorded 55 per cent growth in the first quarter. Mike attributes this growth to repeat walk-in traffic, because 'consumers are prepared to pay a little bit extra for quality.'[63] At the same time, the exclusive, premium product and pricing strategy restricts its market base, because IGB does not 'sell meat that doesn't perform.' Finding the right way to communicate this new proposition is challenging and yet to be perfected.[64] Consumers are sceptical of the new system and approach and still perceive the store as primarily selling expensive steaks for special occasions. Gaining acceptance and appreciation for better value products across the entire range is difficult, given existing meat-buying and consumption behaviours that will be tough to change. Mike expresses this dilemma:

> 'We still have to get across that this is fantastic value – you can buy something for your price point – it will be good and it will all work with no knowledge and no fear. Increasingly we will have to cook it for them – cook the roast for them, cook the veges and so on.'[65]

This constraint, as well as a failed acquisition of the store and concept by a prominent corporate farm, has limited IGB's ability to expand beyond its single store location.

Conclusions

Several common themes, which can be viewed as features of firms becoming market oriented in Australia's meat retailing sector, are revealed through these cases. Although it is unwise to directly compare their performance, owing to marked differences in size, years of operation, and market focus, similarities in how they have applied the concepts that constitute their market-oriented behaviour can be observed. The next section discusses how these concepts – value-added solutions, consumer understanding, leading the market, flexible responsiveness and whole-of-chain coordination – have been implemented by these two businesses and are manifest as key features and consequences of becoming market-oriented meat retailers. Relevant recommendations for managers in this industry are also provided. Finally, a list of issues for further research appears in the final section.

CONTRIBUTIONS AND MANAGERIAL RECOMMENDATIONS

First, innovating to deliver customer-focused value through value-added meat products is an outstanding feature of each retailer and the hallmark of market-oriented behaviour. Both companies have embraced the application of this concept in the agri-food business – meat solutions – through various forms and levels of value adding. In particular, the companies dissected customer value into three main forms: convenience, choice and consistency. Each of these forms resonates with consumers' full and hectic lifestyles. Furthermore, FMR's latest innovation, the store-in-store, recognises the strategic position of supermarkets as one-stop shops, without losing independence or control over its unique value proposition. Increasingly these and other retailers concede that consumers are willing to outsource the time- and labour-intensive value added step of preparing home meals.

Second, the capability to deliver solutions cannot be achieved without in-depth and ongoing consumer understanding or empathy. Even though the major supermarket retailers have access to consumer spending patterns, they often do not cultivate close relationships with regular shoppers, owing to the distance between in-store butchers, counter staff, and checkout operators who tend to work independently and casually and lack the same commitment to their clientele or stake in the success of the business. In contrast, the two meat retailers use owner-operated models, with the bulk of added value occurring in-store and in response to observed consumer preferences rather than, as in large supermarket chains, to centrally organised purchase and merchandising decisions.

Third, both retailers demonstrate at least an awareness of an application of the market-oriented philosophy of leading, or driving, the market. IGB appears at the forefront of its niche gourmet segment and, as an undesirable consequence of this orientation, may have limited ability to expand. Cutting-edge technology and innovation in meat quality assurance and traceability widens the gap between consumer expectations of meat, especially beef, as Australian consumers are unaccustomed to high quality, value-added products. Therefore, their scepticism, and thus willingness to pay, has restricted IGB's ability to carve out a reasonable sized segment, as its positioning only overlaps with a small portion of the population. Meanwhile, FMR's positioning is closer to the mass markets; after initially leading the majority of consumers, it now is in catch-up mode, trailing behind a growing number of innovative and increasingly demanded consumers. Supermarkets are close to consumers but have not led the market and were slow to adopt industry innovations, such as only partially adopting MSA. Instead, they have relied on smaller, branded suppliers to do so and thus have remained sales oriented in their meat categories.

Fourth, independence grants these firms the flexibility to decide their range and cater to customer needs on a daily basis, as evidenced by efficient just-in-time inventory systems of both meat retailers.

Fifth, the two cases exemplify the use of a whole-of-chain approach to coordinating the creation and delivery of value. The effective implementation of such an approach offers some positive outcomes for both, though of different scales. Underlying both approaches is a consumer-focused quality assurance system that substantiates and reinforces marketing communications. The success of FMR's whole-of-chain system is evidenced by its franchise growth. In contrast, the exclusivity of IGB's system may limit or delay its expansion until the market matures.

FURTHER RESEARCH

Researchers in Australia and overseas have signalled the transition of many agri-food marketing systems toward market orientation.[66] Because knowledge is limited about how to achieve this conversion quickly and effectively, greater appreciation is needed of the features of market-oriented agri-food firms. In particular, further research is needed to explain the industry-wide impacts or outcomes of market-oriented firms.

In the case of Australian meat retailing, anecdotal evidence suggests that the transition of these firms has driven the expansion of the major supermarket chains' product ranges, the stabilisation of beef consumption, and enhanced consumer perceptions. Market-oriented behaviours have, in effect, set these firms apart from the pack and raised the standard of meat retailing in Australia. Market followers, or me-too firms, have attempted to emulate these leading-edge meat retailers. However, their success is yet to be demonstrated over the longer term, and research should investigate the underlying systems that they have in place to detect whether they might share common features with the meat retailers examined in this chapter. Similarly, the activities of the major supermarket chains should be studied to determine if they have adopted any of the market-oriented practices documented in this study and to evaluate the consequences for the industry as a whole.

References

1. Insch, A. (2008), 'Triggers and processes of value creation in agri-food supply chains: A study of Australia's chicken meat industry', *British Food Journal*, vol. 110, no. 1, pp. 26–41.
2. Shoebridge, N. (1992), 'Beef and lamb', in Shoebridge, N. (ed.), *Great Australian Advertising Campaigns*, McGraw-Hill Company, Sydney, pp. 97–116.
3. Rabobank International (2002), *Australia and Argentina's Beef Industries: Contrasting Structures and Strategies* (Industry Note 067–2002), Rabobank International, Sydney.
4. Kohli, A.K., and Jaworski, B.J. (1990), 'Market orientation: The construct, research propositions, and managerial implications', *Journal of Marketing*, vol. 54 (April), pp. 1–18.
5. Barksdale, H.C., and Darden, B. (1971), 'Marketers' attitude towards the marketing concept', *Journal of Marketing*, vol. 35 (October), pp. 29–36; McNamara, C.P. (1972), 'The present status of the marketing concept', *Journal of Marketing*, vol. 36 (January), pp. 50–57.
6. Felton, A.P. (1959), 'Making the marketing concept work', *Harvard Business Review*, vol. 37 (July/August), pp. 55–65.
7. McNamara, op. cit.
8. Slater, S.F., and Narver, J.C. (2000), 'Intelligence generation and superior customer value', *Journal of the Academy of Marketing Science*, vol. 28, no. 1, pp. 120–27; Slater, S.F., and Narver, J.C. (1994), 'Market orientation, customer value, and superior performance', *Business Horizons*, March–April, pp. 22–8.
9. Kohli and Jaworski, op. cit.; Slater and Narver 1994, op. cit.
10. Kohli and Jaworski, op. cit., p. 6.
11. Narver and Slater 1990, op. cit., p. 21.
12. Shapiro, B.P. (1988), 'What the hell is "market oriented?"', *Harvard Business Review*, vol. 66 (November/December), pp. 119–25.
13. Jaworski, B.J., and Kohli, A.K. (1993), 'Marketing orientation: Antecedents and consequences', *Journal of Marketing*, vol. 56 (July), pp. 53–70; Kohli and Jaworski, op. cit.
14. Webster, Jr., F.E. (1992), 'The changing role of marketing in the corporation', *Journal of Marketing*, vol. 56 (October), pp. 1–17.
15. Grewal, R., and Tansuhaj, P. (2001), 'Building organizational capabilities for managing economic crisis: The role of market orientation and strategic flexibility', *Journal of Marketing*,

vol. 65 (February), pp. 67–80; Slater, S.F., and Narver, J.C. (1995), 'Market orientation and the learning organisation', *Journal of Marketing*, vol. 59 (March), pp. 63–74; Slater, S.F., and Narver, J.C. (1998), 'Customer-led and market-oriented: Let's not confuse the two', *Strategic Management Journal*, vol. 19, no. 10, pp. 1001–1006.

16. Slater, S.F., and Narver, J.C. (1993), 'Product-market strategy and performance: An analysis of the Miles and Snow strategy types', *European Journal of Marketing*, vol. 27 (10), pp. 33–51.

17. Atuahene-Gima, K. (1996), 'Market orientation and innovation', *Journal of Business Research*, vol. 35, pp. 93–103; Jaworski, B.J., and Kohli, A.K. (1996), 'Market orientation: Review, refinement and roadmap', *Journal of Market Focused Management*, vol. 1, no. 2, pp. 119–35; Slater and Narver 1995, op. cit.

18. Hamel, G., and Prahalad, C.K. (1994), *Competing for the Future*, Harvard Business School Press, Boston, MA; Macdonald, S. (1995), 'Too close for comfort? The strategic implications of getting close to customers', *California Management Review*, vol. 37, no. 4, 8–27.

19. Hamel and Prahalad, op. cit.; Mcdonald, op. cit.; Slater and Narver 1998, op. cit.

20. Grewal and Tansuhaj, op. cit.; Slater and Narver 1998, op. cit.; Slater, S.F., and Narver, J.C. (1999), 'Market-oriented is more than being customer-led', *Strategic Management Journal*, vol. 20, no. 12, pp. 1165–68.

21. Grunert, K.G., Baadsgaard, A., Hartvig Larsen, H., and Koed Madsen, T. (1996), *Market Orientation in Food and Agriculture*, Kluwer, Boston, MA; Beverland, M. and Lindgreen, A. (2007), 'Implementing market orientation in industrial firms: A multiple case study', *Industrial Marketing Management*, vol. 36, no. 4, pp. 430–42.

22. Insch, A. (2005), The *Effects of Marketing Organisation on the Delivery of Added Value: A Historical Comparison of Australia's Beef and Chicken Meat Marketing Systems*, unpublished doctoral thesis, Department of International Business and Asian Studies, Griffith University, Brisbane.

23. Grunert, op. cit.; Lewis, C., Pick, P., and Vickerstaff, A. (2001), 'Trappings versus substance: Market orientation in food and drink SMEs', *British Food Journal*, vol. 103, no. 5, pp. 300–312; Beverland and Lindgreen, op. cit.

24. Beverland, and Lindgreen, op. cit.

25. Eisenhardt, K.M. (1989), 'Building theories from case study research', *Academy of Management Review*, vol. 14, no. 4, pp. 532–50; Keep, W.W., Hollander, S.C., and Dickinson, R. (1998), 'Forces impinging on long-term business-to-business relationships in the United States: An historical perspective', *Journal of Marketing*, vol. 62 (February), pp. 31–45; Miles, M.B., and Huberman, A.M. (1994), *Qualitative Data Analysis: An Expanded Sourcebook*, 2nd edn, Sage, Thousand Oaks, CA.

26. Schatzman, L. and Strauss, A. (1973), *Field Research*, Prentice Hall, Englewood Cliffs, NJ; Yin, R.K. (1994), *Case Study Research Design and Methods*, 2nd edn, Sage, Beverly Hills, CA.

27. Eisenhardt, op. cit.; Eisenhardt, K.M. (1991), 'Better stories and better constructs: The case for rigor and comparative logic', *Academy of Management Review*, vol. 16, no. 3, pp. 620–7.

28. NMAA (National Meat Association of Australia) (2001), 'Submission to the Parliamentary Joint Select Committee on the Retailing Sector', in Parliamentary Joint Select Committee on the Retailing Sector (ed.), *Fair Market or Market Failure?: A Review of Australia's Retailing Sector*, Parliament of Australia, Canberra, ACT .

29. AMLC and AMRC (Australian Meat and Livestock Corporation and Australian Meat Research Council) (various), *Meat Marketing Trends*, Australian Meat and Livestock Corporation and Australian Meat Research Council, Sydney.

30. Rabobank, op. cit.

31. MLA (Meat and Livestock Australia) (2007), 'Retail', <http://www.mla.com.au/TopicHierarchy/ MarketInformation/Domestic Markets/Consumption/Retail.htm>. Accessed November 25, 2007.

32. ACCC (Australian Competition and Consumer Commission) (2007), Woolworths (2001), *Woolworths Meat Industry Commitment.* <http://www.woolworthslimited.com.au/news/ factsheets/ publicdocuments/06–08–2001_b.asp>. Accessed August 12, 2003.

33. Hearnshaw, H. and Shorthose, W.R. (1994), *Tailoring Beef Yield and Quality to Meet Retail and Consumer Preferences by Nutritional and Genetic Means*, Meat Research Corporation, Canberra.

34. O'Keeffe, M. (2002a), 'FMRs: the secrets to success', *Retail World*, vol. 55 (August), p. 29.

35. FMR (2002), *FMR's Corporate History*, FMR webpage. Accessed November 14, 2002.

36. FMR, op. cit.

37. Anon (2002a, 6 October), 'On a winner with a FMR's outlet', *Sunday Mail*, pp. 99.

38. Anon (2002b, 20 October), 'Business and franchises advertising feature fresh food retail leader', *Sunday Mail*, p. 118.

39. Gordon, C. (1998, 14 November), 'Pig Farmers await report', *The Courier Mail*, pp. 5.

40. O'Keeffe, op. cit.

41. O'Keeffe, op. cit.

42. Anon (2002c, 30 September), 'Support means you're not alone', *Sunday Mail*, pp. 81.

43. Anon (2002d, 3 June), product announcement, *International Product Alert.*

44. Anon (2002c), op. cit.

45. O'Keeffe, op. cit.

46. O'Keeffe, op. cit.

47. Hingley, M. (2005), 'Power imbalance in UK agri-food supply channels: Learning to live with the supermarkets?', *Journal of Marketing Management*, special issue: The marketing imperative for the Agri-food sector, vol. 21, no. 1/2, pp. 63–8.

48. O'Keeffe, op. cit.

49. FMR, op. cit.

50. Anon (2003, 3 February), 'Ritchies' IGA stands out', *Retail World*, p. 6.

51. Cooke, S. (2002, 2 May), 'AACo launches retail plans for Victoria', *Stock and Land*, pp. 23; Salins, C. (2002, 22 May), 'Beef up the industry', *Canberra Times*, p. 3.

52. AAC (Australian Agricultural Company) (2003), *AACo Takes First Steps to Implement Branding Strategy*, presentation. Australian Agricultural Company, <http://www.aaco.com.au/ html/ press27nov.htm>. Accessed August 12, 2003; Hopkins, P. (2003a, 13 August), 'AACo shares reach record high on post-drought earnings forecast', *The Age*, p. 14.

53. O'Keeffe, M. (2002b), 'IGBs: The art and science of meat retailing', *Retail World*.

54. IGB (2003), *IGB: Beef at its Best*, IGB, Melbourne.

55. IGB, 2003, op. cit.

56. IGB, 2003, op. cit.

57. O'Keeffe 2002b, op. cit.

58. Hopkins, P. (2003b, 3 January), 'Business – carving out a reputation for quality beef', *The Age*, p. 3.

59. Hopkins 2003b, op. cit.

60. Hopkins 2003b, op. cit.

61. O'Keeffe 2002b, op. cit.

62. Anon 2003, op. cit.

63. Hopkins 2003b, op. cit.

64. O'Keeffe 2002b, op. cit.

65. Champion, S.C. and Fearne, A.P. (2001), 'Alternative marketing systems for the apparel wool textile supply chain: Filling the communication vacuum', *International Food and Agribusiness Management Review*, vol. 4, no. 3, pp. 237–56; Meulenberg, M.T.G. and Viaene, J. (1998), 'Changing food marketing systems in western countries', in Jongen, W.M.F. and Meulenberg, M.T.G. (eds), *Innovation of Food Production Systems: Product Quality and Consumer Acceptance*, Wageningen Pers, Wageningen, the Netherlands, pp. 5–36.

14 Are Consumers Ready for Radio Frequency Identification (RFID)? The Dawn of a New Market Orientation Area

BY LUÍS KLUWE AGUIAR,* FREDDY BROFMAN,† AND
MÁRCIA DUTRA BARCELLOS‡

Keywords

radio frequency identification, RFID, privacy, trust, adoption, loyalty, reward, customer relationship

Abstract

In this chapter, we discuss the far-reaching applications of a wireless tracking technology, radio frequency identification (RFID), and how it could affect customers' relationships with supply chains. Despite gains for many stakeholders in the retailing sector, as the literature review shows, we still know little about consumers' attitudes toward privacy, ethical implications and perceived rewards. An exploratory survey ascertained consumers' attitudes toward RFID adoption regarding the benefits it might provide, by enhancing loyalty schemes and rewards, as well as guarantees of privacy. This survey suggests that the lack of knowledge about what RFID does and what companies do with the information gathered is a problem: Privacy issues are a principal matter of concern. Because the benefits such a technology would bring are unclear, the level of dissatisfaction

---

* Mr. Luís Kluwe Aguiar, Senior Lecturer, School of Business, Royal Agricultural College, Stroud Road, Cirencester, GL7 6JS, England. E-mail: luis.aguiar@rac.ac.uk. Telelphone: +44 1285 652 531.

† Mr. Freddy Brofman-Epelbaum, Research Student, Business School, University of Kent, Wye, Ashford TN25 5AH. E-mail: fmb7@kent.ac.uk. Telephone: +44 0207 594 2973.

‡ Dr Márcia Dutra Barcellos, Aarhus School of Business, University of Aarhus, Denmark, and Business Administration, Pontifícia Universidade Católica, Brazil. E-mail: marcia.barcellos@pucrs.br. Haslegaardsvej 10, 8210 Århus V, Denmark. Telephone: +45 8948 6486.

about the rewards from loyalty schemes is high. In our opinion, more research exploring consumers' beliefs about the possible adoption of RFID would benefit our understanding of consumers' attitudes.

We will briefly review literature detailing RFID technology and its applications, as well as issues associated with its adoption; discuss privacy, trust and risk-perceptions in connection with consumer acceptance; analyse and discuss the findings of a consumer survey in the United Kingdom; and provide conclusions and recommendations for those stakeholders involved with RFID technology.

Introduction

In recent years, the introduction and adoption of new technology developments have dramatically changed the way supply chains operate. Although some technologies have been quickly adopted, others have been strongly rejected.[1]

Radio frequency identification (RFID) is a good example of an emerging technology with potential to streamline supply chains by increasing the identification of products, whether they are static or in transit. Although in its infancy, RFID's great potential lies beyond its tracking and tracing capabilities; it also may provide a tool to strengthen customer relationships.

In particular, RFID technology is cited as the technology of the future, because it possesses a wide range of auto-identification capabilities that incorporate elements of bar coding and biometrics. An improvement on bar coding, it enables the recording of data onto a tag without the need for manual input, so the applications for businesses are vast. The uses of RFID technology range from manufacturing, supply chain management, retailing, payment systems, security, and access control to the most important application: information sharing.[2] Moreover, its possibilities for tracking, tracing, and recall of, for example, any food product are undoubtedly impressive. It has improved stakeholders' capacity for collaboration within a supply chain, which has reduced the length of time required to share information about the products exchanged. Furthermore, sharing information about the location and physical condition of products aids in the co-ordination of actions when the tracking and tracing products is required. As the recent examples of a contamination scare concerning lettuce and spinach in the United States showed,[3] the presence of RFID has dramatically improved product traceability during recall. RFID tagging also may help stakeholders by solving issues at specific item level, because until recently, products could only be tracked and traced in batches.[4] In a recall situation, differentiating products by batch, cargo and entire production day has the potential to reduce fatalities.

Nevertheless, some groups have expressed reluctance concerning its adoption because of RFID's far-reaching capabilities. Consumer responses to RFID product item tagging show that the fear, suspicion and public concern they express associated with RFID technology involve the potential interference in the basic civil right to privacy.[5] The possibility of personal data being lost or shared with third parties has caused disquiet amongst consumers. However, it is with the possibility of data being misused that consumers have voiced the most common fear.[6] In the United Kingdom, several high-profile cases have appeared in the press regarding lapses of security where confidential data has been lost. As a consequence, the general public has shown increasing concern about providing personal data. Mistrust is spread, for example, by the possibility of the cross-checking personal information by law enforcement organisations, courts or insurance companies.[7] Numerous instances of lost data from government departments (e.g., Home Office, Justice, Revenue, Customs, Health) reveal how government bodies, which should have strict procedures for handling thousands of pieces of personal information, can fail.

Carelessness in the management of information has resulted in opposition to the widespread adoption of technologies that might be perceived as too intrusive. Questions such as 'why is this data being collected?' challenge a system's objectives.[8] The benefits of RFID also could seem smaller owing to consumers' perceptions of the use of such an innovative technology.[9]

Yet despite these forms of opposition, RFID technology has abundant potential. Such a technology, when managed well, could provide consumers with valuable information regarding products, pricing and promotion strategies. Such a change in orientation would transform the food supply chain. Until now, bar coding has aided in identifying consumers' shopping habits, which feed into consumer loyalty schemes. However, consumers often receive inappropriate or useless rewards based on data collected about their shopping. For example, many UK loyalty cards offer promotions to encourage the purchase of goods, yet those goods do not match the consumers' shopping profile. Hence, the dissonant rewards of loyalty schemes might have questionable validity. Consequently, it appears that the major problem is the way the data are interpreted by a production-orientation mentality, common to many food manufacturers and retailers. The way RFID could contribute to convenience, accurate pricing and promotion campaigns thus should have a direct effect not only on consumers' attitudes toward that technology but also on how the industry views its use. Consumers' perception of RFID technology will influence market orientations in food retailing. If such a technology were put to good use, in the foreseeable future, food retailing could enter a new era of true market orientation.

This chapter considers risk perceptions and how they might determine the atmosphere of mutual retailer–consumer trust that can support a market orientation. We also ascertain the extent to which consumers are aware of this new technology in food retailing and explore their perceptions of the implications of RFID technology. To achieve these goals, this chapter consists of four sections. First, we provide an introduction to the research question and a brief review of literature related to RFID technology and the applications and issues with its adoption. Second, we review privacy, trust, and risk perceptions, in connection with consumer acceptance. Third, we provide a methodology section, followed by an analysis of the results of a consumer survey in the United Kingdom. Fourth, we offer conclusions and recommendations for those stakeholders involved with RFID technology.

What is RFID and How does it Work?

RFID relies on the Automatic Identification and Data Capture (AIDC) technology that, similarly to bar coding, is used to identify objects and living organisms such as plants, animals and humans.[10]

A wireless tracking technology, RFID that allows a reading device to exchange information with a tag (or transponder) operating via radio frequency. Such a tag may be attached to or embedded in an item, which allows the reading device to either remotely read or write data on the tag.[11] An RFID tag consists of two main components: a tiny silicon computer chip (integrated circuit) that contains a unique identification number, and an antenna, typically a metallic coil arranged flat around the chip, Figure 14.1. The combination of chip and antenna forms a transponder or tag.[12]

The exchange of information is effected by radio waves, silent and invisible electromagnetic energy that is able to travel through solid objects. In a sense, RFID may be considered an extension of bar coding technology, but unlike bar coding technology, which requires the use of an unobstructed laser beam to convey information between the bar code itself and a reader, items with RFID may be located and identified through obstructions.[14] Bar code technology also requires the reader to be visible, so it is very difficult to read a bar code without anyone noticing. With respect of size, tags can be as small as 0.25 sq mm, with some around 0.4 sq mm. Such small dimensions – half the size of a grain of sand – enable companies to place tags into layers of cardboard, plastic, rubber, clothing or labels. Flexible RFID tags also can be used in clothing, in which fine metallic threads sewn into the fabric can act as antenna without the customer ever perceiving their existence. Even some paint can act as antenna for tags, and developments of edible RFID tags have been introduced for some fruit and vegetable applications. Furthermore, unlike the bar code reader, a RFID reader (on the right hand side of Figure 14.2) can be concealed anywhere: doorways, carpets, mats, under floor tiles, ceilings, shelves or behind displays.[15]

As a system, RFID comprises three concepts that may have an impact on business processes: automation, identification and integration.[16] It may be classified according to its tag, reader and application. Tags can be active or passive, according to their capacity;

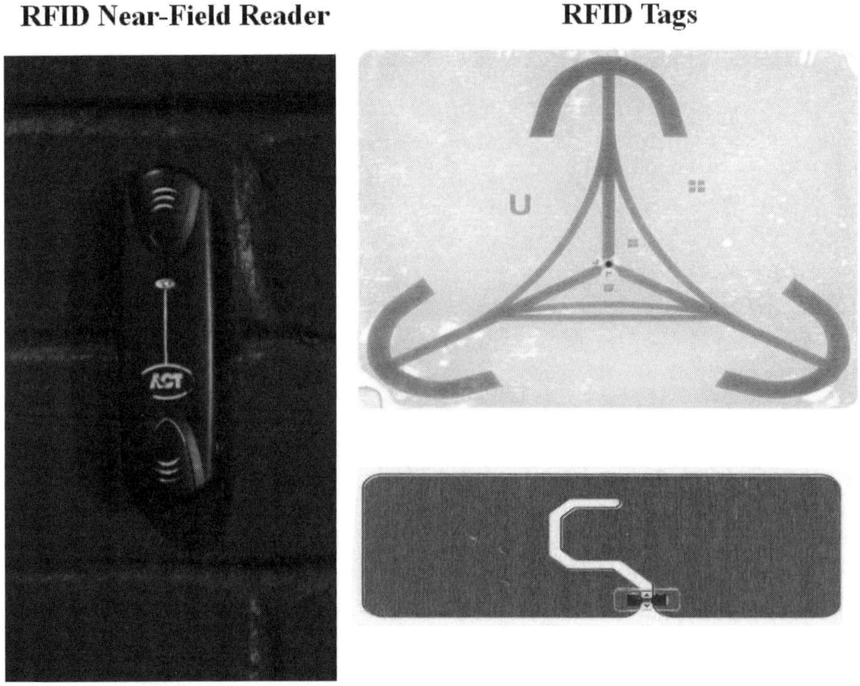

RFID Near-Field Reader   **RFID Tags**

Figure 14.1 Components of a passive RFID system[13]

a passive tag is one in which a tag's antenna picks up a reader's energy, amplifies it and directs it to a chip.[17] Such energy is enough for the chip to beam back its unique identification number, along with whatever other information it has been programmed to relay.[18] Passive tags are usually smaller and cheaper than active tags because they do not carry a battery and have the ability to operate indefinitely.[19] Companies that produce high-value, high-rotation goods, such as clothes, pharmaceuticals, electronics, and entertainment items, are most likely to use RFID tags at the item level. Active tags have a battery and can actively transmit information at longer distances and with greater data capacity. Active RFID systems also may be 'awoken' when they receive a signal from a reader or be programmed to emit signals at intermittent intervals. Active RFID systems can be classified according to the tag's signal emission, as in Table 14.1. Although the majority of tags in the food industry are passive, because of the nature of the products, which vary from dry to wet to humid, a mixture of passive and active RFID systems exist. In dry foodstuffs, despite the use of high and ultra-high frequencies,[20] moisture, the presence of metal, noise originating from electric motors, and fluorescent lights can distort the way radio waves are conducted.[21]

THE SECRETS OF A CHIP

Cost reduction is vital to product technology adoption and dissemination. To keep the cost of RFID tags down, the amount of information stored inside a tag has been kept to a minimum. In a similar way to bar coding, the application is constrained by the amount of

Table 14.1 Categories of tags[22]

Classification	Passive Tags				Active Tags	
Band wave	Low Frequency	High Frequency	Ultra High Frequency	Micro Wave	UHF	MW
Frequency	100–135 KHz	13.56 MHz	868–960 MHz	2.45–5.8 GHz	868–960 MHz	2.45–5.8 GHz
Reading range (m)	0.33	1.00	3.00s	5 to 10	N/A	N/A
Tags communication with readers	Inductive coupling		Radioactive or propagation coupling		Own transmitter	
Lowest price available	U$2–0.50	U$1.15–0.30	U$0.49–0.05	N/A depends on application	Top price range	
Item-level use	Animal tracking	Variety: mostly used for human recognition cheese and wine tracing	Case and pallet tagging	Bottle cap tagging	Asset management consignment management	

numbers a tag can carry. Unlike the Uniform Code Council, whose task is to standardise and regulate bar coding, RFID complies with the global Electronic Product Code (EPC) standard,[23] The EPC tags allow for some 96 bit code combinations, which enable them to carry information about the manufacturer, the product, and a specific serial number for each item, which means they can identify uniquely all objects manufactured for the next thousand years.[24] Therefore, RFID technology has made bar coding look primitive.

As Figure 14.2 shows, EPC coding consists of four categories. First, the Header defines the whole number, type, and length of all subsequent data partitions. Second, the EPC Manager set of numbers may be used by a specific company or entity to which a number has been assigned. The EPC Manager is responsible for sharing the details of a product when it passes through a site data collection or network. Third, Object Class is used to identify stock-keeping units or a group of products, for example, a case, pallet or lorry. It is assigned and managed by each individual company. Fourth, the Serial Number can be used for different purposes as well, such as the date, location or stage of processing of a product. For the serial number alone, there are 68 billion possible combinations of numbers.[25]

1	0000A88	10030G	TL0988333
Header	EPC Manager	Object Class	Serial Number
0 - 7 bits	8 - 35 bits	36 - 59 bits	60 - 95 bits

Figure 14.2 Electronic Product Code categories[26]

RFID DATA USES

EPC data enables many applications,[27] which may be retrieved in different ways to improve the decision-making process and business efficiency and to gain customer loyalty.[28] Unlike bar coding, RFID offers a source of information that is very dynamic and whose location, time and unique identity may be stored inside a common database to be analysed or shared with partners.[29]

Furthermore, RFID technology enables improvements in supply chain management. Being able to manage the flow of information from physical goods at every stage in value chains offers clear gains in logistics.[30] As a cost reduction tool, RFID technology has been put to good use in managing information about inventory control, distribution and sales, which may be made available to partners in the value chain for the benefit of the consumer. In its ability to count or read multiple objects, an RFID system acts as a safety measure in reducing human error.[31] The introduction of RFID tags has helped retailers take more accurate accounting of inventory, thereby facilitating immediate restocking,[32] for entities such as Marks and Spencer, Walmart, CarMax, FedEx, Gillette,[33] Metro,[34] Proctor and Gamble,[35] and Tesco,[36] as well as for farm animals, refugees and prisoner tracking.[37] The benefits also include warehouse rationalisation through the efficient use of space, assets and the labour force.

Despite its current use in identifying objects and living organisms, more is expected to happen in future RFID deployment at the item level. Much of its potential in aiding true market orientation will be derived from item-level identifications that, at the moment, are limited and stop instead at the case, batch and/or pallet level. An enhanced market orientation should derive from RFID's ability to build customer relationships.

When a customer buys a product with an RFID tag on it, he or she is not only buying a product but a history of the processes related to that product. The information that RFID tags offer businesses has the power to alter relationships and business processes, thereby enhancing the marketing orientation in the value chain. Retailers currently are attempting to further reduce transaction costs by integrating the supply chain with RFID technology. Benefits such as reduced internal and external transaction costs, more visible inventory control, delivery compliance, and waste management are already a reality, not to mention the improvements in retail logistics[38] and the savings from electronic proofs of delivery.[39] In the United States, retailers' requirements have focused more on the area of cost reductions (see Table 14.2) and increased efficiency (i.e., the reason for the widespread use of RFID technology at the pallet control level); in Europe, retailers face a different scenario. With less storage space, European companies like Marks and Spencer have sought the advantages of item-level tagging.[40] The European experience could lead to structural changes in food retailing and production: the dawn of a new market orientation era appears just on the horizon.

Table 14.2   RFID cost-saving attributes for retailers at the item level[41]

Product availability	When RFID is deployed on item level, it can help retailers improve the availability of the product by identifying out-of-stocks in real time, increasing the efficiency of inventory control and automatic replenishment systems, and improving the replenishment process from the store backroom.
Product control	RFID will increase the efficiency of tracking and tracing products inside the store, thus, less time will be spent collecting misplaced products and damaged or out-of date products.
Less queue time/ streamlined checkout and payment process	Perhaps the significant RFID requirement for retailers: no more queues at checkouts, and no checkouts at all, because readers can collect information from the shopping basket and charge payment cards, as well as credit loyalty points. In the near future, consumers might waste no time at a check-out counter, because they do not need to even remove a payment form from a wallet.
Price tag replacements	By deploying RFID on item level, smart shelves would automatically recognise a product and display its price. This already has become a reality in recent years.
Loyalty data measurement	Loyalty reports would be able to assess a customer's location in a shop and the shop of purchase. RFID can increase the accuracy of the data, though it also raises privacy issues.
Customer satisfaction	RFID can reward customers based on information they provide inside the store in real time.
Labour use	RFID can aid the restocking process by showing the type of product that should be placed on the shelf and the quantity that needs to be replaced. It can also help ensure the product is properly maintained, for example, the temperature of perishables could be controlled for at all times. RFID is definitely a labour-saving technology. Repetitive bar code scans will be replaced, changing the workforce composition and size. Instead of labour deskilling, it is likely that RFID technology will require complementary skilled labour.

PROBLEMS WITH RFID ADOPTION

Whenever there is a new technology, deployment issues arise. The location context and silent commercial applications in supply chain management that make RFID a key tracking technology are not without problems.[42] The adoption of new technologies is sometimes unclear, as in the case of a Metro Future Shop outlet in Germany. Consumers reacted against the retailer's privacy intrusion. Not only were product items tagged, but the shopping trolleys were too, as were Metro's customer loyalty cards. In this sense, managers could effectively trace every shopper's movement within the store;[43] this ability was considered too intrusive.

Privacy may be understood as an individual's ability to control the terms by which his or her personal information may be acquired and used,[44] or it could be understood as the right to be left alone.[45] In the digital era, informational privacy is of growing concern, because it refers to the right of individuals to retain control over the collection and use of personally identifiable facts and information about their daily lives, as in the case of Internet spyware and cookies tracking people's movements.[46]

The most common problem with RFID relates to privacy and security, including undesirable intrusions into the privacy of individuals and a fear of the misuse of data.[47] Ethical issues have been raised, because many people do not know what is behind such a technology and how it could interfere with their individual privacy.[48] Because passive RFID tags are small and relatively cheap, they are more likely to be ubiquitous in society. Small chips, which could be spying on consumers' habits, have been the main concern of privacy regulations. Yet there might be confusion between active RFID and passive RFID systems, in terms of perceptions of the creation of an 'Orwellian state.'[49] Nonetheless, as the use of RFID systems have removed 'architectural barriers,'[50] the lower costs of the tags have paved the way for privacy intrusions. Their use is diverse, and tags, which can be as thin as paper, are easy to hide, making them visually undetectable.[51] When readers inside a store are installed at strategic locations, all the tags in a basket or trolley may be read. Electromagnetic waves have no problem passing through items that may be considered private, such as purses, wallets, bags, and items of clothing.[52] This ability creates privacy and security problems, and solutions to these problems are necessary. However, due to the small distance required to read passive tags (10 m maximum), it is unlikely that privacy would be a major issue for the food industry.

Invasion of privacy issues also have increasingly attracted society's attention, especially since loyalty cards started gathering enormous amounts of longitudinal information about consumers' purchases and eating habits.[53] The case of British Telecom spyware, which attempted to pilot a behavioural advertising system, illustrates these concerns well.[54] British Telecom, by monitoring Web users' habits without their consent, crossed many legal and ethical boundaries.[55] When consumers realised they were being monitored, they felt their right to privacy had been breached.

The deployment of RFID technology similarly raises privacy issues because anyone who could read any customer's tags would gain immediate knowledge of the basket of products. Thus, it is possible for anyone to be aware of a person's likes and dislikes, as well as some personal and more intimate information. Once the information is associated with a person's profile, he or she could then be targeted by companies – not dissimilar to the BT intrusion. However, it is not clear to what extent consumers welcome the indiscriminate bombardment of promotional offers.

Furthermore, if tagged products keep emitting signals in a place where various readers are located, a person's movements could be also tracked. Therefore, security issues arise if RFID tags are left to open reading. In this case, a competitor retailer could read a consumer's shopping pattern and collect information, to the detriment the original retailer's loyalty or marketing relationship schemes. In an open reading environment, unethical people could also harm retailers and consumers by deactivating tags. To control access to information, companies could deploy patented reading tag protocols, which might create dependence on a single protocol, reducing the benefits of sharing data with suppliers, but also would increase security. Companies and consumers likely will have to face this trade-off in the future.

Moreover, privacy and perceived security may be associated with consumer trust.[56] Because data collection through RFID is invisible to consumers and can happen without the consumer's knowledge, beyond gathering information at a more granular level (item identification), this a technology adds layers of complexity to data handling, sharing, and centralisation. To ensure trust, consumers or users must believe that the benefit derived from the technology is far greater than the costs associated with it.

Trust refers to the belief by one party about another party's ethical behaviour.[57] Trust relates closely to risk, because the need for trust only arises in a risky situation. Yet trust could be considered as part of consumers' 'privacy calculus,'[58] which consists of their relative privacy concerns. In a study of German shoppers, consumers note that they felt powerless in the face of an 'intelligent framework,' because they had no alternative but to succumb to the system.[59] In that study, respondents, despite acknowledging the advantages of RFID, felt that their privacy was being jeopardised. The more educated the respondents, the more helpless they felt in the face of ubiquitous RFID capabilities.

As such, privacy and trust, which can help ensure loyalty, are associated with the perception of risk. However, consumers may engage in a less rational risk perceptions instead of rationally calculating the risks.[60] Therefore, risk perception could be both a psychological and a social construct that, in light of ethical problems, the media, and consumer groups, could exacerbate distrust. Usually the media and pressure groups demonise the technology to alert consumers, such as when they call RFID technology 'spychips,' 'big brother bar code,' or 'tracking devices.' Terms such as 'radio barcode' (Tesco) or 'near field communication' (Nokia) attempt to counteract these negative images.[61]

PRIVACY PROBLEMS IN RETAILERS' ENVIRONMENT

In addition to these more general concerns, RFID technology invokes many other issues that require consideration from the retail perspective. Dynamic pricing is problematic, in that RFID technology can enable retailers to profile consumers and correlate a person's identity with specific products. Retailers could price items according to a customer's preference or loyalty to a shop. If dynamic pricing is possible, an equal pricing policy would be eliminated,[62] and the likely impact on market efficiency would require consideration. It is also worth bearing in mind that a retailer's capacity to offer a greater range of prices to consumers might work to the retailer's detriment. Some customers perceive an advantage, whereas others may feel disadvantaged, which clearly creates a market imbalance,[63] because the increased technological possibilities in the market would allow for 'commodified privacy.' Consumers might be willing to trade some privacy in exchange for a promise of some benefits, such as security, but they more likely prefer cheaper processes. Conversely, a positive attribute of dynamic pricing could relate to food retailers' ability to update prices according to the sell-by date, which might save food retailers the cost of wastage and floor space (i.e., no more 'reduced items' sections in supermarkets) and also directly benefit consumers. In a survey in Greece, consumers expected dynamic pricing by retailing companies to happen in the near future.[64]

Regarding the use of information by less careful marketers, concerns linked to RFID systems have been raised. Marketers could misinterpret consumers' habits, and loyalty programs could backfire.[65] The fear of invasion or breach of privacy is, logically, due to the nature of the tags. Consumers are likely to become annoyed if they perceive they are under surveillance and losing rights without accruing any benefit. A loss of loyalty might result in lost customers. A clear benefit emanating from the RFID technology might involve appropriate rewards to reduce customer dissonance.[66] But with in-store tracking capacities, retailers could really increase advertising efficiency by customising their promotions to specific individuals. Furthermore, linking of RFID with the EPoS communication protocol, as a customer approaches the checkout, could reduce or eliminate waiting times and queues, though this capability might create retailers' overdependence on surveillance systems that, in combination with CCTV surveillance, might seem too intrusive. In the 'spychipped future,' many believe that consumer and individual liberties will be at risk.[67]

Methodology

The literature review pertaining to issues associated with the adoption of RFID reveals that this technology raises consumers' privacy and ethical concerns. To understand many of these concerns, we conducted a survey questionnaire, devised to capture consumers' attitudes towards RFID.

The questionnaire was designed to measure the attitudes of respondents who were or were not acquainted with RFID technology. The questionnaire consists of two main parts. Those who were acquainted with RFID responded to 53 questions in total, and those who were not were invited to answer a different set of 49 questions. Both parts consisted of six subsections, and the final section contained some demographic profiling questions.

The questions explored the respondents' perceptions of issues such as loyalty, trust, privacy, and the use of personal information collected. Questionnaires were sent electronically to the distribution lists of support, clerical, and teaching staff and students at three business schools in three UK universities: the Open University Business School (approximately 250 persons, part of the module B120), The Royal Agricultural College (approximately 820 persons), and the University of Kent–Wye College (approximately 125 persons).

The questions relied on five-point Likert response scales, anchored by (1) totally disagree and (5) totally agree, with (3) corresponding to the neutral position, neither agree nor disagree. In addition, participants could choose option (6), I don't know, if they were unsure about the meaning of the question. In this case, answers would be treated as missing values.

Some 103 questionnaires were returned by the closing deadline of June 2008. Of the total responses, only 92 were valid. The rate of questionnaire returns is low; however, the sample provides a fair representation of a cross-section of the UK society. A larger sample would certainly improve the validity of the results and our ability to extrapolate the findings to the general population. We attempted to establish correlations between the two subsets of responses, namely, those who were or were not acquainted with RFID technology, using Chi-squares. The responses also were tested using t-tests to establish whether differences appeared between the means of the subsets. Due to the limitations of the sample, this analysis necessarily has only a partial character. Nevertheless, it provides an interesting indicator of consumer reactions to RFID issues.

Results and Discussion

The sample of the respondents (92) is approximately evenly distributed in terms of gender (see Table 14.3). The sample also represents modern British demographics well, with smaller households of people living alone or with a partner or friends. However, with regard to educational attainment, the sample was skewed to higher levels, due to the university environment from which we drew the sample. As identified among German consumers,[68] the more educated the respondents were, the more helpless they felt in face of the ubiquitous RFID capabilities. The calculated r^2 coefficient of correlation shows it is positive at 0.194823.

Most respondents visit a supermarket twice a week; none of them did all their shopping online. Consumers used supermarkets as their most common outlet for food shopping.

The respondents possessed two loyalty scheme cards on average. However, their answers ranged from neutral (neither agree nor disagree) to dissatisfaction with the loyalty schemes to which they belonged. This finding should be a matter for concern for managers of relationship schemes. All respondents who were dissatisfied disagreed with statements such as, 'loyalty schemes offer me good deals' and 'loyalty schemes point me to products that I would not normally consider.' Nevertheless, no respondents indicated they had given up their loyalty programmes or felt they were a total nuisance.

With respect to those respondents who belonged to a loyalty scheme, they generally indicated they were neutral in response to the statement: 'I am satisfied with the loyalty

Table 14.3 Sample profile

Gender	Male	Female				
No. of respondents	44	48				
Age	18–29	30–39	40–49	50–59	60+	
	24	31	21	11	5	
Highest education	GCSE	A–Level	Undergraduate	Graduate	Post-graduate	
	5	16	19	25	27	
Household situation	Alone	Both parents	With partner	With friend	In halls	Single
	23	9	43	12	2	3

scheme I belong to.' Their attitude does not disagree entirely with results in the literature, but there is a good indication that their levels of satisfaction are not high. However, when asked whether loyalty programmes prompted them to consider goods they had not considered buying before, most of them rejected the statement. The same question asked in a different way – 'loyalty programmes match goods exactly suited to my taste' – prompts a consistent level of dissatisfaction, though the r^2 at 0.877937 was not significant. However, it may seem illogical for a loyalty scheme or retailer to offer goods that do not suit the consumer's tastes and needs; therefore, we might infer that a product orientation is too deeply embedded into retailers' loyalty schemes.

Some 40 per cent of the sample indicated dissatisfaction with their loyalty schemes. The survey design also captured the main reasons for this dissatisfaction: 'loyalty schemes do not offer me good deals and discounts.'

Moreover, the vast majority of the respondents did not know what RFID meant. The responses indicate an imbalance between those with some knowledge and the vast majority without RFID knowledge, which creates difficulties in establishing strong statistical correlations between the two subsets. All values were half of the expected $t = 1.658$ at 0.05 level of significance. Therefore, it was not possible to compare like to like, as was our initial intention. Nevertheless, these limited results should not be discarded, because they offer important evidence of the low level of knowledge about what RFID means, even in a highly educated sample. With this high level of ignorance about the topic, pressure groups that adopt negative and alarmist approaches to this technology likely have the ability to influence public opinion, to detriment of RFID deployment. The implications of consumers' ignorance about the gathering and use of personal information certainly suggests the need for further investigations into the matter. The phenomenon should be a concern for not only consumers but also those behind the deployment of RFID as a tool to enhance relationships through loyalty schemes.

Despite the sample limitations, it is possible to identify some trends, similar to those noted in prior literature, that confirm consumers' dissatisfaction with regard to the validity and reliability of personal data collection and usage by companies. Some respondents indicated positive ideas about data being collected and fed into a loyalty scheme, but they still perceived the return as rather limited. Respondents who were negative about loyalty schemes strongly oppose the idea that such reward schemes offered benefits at all. Instead, these respondents voiced their opinion that 'loyalty schemes rarely match goods that suited their taste.' Perhaps the industry has not recognised how disaffected these respondents are. Manufacturers and retailers, by providing promotions for goods that appear to suggest a product orientation rather than a market orientation, may be causing themselves harm.

The survey attempts to measure the extent to which consumers were concerned about what companies do with the personal information the collect. The level of apprehension exhibited supports previous literature that notes the general disquiet of people who share information about their lives. The survey also revealed that respondents did not agree about what companies did with their information. The interesting spread of answers might reflect the overall environment of trust that consumers feel toward retailers that collect personal data. It also could also demonstrate how legislation at the European Union level has provided enough assurance that personal data will not be shared without the consumer's consent.

The respondents' views about the possibility of data being collected and analysed for their own benefit revealed that they disagreed with this premise. They clearly believed that the retailer benefited more from the data collected, but the responses from those who were acquainted with RFID technology also did not indicate that they believed it would be an invasion of privacy. In the subset not acquainted with RFID technology, most were concerned with its applications. Nevertheless, the correlation between the two subsets of answers was not significant at -0.4784.

Finally, respondents demonstrated that they were confident data gathering would result in more choice, in line with consumers' privacy calculus.[69] Consumers are willing to trade some privacy in exchange for the promise of some benefits.[70] Provided consumers obtain something in return, there is a trade-off between their relative privacy concerns and trust, and what the loyalty scheme can offer.

Conclusions

With regard to consumers' awareness of RFID technology, in general, there is a lack of knowledge. Consumers with some knowledge of it were a minority in the survey. Despite being aware of its existence, their level of scepticism about its benefits was considerable. Privacy issues and the extent that such a technology would directly benefit consumers were the main concerns expressed by this group.

In terms of overall technology adoption, the introduction and acceptance appear to depend on the consumers' readiness to use such a technology. For those not acquainted with RFID, the lack of knowledge and resulting suspicion about what companies would do with the information represents a constraint. This suspicion generally relates to issues of privacy, for which, according to previous literature, the possibility of such data being misused is the greatest fear. Information gathering is perceived as an invasion of consumers' privacy, without any real benefit. Evidence of this failure to reward appears in the dissatisfaction with loyalty schemes that do not provide members with meaningful promotions.

RFID technology is an extraordinary tool whose application possibilities are just beginning. Thus, it may be a catalyst for how relationships will be structured in the food chain in the future. Its data gathering at the item level, to track, trace, and survey goods and monitor consumption patterns, would be the determining factor in influencing market orientations in food retailing. Hence, the consumer society appears to be in the dawn of a truly new era.

Nonetheless, consumers' awareness of RFID technology in food retailing is low and their perceptions generally negative. Behind the deployment of such a technology, there are fundamental ethical implications that also need to be resolved. However, the questionnaire design did not allow us to capture respondents' perceptions of the ethical implications of RFID technology.

Overall, risk perceptions were not very high, perhaps because the respondents trusted the retail environment. Food retailers fulfil many of the consumers' needs, which implies they might be truly market oriented. However, the results of the survey indicate that the food industry has failed to match consumers' shopping profiles with products on offer. Consumers are dissatisfied and to some extent disillusioned with rewards that never match their profiles. This point raises issues for managers of loyalty programmes, pertaining not only to their present dissonance but also to the introduction of a third-generation data gathering system. Managers in the food industry are likely to face a great deal of resistance.

The constant trade-off between privacy and the infinite possibilities of enhancing market orientation likely require specific guidelines from the industry. If safeguards are in place and communicate enough guarantees, RFID could be key in enhancing customer relationships and market orientation in the food chain. One potential solution might be to create privacy structures that allow consumers to see the advantages of buying RFID tagged products, because they benefit from RFID-enabled loyalty programmes. Overall, the advantages to retailers include improved inventory control, data on consumer preferences, reduced time at the checkout, better control of product expirations, and cost reductions. The respondents understood that the retailers might benefit more from RFID. Although they were sceptical about it providing lower prices, they were confident that it could result in more choice.

In an attempt to address issues of technology adoption, privacy, risk, and trust, consumer groups often highlight a negative image of RFID as a 'Big Brother' technology, enabling an increasingly controlled society. In the food retailing industry, those behind the loyalty programmes need to convey positive and reassuring messages that encourage consumer attitudes that enable the acceptance of RFID technology and thus minimise rejection.

This study suffers from several limitations with regard to the size of the sample and a bias in favour of respondents with more education. The sample also was not sufficiently large to provide strong correlation between the two subsets, which was the intention of the questionnaire design. Despite attempts to balance out the responses, the subset of respondents who were aware of RFID technology was too small and hence not representative. In turn, we could not compare the attitudes of those who considered themselves knowledgeable to those who claimed ignorance about RFID systems. Despite their higher educational attainment, the sample was not better informed about issues surrounding RFID and privacy. Although their general awareness of RFID technology applications in food retailing was minimal, they were not necessarily being averse to such new technology.

For those considering using RFID in retailing as a tool to enhance customer loyalty, we suggest that consumers' perceptions of risk need to be carefully considered, along with how to minimise that risk by creating an environment of mutual retailer–consumer trust that supports a market orientation.

The wider deployment of RFID is an imminent reality, as initiatives such as the tagging of rubbish and recycling bins are occurring in the United Kingdom. In such a future, consumer and individual liberties might be risk, so more investigation should consider the ethical dimensions surrounding RFID technology. The implications of consumer ignorance about the use of personal information also requires further investigations. This phenomenon should be only a concern not just for consumers but for anyone interested in deploying RFID to enhance loyalty schemes.

References

1. Cantwell, B. (2002), *Why Technical Breakthroughs Fail: A History of Public Concern with Emerging Technologies*. MIT Auto-ID White Paper, November. 016:3–18. Auto-ID Center.

2. Brofman, F. (2006), 'Loyalty programs redesign among RFID lines, some examples from industries', *RFID Journal*. Available at http://www.rfidjournal.com (accessed March 29, 2007).

3. ABC News (2007), *Tainted Californian Spinach*. Available at http://www.abcnews.go.com (accessed November 23, 2007).

4. EPCglobal (2006), *Architectural Standards Framework*. Available at http://www.epcgloballink.org (accessed December 10, 2006).

5. Cantwell, op. cit.; Angeles, R. (2007), 'An empirical study of the anticipated consumer response to RFID product item tagging', *Journal of Industrial Management and Data Systems*, vol. 107, no. 4, pp. 461–83.

6. Thiesse, F. (2007), 'Privacy and perception of risk: A strategic framework', *Journal of Strategic Information Systems*, vol. 16, no. 2, pp. 214–32.

7. Vermesan, O., Grosso, D., Dell'Ova, F., and Prior, C. (2006), 'Quo vadis RFID technology?' *RFID Journal*, March-April. Available at http://www.rfidjounal.com (accessed January 6, 2008).

8. Brofman F. and Aguiar, L. (2006), 'Customer's benefits inside retail store with RFID technology', *RFID Journal*. Available at http://www.rfidjournal.com/whitepaper6 (accessed March 29, 2007).

9. Theotokis, A., Pramatari, K., and Doukidis, G. (2007), 'The consumer perspective: An innovative retail service', in *Conference Proceedings of the 4th RFID Academic Forum*. March. Brussels, Belgium. Available at http://www.rfidconvocation.eu/Agenda.htm (accessed April 29, 2007).

10. Brofman, F. (2006), *Loyalty Programs Redesign Along RFID Lines, Some Examples from Industries*. Available at http://www.rfidjounal.com (accessed September 6, 2006).

11. Curtin, J., Kauffman, R.J., and Riggins, F.J. (2007), 'Making the most of RFID technology: A research agenda for the study of the adoption, usage and impact of RFID', *Journal of Information Technology and Management*. vol. 8, no. 2, pp. 87–110.

12. Albrecht, K. and McIntyre, L. (2005), *Spychips: How Major Corporations and Government Plan to Track Your Every Move with RFID*. Nelson Current.

13. Brofman, F. and Aguiar, L.K. (2007b), 'Tracking and tracking food products with RFID technology: An application for agricultural commodities?' In *Conference Proceedings of the 17th Annual Forum and Symposium of the International Agribusiness Management Association*. June. Parma, Italy.

14. Curtin et al., op. cit.; Albrecht and McIntyre, op. cit.

15. Albrecht and McIntyre, op. cit.

16. Langheinrich, M. (2007), *RFID and Privacy*. In Petkovic, M. and Springer, W.J. (eds), *Security, Privacy, and Trust in Modern Data Management*. Heidelberg, New York, pp. 82–94.

17. Finkenzeller, K. (2003), *RFID Handbook: Fundamentals and Applications in Contactless Smart Cards and Identification*. 2nd edn, John Wiley and Sons, New York.

18. Curtin et al., op. cit.

19. Curtin et al., op. cit.; Albrecht and McIntyre, op. cit.

20. Brofman, F. and Aguiar, L.K. (2007a), 'Customers' own information in-store', in *Conference Proceedings of the 4th RFID Academic Forum. The European Parliament*, March 2007. Brussels. Available at http://www.rfidconvocation.eu/Agenda.htm (accessed March 29, 2007).

21. Hingley, M., Taylor, S., and Ellis, C. (2007), 'Radio frequency identification tagging: Supplier attitudes to implementation in the grocery retail sector', *International Journal of Retail and Distribution Management*, vol. 35, no. 10, pp. 803–20.

22. Brofman and Aguiar, 2007b, op. cit.

23. Curtin et al., op. cit.

24. Finkenzeller, op. cit.

25. Brock, D. (2001), *The Electronic Product Code (EPC). A Naming Scheme for Physical Objects.* Available at http://www.autoidlabs.org/page.html (accessed March 29, 2007).

26. Ibid.

27. Harrison, M., Moran, H., Brusey, J., and McFarlane, D. (2003), *PML Server Developments.* Available at http://www.autoidlabs.org/page.html (accessed March 29, 2007).

28. Wolfram, G. (2004), 'Metro future store', *RFID Journal.* Available at http://www.rfidjournal.com (accessed January 31, 2007); Wong, C.Y. and McFarlane, D. (2005), *RFID Data Capture and its Impact on Shelf Replenishment.* Available at http://www.autoidlabs.org/page.html (accessed March 29, 2007); Tellkamp, C., Angerer, A., Fleisch, E., and Corsten, D. (2005), *From Pallet to Shelf: Improving Data Quality in Retail Supply Chains Using RFID.* Available at http://www.autoidlabs.org/page.html (accessed March 29, 2007).

29. Brofman and Aguiar, 2007b, op. cit.

30. Angeles, op. cit.

31. Hingley et al., op. cit.

32. Swedberg, C. (2006), 'MIT and IESE study shows RFID's value', *RFID Journal.* Available at http://www.rfidjournal.com (accessed March 29, 2007).

33. Kelly, E.P. and Erickson, G.S. (2005), 'RFID tags: Commercial applications v. privacy rights', *Journal of Industrial Management and Data Systems*, vol. 105, no. 6, pp. 703–13.

34. Albrecht and McIntyre, op. cit.; Gunther, O. and Spiekermann, S. (2005), 'RFID and the perception of control: The consumer's view', *Communications of the ACM*, vol. 48, no. 9, pp. 73–6.

35. Thiesse, op. cit.

36. Brofman and Aguiar, 2007b, op. cit.

37. Albrecht and McIntyre, op. cit.

38. Brofman and Aguiar, 2007b, op. cit.

39. Swedberg, op. cit.

40. Angeles, op. cit.

41. Curtin et al., op. cit.; Tellkamp et al., op. cit.; Brofman and Aguiar, 2006, op. cit.

42. Angeles, op. cit.; Hingley et al., op. cit.

43. Angeles, op. cit.

44. Westin, A.F. (1967), *Privacy and Freedom.* Athenaeum, New York; Angeles, op. cit.

45. Kelly and Erickson, op. cit.

46. Ibid.

47. Thiesse, op. cit.

48. Albrecht and McIntyre, 2007b, op. cit.

49. Langheinrich, op. cit.

50. Lessig, M. (1999), *Code and Other Laws of Cyberspace.* Basic Books, New York; Thiesse, op. cit.

51. Albrecht and McIntyre, 2007b, op. cit.

52. Ibid.

53. Ibid.

54. *The Guardian* (2008), 'Phorm fires privacy row for ISPs'. Available at http://www.theguardian.co.uk (accessed March 6, 2008).

55. The British Broadcasting Corporation (2008), *Behavioural Advertising.* Daren Waters. Available at http://bbc.co.uk (accessed March 6, 2008).

56. Angeles, op. cit.

57. Thiesse, op. cit.

58. Angeles, op. cit.
59. Gunther and Spiekermann, op. cit.
60. Thiesse, op. cit.
61. Ibid.
62. Brofman and Aguiar, 2006, op. cit.
63. Cantwell, op. cit.
64. Theotokis et al., op. cit.
65. Albrecht and McIntyre, 2007b, op. cit.
66. Brofman and Aguiar, 2007b, op. cit.
67. Albrecht and McIntyre, 2007b, op. cit.
68. Kelly and Erickson, op. cit.
69. Angeles, op. cit.
70. Cantwell, op. cit.

15 *Interrelationship Between Ethnicity and International Trade of Greek Virgin Olive Oil*

BY GEORGES VLONTZOS* AND MARIE-NOËLLE DUQUENNE†

Keywords

olive oil, trade, geographical indications, ethnicity

Abstract

In this chapter, we examine the interrelationship between ethnicity and the evolution of the Greek virgin olive oil trade, as well as specific factors that have direct impacts on the international competitiveness of this traditional Greek product.

We will consider the internal and external environment of the Greek olive oil as a typical example of ethnic food; detect, by implementing an empirical gravity model, the role and impact of noneconomic factors, especially cultural and institutional ones; and suggest some recommendations and proposals for the outline of a marketing mix.

Introduction

This research examines the relationship between ethnicity and the evolution of the Greek virgin olive oil trade globally. The findings resulting from a strengths–weaknesses–opportunities–threats (SWOT) analysis, in accordance with the findings from an implementation of a gravity model, demonstrate the significance of worldwide trade-flows for this Greek agricultural product.

* Dr Georges Vlontzos, University of Thessaly, Agricultural Sciences School, Fitokou Street, 38446 Nea Ionia, Magnisia, Greece. E-mail: georgevlontzos@yahoo.gr. Telephone: + 30 24280 78712.

† Dr Marie-Noelle Duquenne, University of Thessaly, Polytechnic School, Department of Planning and Regional Development, Pedion Areos, 38334 Volos, Magnisia, Greece. E-mail: mdyken@prd.uth.gr. Telephone: + 30 24210 74438.

This article also offers a list of proposals and suggestions that focus on increasing the competitiveness of the sector and armouring it with essential quality and safety reassurances. These proposals represent an aggressive marketing plan for gaining market share in both EU and non-EU countries that uses the product's competitive advantage as a protected designated of origin (PDO) good.[1]

The most recent Common Agricultural Policy (CAP) reform, in accordance with the World Trade Organisation (WTO) negotiations agenda, raised the importance of ethnic foods for international agricultural trade to a top priority. Greece has participated in this new area of interest by trying to establish increased protection and exclusive rights of production and trade for a series of products identified as PDO or protected geographical indication (PGI) goods. Various virgin olive oils have been included in this group in an attempt to end the period of their anonymity. Olive oil has long been a traditional Greek product, produced in every coastal area of the country and an integral part of Mediterranean diet. Recent changes in the EU CAP, in accordance with the ongoing WTO negotiations, and the impact of new nutritional trends have initiated a new global trading environment, capable of adjusting the recent production and trading status quo.

Background

The cultivation of olive trees has been subsidised by the EU since 1980, when Greece became a member. These subsidies were tailored to the production of olive oil and edible olives, such that the EU's olive oil policy was mainly to subsidise farmers and stabilise production by establishing penalties for overproduction. The previous revision of this policy established the maximum guaranteed quantity (MGQ) for every olive oil producer country; if national production exceeded this limit, a gradual reduction in subsidies was implemented. This policy had two targets: to keep the EU budget stable with regard to this subsidy payment and to stabilise the supply of olive oil in the EU market and thus decrease downward pressures on the market price. In contrast, the new agricultural policy, by decoupling subsidy payments from production, provides a secure income to producers and ends an approach that failed to take into consideration the global trends regarding olive oil's great potential.

Methodology

This research implements a SWOT analysis that considers all necessary information about the internal and external trading environment of olive oil. This methodology and analysis of internal and external trading environment recognizes that each production procedure possesses strong and weak characteristics that form the internal environment and experiences opportunities and threats that form the external environment. By applying this methodology, we can examine both economic and noneconomic factors. In the internal environment, the key competencies are marketing, finance, manufacturing and organisational behaviour. We rate each factor as a major strength, minor strength, neutral factor, minor weakness or major weakness. By connecting the ratings vertically for each specific business, we can profile its major strengths and weaknesses. We also list and classify opportunities according to their attractiveness and success probability, which depends on whether the business strengths match the key success requirements for operating in the target market and exceed those of its competitors. Finally, the threat classification considers their seriousness and probability of occurrence. By combining these data, we develop a picture of the major threats and opportunities and thereby estimate overall attractiveness.[2]

Yet we also consider the findings from a gravity model, which provides a reliable instrument for demonstrating the parameters that affect bilateral flows.[3] A gravity model represents a generic name for various quantitative models based on the Newtonian law of gravitation. In the field of international trade, the model developed by Tinbergen[3] and improved by Linneman[4] is considered one of the most fruitful ways to formalise and explain trade flows between geographical entities; its specification is not based solely on quantitative variables that reflect the economic size and characteristics of the partners but also on a set of quantitative and instrumental (dummy) variables, relative to positive or constraining factors, such as cultural or geographical proximity. By implementing the gravity model, we attempt to capture the impact of noneconomic factors on the external trading environment of Greek olive oil. Among these factors, we emphasise the origin of tourist flows in Greece, consumers' preferences for olive oil, institutional proximity (EU integration), and relational proximity (presence of Greek Diaspora in partner countries). More precisely, the gravity equation we use for this study takes the following form:[6]

$$lX_{G,i} = b_0 + b_1.lY_i + b_2.lP_i + b_3.lDis_{G,i} + \sum_k c_k.D_{i,k} + \varepsilon_{G,i} \tag{1}$$

where l denotes the log form, Y_i = economic size of partner i, P_i = population size, $Dis_{G,I}$ = geographical distance between Greece and partner i, and the k variables $D_{i,k}$ are instrumental variables, relative to noneconomic factors. The study covers a sample of 77 trade partners of Greece during 1991–2005. Data were extracted from FAOSTAT trade matrix, the IMF's World Economic Outlook Database, and the General Secretariat of National Statistical Service of Greece. Finally, by estimating equation (1), we capture positive and negative impacts of the selected noneconomic factors.

Findings and Discussion: Internal and External Environment of Greek Olive Oil Sector

To study the virgin olive oil production scheme, we must gather information about the internal and external environment of this sector. The internal environment is characterised by positive and negative aspects with direct influences on the trading performance of the product, in both internal and external markets, whereas the external environment consists of opportunities and threats that affect the entrepreneurial performance of the sector.

INTERNAL AND EXTERNAL ENVIRONMENT OF GREEK OLIVE OIL SECTOR

Strengths

The most important competitive advantage of Greek olive oil is its superior quality, compared with olive oils produced in other Mediterranean countries. This superiority reflects that the percentage of virgin olive oil produced in Greece is the highest in the Mediterranean area, reaching nearly 80 per cent.[7] This element is crucial in the effort to increase the added value of olive oil and shift it gradually into a distinguishable product, rather than the commodity it is today (Table 15.1).

Table 15.1 Greek olive oil sector: internal environment

Factor	Major strength	Minor strength	Neutral	Minor weakness	Major weakness
Superior quality characteristics	✔				
Integral part of Mediterranean diet	✔				
PDO, PGI product		✔			
Positive interrelationship with tourism		✔			
Increased cost of production					✔
Small size of holdings				✔	
Exports in bulk					✔

Since ancient years, olive oil has been an integral part of the Mediterranean diet. This tradition persists and even is strengthening owing to consistent scientific findings supporting the view that the product offers high nutritional value compared with other vegetable oils and fats. These findings provide considerable evidence for increased usage and consumption of olive oil internationally.

Many Greek olive oils have been named PDO and PGI products, according to the EU legislation; specifically, 26 olive oils have been included in this list.[8] These products have a competitive advantage through differentiation and face less competitive intensity from other firms or similar products. As a positive outcome of WTO negotiations, inclusion in these lists can decrease the intensity of rivalry among competing firms and countries and provide more accurate information to consumers about the quality characteristics of the products.[9]

One of the findings derived from the implementation of the gravity model was the significant relationship between the consumption of olive oil and the increase of tourism in Greece. This trend implies a positive and promising precondition for the noteworthy increase of exports, because all parties participating in this production and trading scheme should effectively employ this tendency (Table 15.2).

Weaknesses

One of the most important obstacles for Greek agriculture in general is the high cost of production, in accordance with the general lack of infrastructure. The olive oil sector

Table 15.2 Greek olive oil sector: external environment

Opportunities	Attractiveness	Probability
New subsidy scheme	High	High
Increased adoption of Mediterranean diet	High	High
Increase of tourism	High	High
PDO, PGI wish list at the WTO negotiations	High	Low

Threats	Seriousness	Probability
$–€ parity of exchange	High	High
Geographical expansion of cultivation	High	High
Climate change	High	High
Land uses	Low	High

is not an exception, and this factor simultaneously represents an obstacle to greater competitiveness. In particular, old olive orchards lack irrigation and experience higher costs of inputs, compared with other EU Mediterranean countries, along with gradual increases in the wages of immigrant farm workers.

The other crucial factor that limits competitiveness is the small size of the agricultural holdings specialising in olive trees and the large number of them. This status increases the costs of production and also reduces the bargaining power of producers when they negotiate the selling price of their product. The situation worsens as a result of the poor performance of most cooperatives, which allows traders to offer low prices to producers who have no alternative but to agree and sell their yield.

Most quantities of olive oil are exported in bulk and packaged mostly in Italy or other countries too. This management approach cannot incorporate any added value to the product but instead leaves this significant potential to processors and final distributors. The status quo clearly cannot improve farmers' revenues, and the product cannot gain market shares globally with a Greek identity.

Opportunities

The total decoupling of EU subsidies for the cultivation of olive trees ended a long period of complicated schemes for distributing subsidies to farmers. The new subsidy scheme provides a simpler method to distribute subsidies and secure incomes for farmers. The most important issue though is the abrogation of a series of restricted measures regarding produced quantity on a national level for EU countries. The new CAP moves forward to subsidise actions related to the development of marketing plans and gaining global market shares, in an attempt to improve the economic performance of the sector not through subsidies but through the market.

Another tendency with considerable impact is the increasing adoption of a Mediterranean diet. Consumers in developed countries experience increased concerns about their nutritional habits and gradually are including more food ingredients from the Mediterranean area. Olive oil is the most representative part of this diet, because it accompanies almost every food in this region. Therefore, this parameter must be taken into account by exporting firms hoping to gain market shares globally.

Recent years also have seen a significant increase of tourism into Greece. During their holiday, tourists have the opportunity to taste olive oil and become familiar with it. They therefore represent an opportunity, because when they travel back home, these tourists may remain consumers of the product. This efficient and low-cost promotion strategy offers access to tourists who would not have had such easy access to the product in their own country.

Continuous efforts by the EU during ongoing WTO negotiations have focused on a wish list of PDO and PGI products produced in specific regions of Europe and earning increased protection. This list might be the starting point for increased protection of the 26 PDO and PGI Greek olive oils. Such a development would decrease rivalry among competing firms and provide added value to the product for producers and traders.

Threats

The gradual revaluation of Euro against the dollar creates a negative environment for exports to the United States and other countries whose currencies are pegged to the dollar. In contrast, Mediterranean countries that are not EU members can exploit this situation and gain a competitive advantage regarding price. The revaluation has been stable for some time, which creates scepticism about which alternatives might ease these negative consequences.

The growing interest in olive oil internationally has motivated farmers in other regions, with climates similar to that of the Mediterranean area, to plant olive trees, despite their lack of tradition or experience in olive tree cultivation. Olive tree orchards now produce olives across the American continent, in South Africa, and throughout countries in the Far East.[10] Further increase in the production of olives might partially fulfil the demand for olive oil and reduce the demand for imports from traditional olive oil producing counties.

Although olive trees are hardy plants, resistant to drought and able to use low-quality water, climate changes can cause serious damages in terms of both yields and trees. Extreme weather conditions, such as frost, heat waves, fires and floods, have damaged millions of trees in the past decade and created considerable supply fluctuations. These changes, in the long run, may cause even more serious catastrophes and decrease the quantity and quality olive oil.

Owing to their sensitivity to low temperatures, olive trees generally are planted near the sea, in coastal areas in Greece. These same areas offer the best economic development of the country related to other activities, such as real estate or tourism. The diversification of income, in accordance with pressures for changing land use, has led many olive orchard owners to abandon them. This tendency grew stronger after the last CAP reform, which completely decoupled subsidies from production. A priority therefore must be to motivate olive farmers to continue to cultivate to avoid significant reductions in produced quantities.

The most important results derived from the gravity model we use to examine the export flows of Greek olive oil are as follows:

• Exports of olive oil to the trade partners of Greece are positively affected by their economic and demographic size. Export flows of olive oil increase in greater proportion with the change in gross domestic product (GDP) per capita than with the change in population size. This result is quite logical if we take into account the nature of the product. For most countries, olive oil is not a common product, and the result confirms that in countries with higher living standards, consumers are not oriented merely toward subsistence products but also have the economic capacity to develop consumption of 'ethnic' products.

- The role of distance is not obvious, even if in most cases it represents a 'resistance factor.' Olive oil exports are not always negatively affected by geographical distance. Rather, the negative impact of geographical distance may be due not only to transport costs – though this aspect is critical for fresh products such as olive oil – but also indirectly reflects a 'cultural distance' and a lack of information. Because most Greek exports are absorbed by countries that traditionally consume olive oil, it becomes logical that 'cultural distance' does not play a determinant role. Nevertheless, the impact of geographical distance is not absolutely evident. This result is quite interesting and may suggest that the development of ethnic product markets, especially in developed countries, contributes to the reduction of such cultural distances.
- With regard to the selected dummy variables that attempt to capture institutional and geopolitical characteristics, it appears that the impact of the EU is not always significant for exports of olive oil. The fact that a trade partner is member of the EU does not seem to be an absolute determinant.
- The existence of a Greek Diaspora, as in the case of countries like Germany, Australia or the United States, has a positive and significant impact on trade flows that is determinant across all considered periods. This point is particularly interesting because it shows that even if Greek migration is a relatively old phenomenon, the attachment and links that the Diaspora maintains with the country of origin remain quite strong; moreover, this population appears to promote 'typical' Greek products.
- Finally, the Mediterranean characteristic of the country and the role of tourism generally present a positive impact for the development of oil olive exports. Countries with significant tourist flows to Greece seem to develop olive oil greater imports.[11]

Conclusions

The findings from both the implementation of the gravity model and the SWOT analysis provide useful information about the future of the cultivation and trade performance of olive oil. The first priority is to avoid further reductions in production and establish motives for improving and certifying its quality. Despite the perception of an oversupply of olive oil, which decreases producer prices, the gradual increase of demand globally may ease this pressure. Olive trees produce olives every two years, so an effective stock management scheme must be implemented to balance supply and demand and reduce price fluctuations. The progressive independence of Greek olive oil exports from Italy can significantly improve trading performance and provide a basis for a differentiation strategy that will add value. Another issue pertains to establishing and implementing an aggressive marketing strategy in countries with considerable potential for olive oil consumption, such as those in northern Europe, northern America, and eastern Europe, as well as in the Far East. Such projects are eligible for funding from the second pillar of the new CAP and should be undertaken to increase the competitiveness of the sector.

The only means to improve economic results and profitability is to increase the added value of the product and implement a successful differentiation strategy based on its unique quality and cultural characteristics. Previous subsidy schemes did not motivate such projects. Therefore, the new CAP can provide a starting point for gaining global market share.

Furthermore, the results of the gravity model and the SWOT analysis support the formation of a new marketing mix that can be enriched in the future by newer findings from research into consumer behaviour toward ethnic foods. The well-known marketing mix consists of four elements: product, place, price and promotion. For each element, we provide recommendations aimed to make the most out of the strong points of the production sector, take advantage of the opportunities of the external environment, and minimise the negative effects and importance of the weaknesses and threats.

PRODUCT

The powerful link between olive oil and Greek immigrants worldwide is very important, because this critical mass of consumers can already create demand in markets where the product is not currently negotiable. This status quo will not be satisfactory for long-term marketing performance of Greek olive oil in these markets though because of the marketing policies of competitors. Greek producers and processors must move fast in two directions: rapid increase of 'exportable' quantities and segregation of the product from imitations in consumers' minds. Increasing exportable quantities requires an essential mechanism of quality and safety reassurance that must be established and continue to operate. Although the HACCP protocol, combined with the ISO series, can provide necessary reassurances for olive oil processors, a gap exists for quality reassurances about the production of olive oil. Competitors are aware of this weakness and may use it in the future if they suspect they are losing market share. Consumers of ethnic foods abroad are quite sensitive about such issues, and it would be difficult to defend the reliability of the product against such accusations. Therefore, exported olive oil must have quality reassurances. In this situation, the most feasible scenario is separating olive growers into two categories: (1) those that can produce certified olive oil and (2) those that produce olive oil for consumption in the internal market. This scenario has been implemented with success for other products from the horticulture sector, such as 'Sultanina' table grapes.

PLACE

The implementation of the gravity model provides considerable hints regarding which markets should be top priority targets for Greek olive oil. Countries with Greek immigrants reduce the business risk associated with expanding to foreign markets. The task to increase consumer loyalty in this specific target group represents the easy step; the difficult task is expanding such loyalty among consumers with nutritional lifestyles other than Mediterranean. Market share in markets without Greek immigrants will be more difficult to achieve, because consumers are not familiar with such foods, or their income prevents them from buying products that are not essential or integral parts of their diet. Therefore, it would be judicious to plan such an attempt with a long-term focus, accompanied by a well-organised promotion mix focused on introducing and displaying the distinct differences between the original product and imitations.

Another serious issue involves the gradual decrease of dependence on Greek exports from Italy. Table 15.3 provides useful information about such dependence, which has been dominant for the past 15 years. An encouraging element of this table is the significant increase in the number of trading partners, which has doubled over this period. This tendency proves that an autonomous attempt to gain international market shares is feasible; the only precondition is the will to proceed among Greek production and trading initiatives.

Table 15.3 Geographical diversification trends

Year	Number of trade partners	Exports in tonnes		
		Total	**Italy**	
1990	30	85.556	70.718	83%
1991	33	51.750	43.572	84%
1992	42	155.972	129.455	83%
1993	46	98.935	84.521	85%
1994	46	98.731	79.465	80%
1995	48	133.894	93.024	69%
1996	48	124.209	94.160	76%
1997	48	97.732	87.376	89%
1998	52	107.339	92.857	87%
1999	65	152.706	109.578	72%
2000	60	104.106	84.065	81%
2001	59	178.122	93.948	53%
2002	60	73.474	60.221	82%
2003	60	96.704	78.177	81%
2004	56	63.294	48.327	76%
2005	66	98.825	79.012	80%

Source: FAOSTAT, detailed trade matrix, http://faostat.fao.org.

PRICE

The determination of a final consumer price is not an easy task, because many controversial parameters must be taken into account. Owing to the internal problems the olive oil production sector faces, it is almost impossible for it to adopt a price leadership strategy. The cost of production is higher, compared with competitive products produced in the

Mediterranean area, so a differentiation strategy based on product authenticity is more appropriate. A price level correlation must exist among similar products, because of the comparisons made before the final purchase decision. Research in these target markets should evaluate and quantify the maximum difference in price between Greek olive oil and others that will motivate consumers to include it in their diet. Similar attempts generally minimise this difference in the short term, to encourage consumers to purchase, then increase the difference in the long term, when the trading environment is safer, to achieve profit maximisation.

PROMOTION

The success of an expansion into foreign markets in which olive oil is largely unknown requires a well-planned and implemented promotion mix. The target consumer group should be informed about the product's characteristics that make it distinctive. Such a campaign can achieve increased loyalty to the specific product and a low level of competition. A successful promotion also will reduce misunderstandings and the entrepreneurial risk that usually accompanies such attempts.

One of the most effective ways of introducing a product to consumers is through television. Commercial advertisements that provide information about original olive oil should persuade significant consumer groups to try the product. Furthermore, ads that emphasise the origin and traditional processing procedure should motivate consumers to include olive oil in their diets and increase sales. Another effective promotion relies on offers, usually implemented through discounts or extra product at no additional charge. Information about how to use olive oil also should be accessible to consumers in written recipes on packaging or leaflets with relevant information distributed in places the product is sold. Participation in food exhibitions and advertisements or articles in food-oriented magazines can be helpful as well. Finally, producers should cooperate closely with tourism experts, because tourism and sales of traditional products are significantly interrelated parameters.

As this analysis reveals, increased protection in the EU market does not improve the trading performance of olive oil. Much remains to be done at both production and market levels to reduce the intensity of rivals and gradually increase market shares globally.

References

1. Babcock, B. (2003), 'Geographical indications, property rights and value added agriculture', Review paper (IAR 9:4:1–3). Centre for Agricultural and Rural Development, Iowa State University, Ames, Iowa.
2. Kotler, P. (1994), *Marketing Management: Analysis, Planning, Implementation and Control*, Eighth edn, Prentice Hall, Englewood Cliffs, NJ, US.
3. An interesting review of literature pertaining to the gravity model appears in Martinez-Zarzoso and Nowak-Lehmann (2003), 'Augmented gravity model: An empirical application to Mercosur-European Union Trade Flows', *Journal of Applied Economics*, vol. 6, no. 2, pp. 291–316. See also Bun, M. and Klaassen F. (2002), 'The importance of dynamics in panel gravity models of trade', working paper, University of Amsterdam, Faculty of Economics and Econometrics.

4. Tinbergen, J. (1962), *Shaping the World Economy. Suggestions for an International Economic Policy*, New York.

5. Linnemann, H. (1966), *An Econometric Study of International Trade Flows*, Amsterdam.

6. The model was developed by Vlontzos, G. and Duquenne M.N. (2008), 'Evolution of trade flows for olive oil: An empirical model for Greece', http://aede.osu.edu/programs/Anderson/trade/VlontzosOliveoil.pdf.

7. General Secretariat of National Statistical Service of Greece (2007), Primary Sector, Annual Agricultural Survey, Athens, Greece.

8. Commission of the European Communities (2008), 'Protected Designation of Origin (PDO)/Protected Geographical Indication (PGI) list', http://ec.europa.eu/agriculture/qual/en/el_en.htm.

9. World Trade Organisation (2005), TN/IP/W/11 document, Cancun.

10. FAO, (2006), 'Production of olive oil', http://faostat.fao.org/site/567/default.aspx.

11. Vlontzos and Duquenne, op. cit.

16 Organic Wine: Perceptions and Choices of Italian Consumers

BY MARCO PLATANIA* AND DONATELLA PRIVITERA†

Keywords

organic wine, consumer behaviour, monovariate analysis

Abstract

Organic agriculture as a cultural evolution originated in an environmentalist culture. Focus on organic products derived from demand for healthy foods with high quality standards that limited the use of chemical substances. This chapter analyses knowledge about organic product consumption in general and wine consumption in particular. The resulting framework for organic wine shows that the sector is still expanding, though organic wine consumption remains limited to a few consumers. Not only are these consumers few in number, but they seem less affected by the organic certification in terms of their choice; it is just another factor, along with place of origin and brand, to consider. For this type of product at least, the value connected with the organic product concept appears to have unexpressed potential.

Introduction and Theoretical Background

Although organic agriculture has not been well supported by the Italian government, it provides the main drawing power for Italian agriculture. It is no longer a niche product, and its selling points – healthiness and respect for the environment – are widely taken for granted by consumers.

*   Dr Marco Platania, University of Catania, Via Biblioteca 4, Palazzo Ingrassia, 95124 Catania, Italy. E-mail: marco.platania@unict.it.

†   Dr Donatella Privitera, University of Catania, Via Biblioteca 4, Palazzo Ingrassia, 95124 Catania, Italy. E-mail: donatella.privitera@unict.it.

Italy is the first European nation in terms of its number of organic operators, with more than 1 million cultivated hectares, equal to 8 per cent of the usable agricultural area (UAA) (Reg. CEE 2092/91). According to the most recent census returns, 50,000 operators function in this sector – 45,000 farmers and 5,000 manufacturers. Moreover, 17 certifying organisations authorised by the Minister of Agriculture operate in this field.

The market analysis of organic product consumption reveals that it represents roughly 2 per cent of all domestic food. According to some estimates, Italian organic production potential is not confined to the internal market but also includes 33 per cent for export (roughly 400 million Euros) and 10 per cent for the uncertified market, with a significant portion dedicated to animal nutrition.

As a cultural evolution, organic agriculture originated in an environmentalist culture that has been well-established for at least 20 years. Furthermore, the focus on these products reflects the great demand for healthy foods with high quality standards[‡] and limited use of chemical substances. This chapter studies organic consumers to determine their evaluation criteria for such products. In particular, we consider organic product consumption in Italy.

Wine an agricultural product that is emblematic of Italian agri-food production, so we study how consumers choose wine and, in particular, organic wine.[§] Organic wine is appreciated for its taste as much as are conventional types of wine and may be considered more 'genuine.' This latter attribute justifies a 10–20 per cent price premium, which also takes into account the period necessary to obtain organic certification. Interest in these products, however, relates to renewed demand for food and drink that is safe from a hygienic and sanitary point of view, with a qualitatively superior standard, that is less harmful to health, and that uses fewer chemical substances. The results of this study indicate how well the production process of organic wine is known.

Significant research investigates the sustainability of organic agriculture in the Italian and international markets from economic, environmental and health points of view. Many studies, albeit qualitative, focus on consumers and their preferences for organic production;[2] the bibliography of quantitative studies is somewhat shorter.[3] The first consumer studies date from the 1980s and display various points of view. Sandalidou, Baourakis and Siskous deduce that consumers need more information, particularly on the label and packaging, because of the higher price for products considered 'healthy.'[4] According to Poelman and colleagues, the influence and incidence of information is such that it can lead to different perceptions of organic products, though they depend on positive or negative attitudes of consumers toward such products.[5]

Other authors note the elasticity of demand for these products with respect to price and income, indicating that economic factors have greater influence than they do for conventional products, though to varying extents according to the product (e.g., greater elasticity for dairy products than for cereals).[6] Price is an important factor in consumers' choice of foodstuffs, and high prices may lead to negative attitudes toward organic products in particular.[7] Moreover, organic products often are considered less innovative

---

[‡]   We use the generic definition of organic standards, that is, 'ecologically farmed land and a careful eye toward soil fertility maintenance.'[1]

[§]   When we refer to organic wine, we mean wine produced with grapes of biological production. In Italy, it complies with European legislation on biological agriculture (Rule EEC 2092/91). At the international level, there are some differences in legislation between Europe and the United States and the norms Ifoam and Codex Alimentarius. However, consumers rarely understand the true meaning of biological wine.

and more traditional than, for example, cereal-based products and otherwise functional foods and drinks.[8]

Regarding specific products, such as wine in the Italian and international markets, many market and distribution studies attempt to quantify exchanges and their challenges.[9] To a lesser extent, analyses investigate consumer preferences for the products.[10] Many authors also suggest that the country of origin is of primary importance for consumers' purchase choices for conventional wine,[11] but according to Thrupp, the choice of organic wine depends on both an excellent taste and the intrinsic value of safeguarding the environment.[12] This latter attribute justifies the 10–20 per cent higher price that reflects the time necessary to obtain an organic certificate. Consumers cannot taste a wine before purchase and must therefore base their choice on information provided on the label and bottle, including the brand, region, price and awards won.[13]

Moreover, consumers are becoming more and more familiar with the importance of appreciating and protecting typical products such as wine from an insidious globalised market, which respects neither nature's natural rhythms nor the particularities of different areas, as well as associating such products with the richness of biodiversity.[14]

Methodology

The data come from a sample of 1,000 families,[¶] representative of the Italian population. The research consisted of three phases. First, qualitative research based on focus groups with consumers identified the attributes to be used in the analysis of consumer choices. These results led to a greater understanding of some important aspects of organic products in general and wine in particular, which enabled us to choose the critical aspects on which to base the questions. Second, we created the questionnaire and conducted a pre-test with 100 respondents. Third, the last phase consisted of gathering and elaborating the data.

The topics in the questionnaire include both a part about organic products in general and a more specific part relating to organic wine. Both sections collect information about consumers' beliefs and attitudes toward organic products. The respondents viewed a matrix with a set of attributes (material and immaterial)[**] that can be categorised as relating to alimentary, aesthetic, environmental, or comparative values (Table 16.1). They then provided ratings between 1 (not at all in agreement) and 5 (in total agreement).

The section pertaining to organic wine collects information about purchase choices, including conventional wine. The questions relate to consumption and corresponding choice criteria, ordered according to the focus attention to information while buying.

The questionnaires were completed online during March 2006 by the head of the household, who logged on to the data centre of the market research agency every week, in accord with an existing agreement. Respondents took approximately 25–30 minutes to complete the questionnaire. We use descriptive statistics techniques to summarise and

---

¶   The sample comes from a panel run by Customized Research Analysis, a company that specialises in market research. The reference pool consists of 3,500 Italian families (9,520 individuals) that can be divided into samples of various sizes. The study sample is stratified at both family (20 Italian regions; community sizes from fewer than 5,000 to more than 100,000 inhabitants; nuclear family sizes from 1 to 5) and individual (sex, age, education, occupation) levels.

**   The material attributes refer to organolettiches characteristics, whereas immaterial attributes refer to qualitative aspects.

Table 16.1 List of evaluation attributes for organic products

Attribute 1	Safer for health
Attribute 2	More controlled
Attribute 3	Produced more traditionally
Attribute 4	Have taste of the past
Attribute 5	Produced by serious and professional firms
Attribute 6	No preservatives
Attribute 7	More genuine
Attribute 8	Good appearance
Attribute 9	No GMOs (genetically modified organisms)
Attribute 10	Do not damage the environment
Attribute 11	No different from other products
Attribute 12	Too expensive

present the data and adopt a monovariate analysis technique to describe consumers' preference systems and choice processes.

Analysis of Data and Main Results

ORGANIC PRODUCT PERCEPTION

With 864 valid questionnaires collected, the sampling methods made it possible to form a representative sample of the Italian population (Table 16.2). The sample contains more female respondents (63.8 per cent), and the most frequent age range is 35–44 years (32.6 per cent). A significant number of consumers have a medium to high level of education (secondary school and university together represent about 60 per cent of the sample), and the largest occupation groups are white-collar workers (36.1 per cent) and housewives (23.4 per cent). Almost 38 per cent of the respondents earned a maximum monthly income of €1.750, and there were usually four members in the family.

The first results indicate the percentage of consumers who are acquainted with organic products. When asked if they knew these products, 735 consumers (85 per cent) answered that they did. This high response rate indicates that organic products are part of consumers' lives.

For the collection of general information about consumer choices, we considered it useful to find out more about the level of information about products related to organic ones (Table 16.3), such as those produced from integrated, genetically modified organism (GMO)-free, and zero residue agriculture. Even 15 years after organic certification was introduced in the EU, great confusion still exists among consumers with regard to these categories of products.

Table 16.2 Demographic characteristics of respondents

Variables	n.	%	Variables	n.	%
Sex			**Occupation**		
Male	313	36.2	Self-employed	104	12.0
Female	551	63.8	White-collar worker	312	36.1
Total	864	100.0	Blue-collar worker	118	13.7
			Housewife	202	23.4
Age			Pensioner	95	11.0
Up to 34	76	8.8	Unemployed	25	2.9
35–44	282	32.6	Other	8	0.9
45–54	262	30.3	Total	864	100.0
55–64	161	18.6			
65 or over	83	9.6	**Education**		
Total	864	100.0	Primary school	64	7.4
			Middle school	286	33.1
			Secondary school	405	46.9
Social-economic level			University degree	109	12.6
Low	253	29.3	Total	864	100.0
Medium low	167	19.3			
Medium	223	25.8	*Monthly income*		
Medium high	134	15.5	Up to 1250 euros	181	20.9
High	87	10.1	1251–1750 euros	147	17.0
Total	864	100.0	1751–2250 euros	100	11.6
			above 2250 euros	159	18.4
			Not declared	277	32.1
			Total	864	100.0

The results indicate a very low level of knowledge of the meaning of these alternative products: Only 1.5 per cent recognise the meaning of food produced from zero-residue agriculture and 3.6 per cent recognise food from integrated agriculture. There appears to be a greater awareness of GMO-free products (26 per cent, though 23.4 per cent of the respondents said they were not aware of them), perhaps because the controversial debate

Table 16.3 Knowledge of possible competitors for organic products (per cent value)

	Food from integrated agriculture	Food from zero residue Agriculture	GMO-free food
I don't know them	49.2	70.9	23.4
I have heard about them at some time but I have no idea what they are	25.6	16.5	22.9
I've heard about them and know the meaning	11.9	6.4	26.4
I know them but not very well	9.8	4.6	17.5
I know them well	3.5	1.6	9.8

about GMO[††] products is very public. The number of competitors of organic products appears to be increasing the existing difficulty consumers have in understanding these products.

The results of the interviews about the characteristics of organic products indicate that the image of a genuine product with no preservatives is most important to consumers (Figure 16.1). Comparing organic products with conventional ones, they prefer the former, which they consider to have greater intrinsic and extrinsic value. These respondents perceive organic products as 'different' and worthy of greater attention. In other terms, consumers see 'organic' as a synonym for a product that guarantees greater respect for the environment and attention to their health. The consequence of this 'therapeutic' image may be a justification for a higher price.

CHARACTERISTICS OF ORGANIC WINE CONSUMERS

In the study of choice criteria for organic wine,[‡‡] we first wanted to establish the consumption frequency. The results show that the majority of consumers (80 per cent) state they do not consume organic wine; only 2 per cent say they consume it almost exclusively (Table 16.4).

This slight preference reflects the information that consumers consider important to use during purchase. In the interviews, consumers ordered, from the most important to the least, the information they use to make purchase choices. They prefer to choose wine according to its denomination (45 per cent) and the presence of quality certification, which gives the product a recognisable identity, an important factor for this type of product. The organic nature of the product ranks after these considerations (4.3 per cent).

[††] The worldwide debate over GMO products is heated; in Italy in particular, the issue has created great doubt and perplexity – not least because it contrasts with that country's general strategy to promote traditional, small to medium-sized enterprises that typify the Italian food industry and produce 'genuine', traceable products with profound links to Italian territory, history and culture.

[‡‡] As already noted, during the interview, consumers had little understanding of organic wine; most cannot describe how such a product can be realised.

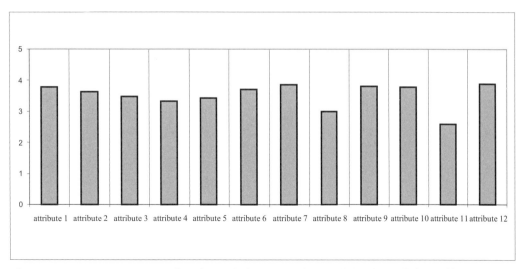

Figure 16.1 Consumer evaluation of the material and immaterial attributes of organic products (average values)

Table 16.4 Frequency of organic wine consumption (per cent value)

I almost exclusively drink it	2.0
I usually drink it	4.0
I drink it sometimes	14.0
I do not drink it	80.0
Total	**100.0**

This attitude also occurs among consumers who are particularly sensitive to the idea of organic products. When we examine the choices of organic consumers, so defined by their purchase frequency (i.e., who state they buy organic products frequently), we find that few of them attribute the greatest importance to the organic nature of the wine (Table 16.5). For the majority, the most important criterion is the name of the species of vine, followed by quality certification. These considerations precede area of origin, the name of the producers' vineyard, organic certification, and, finally, price.

To examine the links between the purchase choice variables and the purchase and consumption of organic wine, we used Kendall's tau-b, a nonparametric measure of association for ordinal or ranked variables that takes ties into account. Kendall's tau is probably the most frequently used ordinal measure of association. It measures 'the extent to which a change in one variable is accompanied by a change in another variable.'[15] The sign of the coefficient indicates the direction of the relationship, and its absolute value indicates the strength, with larger absolute values indicating stronger relationships. If the value of this indicator is 0.7 or above, there is a high relationship, a value between 0.3 and 0.7 indicates a moderate relationship, and a value below 0.3 indicates a modest relationship.[16]

We compared the consumption frequency of organic wine with some variables related to consumption behaviour toward organic overall products and the socio-economic characteristics of the sample. We find no significant relationship with the latter, which

Table 16.5   Procedural behaviour of consumers who often buy organic products

	Choice Frequency (number of consumers)					
Choice Criteria	**1**	**2**	**3**	**4**	**5**	**6**
Designation (e.g., Chianti, Barolo)	14	7	7	5	4	9
Presence of quality certificate (e.g., IGP, DOC)	8	11	3	11	4	9
Area of origin	8	7	14	7	4	6
Producer	4	6	6	12	12	6
Organic producer	6	7	9	7	13	4
Price	6	8	7	4	9	12

confirms that organic wine consumption does not appear linked to socio-economic characteristics but rather to lifestyle.

The most interesting correlations appear in Table 16.6. There are only two results that indicate a moderate relationship; the others indicate modest relationships.

Specifically, we find a positive correlation between the variable relating to the consumption frequency of organic wine and the consumption of organic extra-virgin olive oil, as well as the proportion of food expenditures for organic products, which is to be expected. The correlation between these variables seems to indicate that a consumer who chooses organic alimentation also tends to consume organic oil and wine, products that have notable effects on their total expenditures. More modest values occur for the purchase frequency of organic and GMO-free products.

To determine some of the reasons for organic wine consumption, we group those categories that stated they usually or exclusively bought such products. These consumers represent 6 per cent of the sample, or 46 respondents, and we base the following analyses on this group of 'faithful' consumers. In this group, 63 per cent are women and 41 per

Table 16.6   Kendall correlation coefficients between organic wine consumption and consumer behaviour variables

Questions	tau-b	Significance
Sex	-0.087	0.019
Age of purchaser	0.079	0.015
Organic product purchase frequency	0.190	0.000
Proportion of food expenditure for organic products	0.297	0.000
Purchase of GMO free products	0.202	0.000
Organic extra-virgin oil consumption	0.536	0.000

Notes: The value of the coefficient tau-b ranges from −1 to 1. A value of 0 indicates no correlation, positive values indicate a direct correlation, and negative values indicate an inverse correlation.

cent are white-collar workers. The majority have a medium level of education (41.3 per cent middle school, 43.4 per cent secondary school), and 39.1 per cent belong to the medium socio-economic category (Table 16.7).

We again decided to examine their preference system and choice of organic products in general to assess whether the choice of organic wine linked to their general preference for organic products or depended on other reasons. For this group, the presence of organic products in the shopping basket has contradictory meanings: Some consumers (a little more than 33 per cent) buy organic products at least once a week, whereas others buy them very rarely or never (Table 16.8).

This contradictory attitude also emerges with regard to the 'duration of consumption,' that is, how long these consumers have been buying organic products. More than one-quarter (26.8 per cent) of the consumers say they have bought these products for many years, and just a few affirm that they only started buying them recently

Table 16.7 Main features of 'faithful consumers' interviewed

Variables	n.	per cent	Variables	n.	per cent
Sex			**Occupation**		
Male	17	36.9	Self-employed	4	8.7
Female	29	63.1	White-collar worker	19	41.3
Total	46	100.00	Blue-collar worker	6	13.0
			Housewife	13	28.3
Age			Pensioner	4	8.7
Below 35	3	6.5	Total	46	100.0
35–44	12	26.1			
45–54	15	32.6	**Education**		
55–64	11	23.9	Primary school	3	6.5
65 and over	5	10.9	Middle school	19	41.3
Total	46	100.0	Secondary school	20	43.5
			University degree	4	8.7
			Total	46	100.0
Social-economic level					
Low	8	17.4	**Monthly income**		
Medium-low	6	13.0	Up to 1,250 euros	9	19.6
Medium	18	39.1	1,251–1,750 euros	6	13.0
Medium high	11	23.9	1,751–2,250 euros	4	8.7
High	3	6.5	above 2,250 euros	11	23.9
Total	46	100.00	Not declared	16	34.8
			Total	46	100.0

Table 16.8 Purchase and consumption of organic products by 'faithful consumers'

How often do you personally buy organic products for yourself or your family?

Answers:	Frequency	Percent
More than once a week	5	10.86
Once a week	10	21.73
Between once and three times a month	7	15.21
Less than once a month	7	15.21
Never	9	19.56
I don't remember	8	17.39
Total	46	100.00

How often have you been buying products derived from organic agriculture?

Answers:	Frequency	Percent
A few months	3	6.52
About a year	7	15.21
More than 3 years	7	15.21
Far more than 3 years	12	26.08
I don't remember	8	17.39
I don't personally buy organic products	9	19.56
Total	46	100.00

What proportion of your food and drink expenditure is used for organic products?

Answers:	Frequency	Percent
Zero	1	2.17
Less than 20 per cent	20	43.47
From 20 to 50 per cent	10	21.73
More than 50 per cent	5	10.86
100 per cent	1	2.17
I don't personally buy organic products	9	19.56
Total	46	100.00

Finally, their answers to the question about the proportion of total food expenditures spent on organic products show that few consumers' choice of food is totally governed by avoiding artificial versions of products. Nearly half spend 20 per cent of their food budget on organic products.

Conclusions and Implications

The results obtained in this study contribute to an analysis of organic product perception, particularly that of wine. Organic agriculture products have been in Italian consumers' shopping baskets for several years, but they also have become more and more popular and an integral part of the lives of some categories of consumers, even if their significance and value is not yet entirely clear.

In this empirical survey, in which a sample representative of the Italian population completed a questionnaire, we assessed knowledge of organic products, especially in comparison with other, similar products (e.g., integrated agriculture, zero-residue and GMO-free products). The results indicate a low level of knowledge and a great deal of confusion. In terms of their purchase motivation, consumers seem most interested in the greater genuineness and healthiness of these products, as well as their lower impact on the environment. The high price of organic products is, however, an important factor, though it is mitigated by the healthy characteristics of the product.

This chapter focuses in particular on the appreciation and consumption of an emblematic Italian agri-food product: organic wine. The results of the analysis of the extent to which it is known and the factors governing consumers' purchase choices lead to some brief considerations. First, organic wine consumption is limited to a small number of consumers. Not only are they few in number, but the organic certification does not seem to influence their choice as much as the place of origin or brand. For this type of product at least, the value connected with the organic product concept appears to have unexpressed potential. Mostly, organic wine gets purchased with other products (e.g., organic oil) because of consumers a healthy alimentary philosophy.

Second, in Italy, there are excellent wines on the market from both organoleptic and qualitative points of view, and these wines charge high prices. They also compete with wines derived from organic agriculture, along with other substitutive products. The ineffective information communicated by producers about this type of product thus likely contributes to slowing the development of a sector that remains in expansion but perhaps could expand even faster.

Acknowledgments

This paper was developed within the aegis of the Relevant National Interest Projects (PRIN), 2004, 'Modern distribution, food safety: the perspectives of the organic agriculture in Italy'. The paper is the result of collaboration between the two authors. Marco Platania wrote the 'Methodology' and 'Characteristics of Organic Wine consumers' sections, and Donatella Privitera wrote the 'Introduction and Background' and 'Organic product perception' sections. The 'Conclusions and Implications' section is a joint contribution.

References

1. Dimitri, C. and Oberholtzer, L. (2005), 'Development of the U.S. and EU organic agricultural sectors', Electronic Outlook Report from the Economic Research Service, WRS-05–05, August 2005.

2.   Cicia, G., Del Giudice, T., and Scarpa, R. (2002), 'Consumers' perception of quality in organic food, a random utility model under preference heterogeneity and choice correlation from rank-orderings', *British Food Journal*, vol. 104, pp. 200–213.

3.   Pinton, R. (2007), 'Niente crolli o boom, il biologico viaggia con una crescita costante', *AZBIO*, No. 1/2.

4.   Sandalidou, E., Baourakis, G., and Siskous, Y. (2002), 'Customer' perspectives on the quality of organic olive oil in Greece', *British Food Journal*, vol. 104, pp. 391–406.

5.   Poelman, A., Mojet, J., Lyon, D., and Sefa-Dedeh, S. (2008), 'The influence of information about organic production and fair trade on preferences for and perception of pineapple', *Food Quality and Preference*, vol. 19, pp. 114–21.

6.   Wier, M. and Calverley, C. (2002), 'Market potential for organic foods in Europe', *British Food Journal*, vol. 104, no. 1, pp. 45–62; Mauracher, C. (2007), 'Analisi delle determinanti della domanda di prodotti biologici. Aspetti teorici ed evidenze empiriche', in Cicia, G. and de Stefano, F. (eds), *Prospettive dell'agricoltura biologica in Italia*, Edizione Scientifiche Italiane, Napoli.

7.   Chen, M. (2007), 'Consumer attitudes and purchase intentions in relation to organic foods in Taiwan: Moderating effects of food-related personality traits', *Food Quality and Preference*, vol. 18, pp. 1008–1021.

8.   Huotilainen, A., Pirtila-backman, A-M., and Tuorila, H. (2006), 'How innovativeness relates to social representation of new foods and to the willingness to try and use such foods', *Food Quality and Preference*, vol. 17, pp. 353–61.

9.   Zanoli, R. (1997), 'La filiera del vino di agricoltura biologica', in Cantucci, F. (ed.), *Le filiere del biologico*, Università di Perugia, GRABIT; Sharples, L. (2000), 'Organic wines–the UK market: A shift from 'niche market' to 'main stream' position?', *International Journal of Wine*, vol. 12, No. 1, pp. 30–41; Crescimanno, M., Ficani, G.B. and Guccione, G. (2002), 'The production and marketing of organic wine in Sicily', *British Food Journal*, vol. 104, no. 3–5, pp. 274–86.

10. Cicia, G. and Perla, C. (2000), 'La percezione della qualità nei consumatori di prodotti biologici: uno studio sull'olio extra-vergine di oliva tramite conjoint analysis', in de Stefano, F. (ed.), *Qualità e valorizzazione nel mercato dei prodotti agroalimentari tipici*, Edizione Scientifiche Italiane, Napoli; Cicia et al., op. cit.

11. Dean, R. (2002), 'The changing world of the international fine wine market', *The Australian and New Zealand Wine Industry Journal*, No. May/June, pp. 84–8; Koewen, C. and Casey, M. (1995), 'Purchasing behaviour in the Northern Ireland wine market', *British Food Journal*, vol. 97, No. 11, pp. 7–20; Skuras, D. and Vajrou, A. (2002), 'Consumer's willingness to pay for origin labelled wine: A Greek case study', *British Food Journal*, vol. 104, no. 11, pp. 898–912.

12. Thrupp A. (2004), 'Market opportunities and challenges for organic and sustainably grown wines', available at http://www.vineyardteam.org/pdf/Econ04_thrupp.pdf (accessed March 2008).

13. Lockshin, L., Jarvis, W., d'Hauteville, F., and Perrouty, J.-P. (2006), 'Using simulations from discrete choice experiments to measure consumer sensitivity to brand, region, price and awards in wine choice', *Food Quality and Preference*, vol. 17, pp. 166–78.

14. Prosperi, L. (2006), 'La scelta del consumo biologico, tra passato e futuro: ragioni e suggestioni di una nuova coscienza alimentare', in De Lorenzo, A. and Di Renzo, L. (eds), *Nutrire per prevenire. Quali indicatori di rischio nutrizionale?* INEA.

15. Weisberg, H.F., Krosnick, J.A., and Bowen, B.D. (1996), *An Introduction to Survey Research, Polling, and Data Analysis*, Sage Publications, Newbury Park, CA.

16. Ibid.

PART IV

Market Orientation For Specialty Products

CHAPTER 17

Consumer Values and the Choice of Specialty Foods: The Case of the Oliva Ascolana del Piceno (Protected Designation of Origin)

BY ALESSIO CAVICCHI[*] AND ARMANDO MARIA CORSI[†]

Keywords

theory of consumption values, consumer behaviour, focus groups, specialty food, protected designation of origin

'I have no special talents. I am only passionately curious.'

Albert Einstein

Abstract

This work aims to stress the main values considered by consumers as determinants of their market choice behaviour, with specific regard to the Oliva Ascolana del Piceno, a protected designation of origin of the Le Marche and Abruzzi regions (Central Italy). In this chapter, we will:

- introduce the research topic and the background of the research;

---

[*]   Dr Alessio Cavicchi, Department of Studies on Economic Development, University of Macerata, Piazza Oberdan 3, 62100 Macerata, Italy. E-mail: a.cavicchi@unimc.it. Telephone: + 39 0733 258 3919.

[†]   Mr Armando Maria Corsi, Department of Agricultural and Resource Economics, University of Florence, Piazzale delle Cascine 18, 50144 Florence, Italy. E-mail: armando.corsi@unifi.it. Telephone: +39 055 3288 220.

- describe the Italian market of protected designation of origin and protected geographical indication products, with particular attention to the Oliva Ascolana del Piceno;
- describe the theory of consumption values, optimal stimulation values, and the reasons we focus on the epistemic value of a product;
- describe the methodology used;
- propose and discuss the main findings of the research.

Introduction and Background to the Research

Italian farmers traditionally have achieved a high level of quality in agricultural products. In the past few years, the common agricultural policy, among other factors, has pushed farmers toward a market-oriented production and a greater consideration of consumers' needs.

The concept of market orientation developed in the early 1990s, thanks to works by Narver and Slater,[1] and of Kohli and Jaworski.[2] Taking into account the definition given by these authors, market orientation is 'the organizational culture that most effectively and efficiently creates the necessary behaviors for the creation of superior value for buyers and, thus, superior performance for businesses.'[3] In their view, market orientation embodies three behavioural components: customer orientation, competitor orientation, and inter-functional coordination.[4] Continuous innovation is present latently in each of these elements,[5] and the main objectives are to generate a positive, long-lasting effect on firms[6] and to help them gain a successful competitive advantage.[7]

The new European agricultural policy stimulated a new, entrepreneurial approach to farm management, aimed at the most profitable market segments. From the European standpoint, 2007 was the beginning of a new programming period, according to the rules and norms expressed in the European Community regulation no. 1698/2005. The first of the three facets around which this regulation is structured covers an improvement in competitiveness. In this context, quality has primary importance; consequently, an entire subsection is devoted to the 'quality of agricultural production.' The related measures, adopted at the regional level through the 'rural development plans,' aim at improving the quality of agricultural products by:

- helping farmers adapt to the demanding standards of the European Community legislation;
- supporting farmers who participate in food quality schemes;
- supporting producer groups in information and promotion activities for products within food quality schemes.

Another relevant issue in the evolution of marketing systems is the Leader initiative. Leader is a French acronym for 'Liaison Entre Actions de Développement de l'Économie Rurale,' meaning 'links between the rural economy and development actions.'[8] The Leader initiative was launched for the first time in 1991. Since then, three Leader initiatives have followed: Leader I (1991–93), Leader II (1994–99), and Leader Plus (2000–2007). The most interesting change brought about by European Community regulation no. 1698/2005 to the Leader initiative is that the latter is not subject to a specific communitarian communication anymore, though the project became the fourth axis of the main regulation.[9]

This initiative has reached a level of maturity that has enabled rural areas to implement the Leader approach more widely in mainstream rural development programmes. This endeavour guarantees an organic approach in rural development projects and calls attention to the need for European agricultural funds for rural development to complement national, regional and local actions that contribute to the Community's priorities on one side and ensure the coordination between various funding aids (e.g., European regional development fund, European social fund, Community fund) and other Community financial instruments on the other side. To support this strategic approach, the regulation also covers vocational training and information actions and defines local strategies of development. Moreover, the regulation integrates a specific measure for the diffusion of scientific knowledge and innovative practices (European Community regulation no. 1698/2005 defines four axes, each of which represents one of the four macro areas of intervention[10]). Furthermore, the regulation proposes regional policies aimed at activating and pursuing strategies for the implementation and communication of quality products.

A third topic of discussion is that in a globalised market, the proliferation of different sets of standards for quality certification lead to consumers' confusion, and consequently both public authorities and entrepreneurs must communicate the specifics about each scheme and certified food product, because 'if there are too many differentiated products relative to the degree of consumer heterogeneity, or to the consumer's capacity to understand and evaluate the array of different product types offered, then the outcome will be sub-optimal.'[11] Thus, especially with a differentiation based on 'region of origin,' the proliferation of protected designations of origin, and protected geographical indication labels, there is a need to capitalise on the unique features of regional products. Which consumer segments can be targeted by managers? What motivates consumer choice? Which characteristics of the product can be emphasised? Questions like these are fundamental for a successful marketing approach and an efficient use of public funds.

If we compare the definitions of market orientation with the preceding series of norms and regulations, we find that the most fundamental change in the past few years is the shift from a focus on production to a market-based orientation.

In light of this view, the object of this study is consumers' perceptions of the Oliva Ascolana del Piceno (a product with a protected designation of origin label). As reported by European Community regulation no. 1855/2005, this name denotes the *ascolana tenera* variety of olive, either in brine or stuffed. The area of production for this protected designation of origin covers certain municipalities in the provinces of Ascoli Piceno and Teramo, located in the Le Marche and Abruzzi regions. The original recipe for the stuffing can be traced back to the time of the unification of the Kingdom of Italy (1859–61). The Oliva Ascolana del Piceno is a product underpinned by quality and an excellent reputation, owing to its age-old history and deeply rooted tradition (see the application for the request of the protected designation of origin label no. IT/00331/04.12.2003).

This exploratory chapter aims first to describe the current situation of protected designation of origin and protected geographical indication products in Italy, with a specific emphasis on the Oliva Ascolana del Piceno. A literature review on the theory of consumption values[12] follows. The methodology of the research and the results relating to the main factors driving the behaviour of the respondents appears next. Finally, this chapter suggests implications for practitioners, along with some conclusions, limitations and guidelines for further research.

Italian Market of Protected Designation of Origin and Protected Geographical Indication Products

The EU policy regarding protected designation of origin and protected geographical indication products allows the collective organisation of producers to benefit from a rent that is similar to the reputation rent of a private brand. Therefore, the regulator enables producers to reap the benefits of the reputation rent without incurring the costs that a private company incurs when striving to establish the reputation of its commercial brand name.[13]

For this reason, there has been a proliferation of protected designation of origin and protected geographical indication products in Italy in recent years. In July 2008, Italy led the European ranking unchallenged, with 171 registered products,[14] 16 more than those recognised in December 2006.[15] The fruit and vegetables sector covers 34 per cent of the denominations, followed by oils and fats (22 per cent), cheeses (21 per cent), and meat-based products (17 per cent).[16] For the year ending October 2007, Italian certified products represented 21 per cent of the total Communitarian basket; France followed with 155 denominations (20 per cent), and then Spain with 114 (14 per cent), and Portugal with 104 (13 per cent).[17]

For the year ending December 2007, Italy counted 111 protected designations of origin (67 per cent) and 54 protected geographical indications (33 per cent).[18] In the past decade, protected designation of origin and protected geographical indication products nearly doubled, with the fruit and vegetable sector as the 'denominated' leader, with 58 labels.[19] The Italian Institute for Services in Food and Agricultural Market (ISMEA) reports that in February 2007, 45.8 per cent of the Italian denominations were concentrated in Northern Italy, followed by Southern Italy and the Islands (32.7 per cent) and Central Italy (21.5 per cent). However, there was a considerable increase in the number of protected designation of origin and protected geographical indication products of Central Italy, mainly to the detriment of Northern Italy, while Southern Italy and the Islands remained stable.[20] Moreover, firms producing certified products decreased in Central Italy since 2004 by 13.6 per cent, but they increased in Northern (+12.4 per cent) and Southern (+21.3 per cent) Italy. A similar situation occurred with transformers, which decreased in Central Italy (-15.8 per cent) but increased in Northern (+17.5 per cent) and Southern (+6.9 per cent) Italy. The number of cattle breeders escalated considerably in Southern Italy (+610 per cent); smaller improvements occurred in Northern (+14.5 per cent) and Central (+11.5 per cent) Italy. Finally, the number of hectares dedicated to certified productions increased significantly in Northern Italy (+78 per cent), decreased in Central Italy (-5.7 per cent), and slightly increased in Southern Italy (+9.5 per cent).[21]

Protected designation of origin and protected geographical indication products play a significant role in the Italian economic system. In 2007, these products produced a total turnover of €5 billion and a value of production and consumption of €9 billion, which correspond to increases of 7.9 per cent and 5.4 per cent, respectively, compared with the previous year.[22]

The consumption of Italian protected designation of origin and protected geographical indication products is mainly located in the domestic market, which absorbs 77 per cent of the total production value, with cheeses and meat-based products as the current product leaders. Only major productions, such as Grana Padano or Parma ham, exhibit slightly lower percentages, as their worldwide fame and supplying capacity allow them to cover domestic and foreign demand.[23] These data show that the presence of specialty food abroad is largely underdeveloped, mainly due to the excessive fragmentation of Italian food firms and the scarce presence of Italian grocery chains in foreign markets – contrary to the practices in France (e.g., Carrefour, Leclerc, Auchan) or Great Britain (e.g., Tesco, Sainsbury's, Asda). Beyond these structural deficiencies, we highlight the so-called 'agripiracy' phenomenon, which causes an annual estimated loss of more than €1.5 billion. This term refers to the practice of producing products that try to imitate original protected designation of origin and protected geographical indication products; for example, the most 'imitated' product is Parmigiano Reggiano, whose strange and false relatives are present in Argentina (Regianito), Brazil (Parmesao), and the United States (Parmesan), to name a few. However, beyond the simple though severe damage caused by falsification, 'Italian-sounding' products seem even more dangerous.[24] This threat is more complex and lasting, as it refers to the use of names, images and shapes that refer to the food of and cultural Italian traditions, generating confusion in consumers' minds and convincing them not to buy original Italian protected designation of origin and protected geographical indication products. Italian-sounding items generate a total turnover of €33.4 billion; that is, the total amount of imitation generates a turnover more than three times higher than exported original products (€9.9 billion).[25]

The Oliva Ascolana del Piceno

The Oliva Ascolana del Piceno was granted a protected designation of origin label in 2005, as reported in the European Community regulation no. 1855/2005. As previously mentioned, this name denotes the *ascolana tenera* variety of olive, either in brine or stuffed. The stuffed version is a world-famous example of the Italian culinary style. By the end of 2006, the production of denominated olives accounted for 4230 kg of brine olives and 600 kg of stuffed olives.[26] This small production level was handled by 7 farmers, 2 cattle breeders, 14 olive collectors, 5 firms in charge of the elimination of the bitter taste, and 11 producers dedicated to the filling process for the stuffed version of the olives. Among these various agents, 10 producers adhere to the 'Consortium for the protection of the Oliva Ascolana del Piceno.' Data on the current distribution of these olives in the domestic and international market are scarce due to the product's limited production and the recent denomination of origin acquisition. However, the total added food value of the geographic area affected by the production of these olives is estimated at €639,300.[27] The strengths of these olives include high intrinsic and perceived levels of quality; their weakness is a lack of logistic and distribution knowledge.

The production process for the stuffed version of the olives consists of three main phases. First, the bitter taste of the olives is eliminated, and they are destoned within 48 hours of harvesting. Second, they are filled with a preparation of meat (40–70 per cent beef, 30–50 per cent pork, and a maximum of 10 per cent poultry/turkey), eggs, extra-virgin olive oil, and, eventually, onions, dry white wine, other vegetables, salt, black pepper, nutmeg and additional flavourings. Third, the olives are rolled in wheat flour, eggs and bread crumbs. At the end of this process, the original olive should still make up at least 40 per cent of the entire stuffed olive. The olives are packaged for either short- or long-term consumption. The code of production requires producers to follow special rules for cooking of the meat and filling and breading the olives. Stuffed olives must present certain sensorial characteristics to be recognised as a denominated Oliva Ascolana del Piceno.

Literature Review: The Theory of Consumption Values

To achieve our research objective, we use the theory of consumption values[28] as a theoretical framework, as we show in Figure 17.1. The theory is based on three fundamental axioms: the market choice is a function of different values, these values have different influences according to different situations, and the values are independent. Consumer choice behaviour, depending on the kind of good and the context, is based on one or more of the following values: functional, social, emotional, conditional and epistemic.

Figure 17.1 Theory of consumption values[29]

EMOTIONAL VALUE

Emotional value relates to the ability of a product to arouse desired emotions and feelings.[30] Emotions usually come before cognition when consumers face a choice, because they have a general propensity to seek affective situations.[31] Thus, emotional positioning can be a successful marketing strategy for regional products, because emotions have an important role in consumer behaviour.[32] Moreover, the emotional link between consumers and a product's country of origin increases the perception of the typical food produced in that territory.[33]

FUNCTIONAL VALUE

Regional products also have a functional value when the perceived quality of a product influences beliefs about its overall fitness for use.[34] Functional value relates to the capability of a market choice to satisfy utilitarian or physical purposes.[35] In this sense, if a region and a product match, the regional indication works as a brand, bringing producers more advantages (mainly economic) than would introducing a new brand.[36] However, for durable products, other values such as emotions are able to influence consumers' choices.[37]

SOCIAL VALUE

Social performance refers to the ability of a product or service to portray an image to others that is congruent with the norms of significant others.[38] For example, social values are responsible for the differences in choice behaviour between male and female online purchasers. In particular, social prestige is the main discriminating social value for Web use between male and female purchasers, because female, but not male, consumers often decide not to buy online because of the perceived risk.[39] Social values are also able to differentiate the behaviour of past or current purchasers of organic food versus non-purchasers.[40] These findings are consistent with Sheth's assumption about social value: 'buy versus no-buy decisions are influenced by social values, in that consumers perceive various product classes as either congruent or incongruent with the norms of the reference groups to which they belong or aspire.'[41]

CONDITIONAL VALUE

The conditional value relates to the purchase of a product when a particular situation faced by the consumer alters his or her behaviour.[42] The alternative behaviour acquires conditional value if a physical or social contingency enhances the functional or social value that the alternative does not otherwise possess.[43] Therefore, the conditional value relates to a series of circumstances that are able to modify consumer choices.[44] For example, if a consumer desires to buy buffalo mozzarella, but on the way to the store reads news about a possible contamination of this product by dioxin, the consumer may decide to buy something else.

EPISTEMIC VALUE

Epistemic curiosity denotes a desire for knowledge and mainly applies to humans. Specific curiosity refers to the wish to have a particular piece of information, whereas diverse curiosity links to a more general desire for knowledge (closely related to boredom).[45] A further distinction exists between state and trait curiosity.[46] State curiosity refers to curiosity about a particular situation; trait curiosity refers to the propensity of every human being to explore new things. These definitions do not resolve the causes of curiosity though. In the first half of the twentieth century, curiosity was considered a homeostatic drive (internally driven stimulus),[47] but Berlyne broke with tradition by affirming that curiosity may be externally stimulated. In particular, he defines this external stimulus as 'stimulus conflict' or 'incongruity,'[48] also considered the underlying cause of curiosity by Hebb[49] and Hunt,[50] though they do not agree that curiosity is a drive.

OPTIMAL STIMULATION LEVELS

Together with the concept of curiosity, an optimal level of stimulation is a fundamental concern. Personality traits are the primary drivers of people's exploratory behaviour.[51]

However, to stimulate this explorative behaviour, an optimal individual stimulation level must be found. The levels, also referred to as 'optimal stimulation levels,'[52] vary according to each person.[53]

To acquire a satisfactory level of stimulation, consumers explore their environment. The discrepancy between the actual amount of stimuli consumers receive from the outside and their personal optimal stimulation levels helps researchers identify whether they must augment or reduce stimulation. Moreover, stimulation is a dynamic concept.[54] If repeat purchases of a product lead to a greater familiarity with it, the perceived utility of an unfamiliar item increases with consumption, but after a point, it should start to decline, leading to variety-seeking behaviour. The ability of new products to exist in the market when consumers are seeking variety, and the capability of marketers to find an optimal stimulation level to push consumers to try new products, are fundamental keys of success.

Prior literature suggests general agreement about the tendencies that lead to exploratory behaviour, categorised as curiosity-motivated behaviours, variety-seeking tendencies, and innovative behaviour. However, a fourth aspect links epistemic curiosity to optimal stimulation levels and exploratory behaviour: the wish to collect new information that pushes consumers toward exploratory behaviour and knowledge acquisition.[55] This factor helps consumers reduce the unpleasant feeling of deprivation owing to a lack of knowledge[56] and stimulate a 'feeling-of-knowing,' 'feeling-of-not-knowing,' or 'feeling-of-certainty,' depending on the perceived discrepancy between what they know and do not know.[57] In particular, typical feelings of knowing are associated with the so-called 'tip-of-tongue' state, which reflects a situation in which a person is able to retrieve a target word or object only partially and wishes to fill this gap.[58] In this sense, epistemic curiosity helps people use exploratory behaviour (especially in tip-of-tongue states) to reduce this state of painful deprivation. The exact amount of information required to fill this gap represents the optimal stimulation level that each person needs.

WHY IS IT INTERESTING TO FOCUS ON THE EPISTEMIC VALUE OF A PRODUCT?

In previous research,[59] epistemic value was not specifically addressed during focus-group sessions about consumers' perceptions of regional products. Nevertheless, this value seems important in the Italian context, where several products have appellations, and many differences exist among regions regarding the way a certain product is cooked, prepared, and served. Although regional products do not always seem to differ directly from other new or different products in providing consumers with epistemic value, van Ittersum[60] stresses one aspect that may provide a regional product with more epistemic value than other new products: the perceived match or congruity between a product category and the product's region of origin. There are further reasons to investigate the elicitation of this specific value in consumers' minds in more depth. First, region of origin acts as a quality cue with a twofold impact, one direct and one indirect.[61] The indirect impact refers to quality cues consumers may access before consumption and are perceived in relation to other products.[62] These cues may be intrinsic and/or extrinsic. The former refers to the physical characteristics of a product (e.g., colour, shape, size), and the latter refers to characteristics related to the product but that are not part of its physical description.[63] All these attributes may help consumers reduce the gap between the current and ideal levels of stimulation and therefore reach an optimal stimulation level. Second, regional food products can supply consumers' needs for variety and quality when they shop for food.[64] Third, peculiarity and homogeneity decline as the boundaries of a region expand. As a consequence, recipes or food processes may not be the same in every area. When recalling a processed regional food product produced over a vast area, a tip-of-tongue feeling may be generated.[65] Fourth, two clear trends in food production and consumption emerge: (1) Consumers devote more attention to agricultural products that bear a geographical indication and reflect traditional methods of production and (2) though household expenditure records for products awarded a protected designation of origin or protected geographical indication are difficult to obtain, there is clearly a change in the demand for these kinds of products. This shift shows a tendency to adopt new kinds of products (which are traditional in certain areas), defined as adoptive innovativeness.[66]

Methodology

The importance of these values and their level of interaction have been investigated with a series of focus groups, which elicit people's views, opinions and concerns and provide an opportunity for an in-depth understanding of the dimensions of consumer choice. A protocol for conducting focus groups was prepared and tested through a pilot trial performed with eight students from Le Marche Region who attended courses at University of Florence. Four focus groups were held in the Italian Center of Sensory Analysis (Matelica–Le Marche Region) in February 2006. The questions asked during focus groups, which lasted approximately 90–120 minutes, were an adjusted version of standard protocol proposed by Sheth,[67] pertaining to:

- the benefits and problems associated with the Oliva Ascolana del Piceno and alternative products (functional value);
- groups of people both most and least likely to purchase the Oliva and alternative products (social value);
- feelings aroused by the decision to buy the Oliva and alternative products (emotional value);
- determinants that trigger the decision to buy the Oliva and alternative products (epistemic value);
- circumstances that cause starting or stopping buying the Oliva and alternative products (conditional value).

There were eight participants in each focus group, all sensory panel members of the Center, who represented different social positions. The groups were built according to the homogeneity of their cultural and working dimensions (e.g., housewives, university students, entrepreneurs, cooks and waiters[68]) and demographic heterogeneity within each group (Table 17.1).

The participants came from Le Marche, the same region of origin as the Oliva Ascolana del Piceno. The denomination of origin value cannot be limited only to 'foreign' consumers (and therefore to extra-regional exports, tourist fluxes, especially agri-tourist and cultural ones, or 'Sunday trippers') but must extend to local populations – both rural and those of adjacent urban areas.[69] In this context, the local population should be addressed as a significant segment with a precise marketing approach; a 'one-size-fits-all'

Table 17.1 Focus groups' composition

Groups	Female			Male			Total
	25–34	35–44	45–65	25–34	35–44	45–65	
Housewives	2	3	3	0	0	0	8
University students	3	1	0	4	0	0	8
Entrepreneurs	0	1	1	1	2	3	8
Cooks and waiters	1	1	1	1	2	2	8
Total	6	6	5	6	4	5	32

strategy is not possible in the present day.[70] A study of the plurality of organisational forms in the supply of typical products in Italy[71] shows that, for consumers coming from the area of production, the sense of belonging and shared knowledge and traditions is strong and gives rise to a situation of total trust, as well as to knowledgeable evaluations in the purchasing phase. Nevertheless, when a product is extremely elaborate and is the result of a complex recipe (as in the case of the Oliva), an appropriate educational process aimed at all supply chain actors is important to promote the recovery of historic values and communicate traditions that risk being lost, even in the local community.

Findings and discussions

Textual analysis of transcriptions shows that the regional product Oliva Ascolana del Piceno contains every value described by Sheth's theory. The findings for each of the values, with the exception of the epistemic one, are summarised in Table 17.2.

As noted by van Ittersum,[72] regional products have functional value whenever they provide consumers with attributes that match consumer goals, such as a desire for high quality, good taste and health. The social value is elicited when consumers want to identify themselves with a social group either directly or indirectly linked to the region of origin. The conditional value is related to the usage situation or environment of consumption, and the Oliva Ascolana del Piceno is well-suited to special occasions like parties or 'happy hour' in bars. Consumers also may have different feelings about past experiences in the region (emotional value).

These values, considered as a whole, seem to show the interviewees' preferences for this product to 'reassure' their own identities and preserve their cultural boundaries.[73] As stated by Giddens, regions become more important because, with the enlargement of traditional borders through globalisation, the local citizen/consumer wants to better identify with his or her home region.[74]

Epistemic value is the perceived utility acquired by an alternative as a result of its ability to arouse curiosity, provide novelty, and/or satisfy a desire for knowledge. Of particular interest, epistemic value represents one of the main influences on choice.

The study reveals the need for knowledge about the original Oliva Ascolana del Piceno, now the subject of many industrial imitations in Italy. The local expert population more or less knows the recipe ingredients, but a definite lack of knowledge about the historical

Table 17.2 Most important findings

Functional	• *"First thing that comes to my mind is the wonderful taste of this product."* • *"It's crispy and gold outside and creamy inside."* • *"It's a product of high quality."*
Emotional	• *"I feel happy when I eat it because it's a way to remember the way my grandmother cooked it when I was young."* • *"The Oliva tells the story of passion of farmers in the area of Ascoli."*
Social	• *"It's a sort of flag for people from Marche"; "It's an image of our region."* • *"Oliva and Verdicchio (a kind of wine), are parts of our tradition."*
Conditional	• *"When I prepare a special dinner for friends who come from outside the boundaries of our region I cook the Oliva."* • *"The Oliva is great as an appetizer."* • *"It cannot miss wedding parties."*

origin and production process of the olives exists. When the original process and recipes were declared during the focus group, with a tasting session for the original olive, many comments emphasised a sense of curiosity and desire to learn the traditional methods.

The analysis of focus groups showed that all the respondents were familiar with the Oliva Ascolana del Piceno (see Figure 17.2). Moreover, they all affirmed that the Oliva has belonged to their culinary tradition since their childhood. Hence, a feeling of familiarity toward the Oliva Ascolana del Piceno occurs for our sample. The feeling of familiarity, defined as a long-term recognition memory process that refers to a subjective state of awareness based on judgments of the item's prior occurrence,[75] links directly to a feeling of certainty,[76] which in turn leads to a lack of curiosity and exploration.[77] However, when consumers were asked to list the ingredients necessary to produce the Oliva Ascolana del Piceno or other relevant information about the product – production area, historical background – they were at a loss.

If a feeling of familiarity or certainty continued to prevail, respondents would not have been interested in answering the questions any more, because in this scenario, curiosity would be almost entirely absent.[78] However, this feeling did not ensue; the more the respondents were stimulated to engage in exploratory behaviour to obtain their desired level of knowledge about the Oliva Ascolana del Piceno, the more they felt close to resolving the discrepancy between what they already knew about the product and what they could learn, so epistemic curiosity reached its peak.

If we consider that the focus groups involved people of different generations, but all from the same area, we can easily understand why fully objective knowledge of what ingredients compose the Oliva Ascolana del Piceno, or other relevant product information, might be missing. The focus groups demonstrated that everyone purported to know the product objectively, but this feeling did not derive from a rational cognitive process of data collection. Rather, it was the general feeling of knowing brought on by the environment in

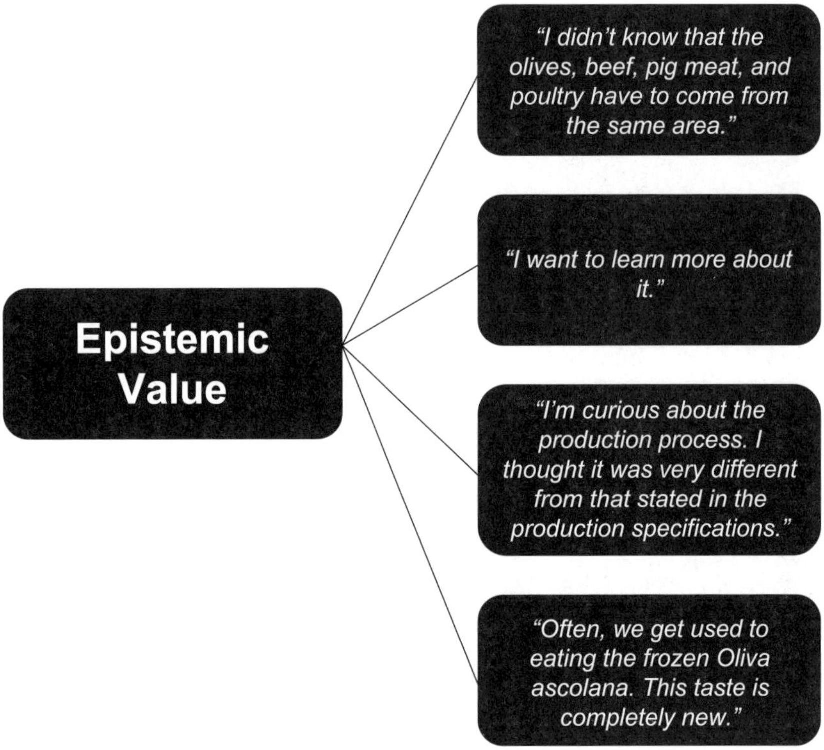

Epistemic Value

"I didn't know that the olives, beef, pig meat, and poultry have to come from the same area."

"I want to learn more about it."

"I'm curious about the production process. I thought it was very different from that stated in the production specifications."

"Often, we get used to eating the frozen Oliva ascolana. This taste is completely new."

Figure 17.2 Brief list of epistemic items generated by focus groups

which respondents have always lived. This effect could be a potential problem if marketers wish to use the epistemic value associated with a product to stimulate consumer choices, because the feeling of familiarity hardly stimulates people's curiosity. However, respondents seemed to become more curious when stimulated to learn something more about the Oliva Ascolana del Piceno. Thus, managers could develop promotional and communication strategies aimed at decreasing feelings of familiarity, thereby awaking a sense of novelty for products like the Oliva Ascolana del Piceno – a product that otherwise would be consumed without thinking about the extrinsic and intrinsic characteristics of the product.

In turn, young consumers (especially university students, young housewives and entrepreneurs) desire more experience and familiarity with these issues. In general, specialty shops are preferred to other shops (supermarkets) to purchase the Oliva Ascolana del Piceno, because trusted local vendors give advice and information to customers about the production process and the cultural and historical origin. Thus, price is not an obstacle; interviewees declared that they were willing to pay more as long as the price reflects a corresponding level of quality and information.

In light of these results, and according to van Ittersum,[79] to benefit from the effects of consumers' sense of belonging to the region and their involvement with the product category on information accessibility, we assert:

- regional product information should be made available at places where consumers expect to find regional and product category information (e.g., ads in regional media, displays in regional retail outlets);

- marketers should communicate the high quality of the regional product to consumers and explain how human expertise and natural and climatic conditions in the region contribute to the product's quality;
- the regional character of the product might best be emphasised by providing consumers with information about the social and emotional value of the regional product;
- for local consumers, the communication of attributes generating the functional, social and emotional value of a product increases the epistemic value through a corresponding arousal of consumers' curiosity.

Conclusions

The findings of this work confirm that to develop market strategies for specialty products, managers must have a deep understanding of consumers' values. Although small, Italian, agri-food firms are attracted by supermarket chains, they rarely understand the real perception and willingness to pay for their product's attributes by specialty stores. Moreover, when they initiate mass market distribution, they must face at least three major risks: the monopsonistic power of the retailer that becomes the exclusive customer of the small firm, the impossibility of delivering huge volumes of product to cover extra-regional point of sales, and the lower added-value for the small firm from the contract.

To develop efficient strategies, SMEs can benefit from the European regulation that proposes many regional policies aimed at activating and pursuing strategies for the implementation and communication of quality products.

The object of this chapter is consumers' perceptions of the Oliva Ascolana del Piceno, a famous variety of olive, underpinned by its quality and excellent reputation, founded on an age-old history and deeply rooted tradition. Consumers' perceptions of this product are investigated on the basis of the framework of the theory of consumption values, through a methodology involving focus groups.

The groups of consumers selected to participate in this research are local people, all sensory panel members of Italian Center of Sensory Analysis in Matelica. The decision to concentrate on this segment is based on the role of local populations, both rural and those in adjacent urban areas, to improve the local agri-food economy. Moreover, local people's regional sense of belonging represents, if properly elicited, a source of rediscovery of traditions that risk being lost. Finally, local consumer loyalty toward the regional product is a potentially sustainable effect of the use of the region of origin as a marketing tool.[80] As stated by Porter, 'paradoxically, the enduring competitive advantages in a global economy lie increasingly in local things – knowledge, relationships, and motivation that distant rivals cannot match.'[81]

From this perspective, the sense of belonging to a tradition and the rediscovery of a natural lifestyle becomes an opportunity for social growth in the local community, especially where it is possible to integrate productive activities with the climate and the history of a population and territory. For a specialty product, all these characteristics are perfectly melded together.[82]

Consumers' desire to support, protect and assert their identity may result in the purchase of a regional product. Therefore, marketers should articulate how the purchase of a regional product enables consumers to strengthen their sense of belonging to the region and supports the local economy.[83]

An appropriate educational process aimed at all supply chain actors can promote the recovery of historic values and communicate traditions that risk being lost, even in the local community. The evidence of this risk is exemplified by the sense of novelty noted by the local population, linked to the description of traditional production process and the taste of the original Oliva Ascolana del Piceno during focus-group sessions. Thus, a training process aimed at a diffusion of knowledge about both intrinsic and extrinsic quality attributes is necessary to maximise complex research activity about historical patterns, social habits, sensory characteristics, and more general quality cues, all of which are needed to obtain the registration of protected designation of origin from the European Community.

Marketers of regional products must particularly target quality-conscious consumers, who appreciate naturalness and craftsmanship and have a strong sense of belonging to the region.[84] This epistemic value becomes especially relevant for processed products such as the Oliva Ascolana del Piceno, for which the complexity of the production process increases the amount of historical, anthropological, environmental and production information available to consumers. Transformed regional products perform better when their relevant characteristics match the consumer's image of the region in terms of human factors[85] and local know-how.

In light of these considerations, it is important for local farmers and small firms to investigate which values drive the decision to buy and which values are predominant in a certain consumer segment. The limitations of this research relate mainly to the choice of consumer segment, which requires further investigation. The local population obviously is more involved in the historical and traditional characteristics of the product than are other consumer segments. However, the epistemic value, if properly stimulated through communication, also may improve local consumers' motivations and buying behaviours.

This strategy is pursued through the joint work of public institutions (interested in developing territory reputation) and economic actors. As the regional product delivered by multiple SMEs, the marketing of a regional product should be centrally organised and coordinated.[86]

References

1. Narver, J.C. and Slater, S.F. (1990), 'The effect of a market orientation on business profitability', *Journal of Marketing*, vol. 54, no. 4, pp. 20–35.
2. Kohli, A.K. and Jaworski, B.J. (1990), 'Market orientation: the construct, research propositions, and managerial implications', *Journal of Marketing*, vol. 54, no. 2, pp. 1–18.
3. Narver and Slater, op. cit.
4. Ibid.
5. Verhees, F.J.H.M. and Meulenberg, M.T.G. (2004), 'Market orientation, innovativeness, product, innovation, and performance in small firms', *Journal of Small Business Management*, vol. 42, no. 2, pp. 134–54.
6. Ibid.

7. Grunert, K.G., Fruensgaard Jeppesen, L., Risom Jespersen, K., Sonne, A., Hansen, K., and Trondsen, T. (2005), 'Market orientation of value chains: a conceptual framework based on four case studies from the food industry', *European Journal of Marketing*, vol. 39, no. 5/6, pp. 428–55.

8. European Commission (2008), *What Does the Acronym Leader Stand for?* Available at http://ec.europa.eu/agriculture/rur/leaderplus/faq_en.htm#37 (accessed September 15, 2008).

9. European Commission (2006), *The Leader Approach: a Basic Guide*. Available at http://ec.europa.eu/agriculture/publi/fact/leader/2006_en.pdf (accessed September 15, 2008).

10. INEA (2006), *La Riforma dello Sviluppo Rurale: Novità e Opportunità*. Available at http://www.inea.it/ops/pubblica/quaderno/workingpaper/quadernoSR1.pdf (accessed September 15, 2008).

11. Burrell, A., Gijsbers, G., Kosse, A., Nahon, D., Réquillart, V., and van der Zee, F. (2006), *Overview of Existing Studies – Preparatory Economic Analysis of the Value Adding Processes within Integrated Supply Chains in Food and Agriculture*, Seville, European Commission Directorate-General Joint Research Centre.

12. Sheth, J.N., Newman, B.I., and Gross, B.L. (1991), *Consumption Values and Market Choices. Theory and Application*, South Western Publishing, Cincinnati.

13. Bureau, J.C. and Valceschini, E. (2003), 'European food labeling policy: successes and limitations', *Journal of Food Distribution Research*, vol. 34, no. 3, pp. 70–6.

14. ISMEA (2008), *Tendenze Recenti del Mercato delle DOP e IGP*. Available at http://www.istat.it/istat/eventi/2008/fierabologna/SANA/finizia.pdf (accessed September 15, 2008).

15. ISTAT (2008), *I Prodotti Agroalimentari di Qualità DOP e IGP*. Available at http://www.istat.it/salastampa/comunicati/non_calendario/20080912_00/testointegrale20080912.pdf (accessed September 15, 2008).

16. ISMEA (2008), op. cit.

17. Qualivita (2008), *Osservatorio Socio-Economico Qualivita*. Available at http://www.qualivita.it/FrontEnd/OSS2007/studio.aspx (accessed September 15, 2008).

18. ISTAT (2008), op. cit.

19. ISMEA (2008), op. cit.

20. ISMEA (2007), *Le Tendenze del Mercato delle DOP e IGP*, Rome, ISMEA.

21. ISTAT (2008), *Evoluzione della Qualità Certificata dal 2004 al 2007*. Available at http://www.istat.it/istat/eventi/2008/fierabologna/SANA/adua.pdf (accessed September 15, 2008).

22. ISMEA (2008), op. cit.

23. Qualivita (2008), *Osservatorio Socio-Economico*, op. cit.

24. Qualivita (2007), *Il Ruolo dei Prodotti di Qualità nelle Esportazioni Agroalimentari Italiane e il Fenomeno dell'Agropirateria*. Available at http://www.qualivita.it/New/studi/studio.aspx?NID=210 (accessed September 15, 2008).

25. Qualivita (2007), op. cit.

26. Qualivita (2008), *Oliva Ascolana del Piceno DOP*. Available at http://www.qualivita.it/GALLERY/PRODOTTI/265/ATLANTE.pdf (accessed September 15, 2008).

27. Qualivita (2008), *Oliva Ascolana*, op. cit.

28. Sheth, J.N., Newman, B.I., and Gross, B.L. (1991) *Consumption Values and Market Choices. Theory and Application*, South Western Publishing, Cincinnati.

29. Ibid.

30. Ibid.

31. Tsai, S. (2005), 'Utility, cultural symbolism and emotion: a comprehensive model of brand purchase value', *International Journal of Research in Marketing*, vol. 22, pp. 277–91.

32.   Giraud, G. and Halawany, R. (2006), *Consumers' Perception of Food Traceability in Europe*, paper presented at the International Food and Agribusiness Management Association World Food and Agribusiness Symposium, Buenos Aires, Argentina, June 10–11, 2006.

33.   Resano, H., Sanjuan, A.I., and Albisu, L.M. (2007), 'Consumers' acceptability of cured ham in Spain and the influence of information', *Food Quality and Preference*, vol. 18, pp. 1064–1076.

34.   Steenkamp, J.B. E.M. (1989), *Product Quality: An Investigation into the Concept and How it is Perceived by Consumers*, van Corgum, Assen.

35.   Tapachai, N. and Waryszak, R. (2000), 'An examination of the role of beneficial image in tourist destination selection', *Journal of Travel Research*, vol. 39, pp. 37–44.

36.   van Ittersum, K., Candel, M.J. J.M., and Meulenberg, M.T.G. (2003), 'The influence of the image of a product's region of origin on product evaluation', *Journal of Business Research*, vol. 56, pp. 215– 26.

37.   Sweeney, J.C. and Soutar, G.N. (2001), 'Consumer perceived value: the development of a multiple item scale', *Journal of Retailing*, vol. 77, pp. 203–220.

38.   Overby, J.W., Fisher Gardial, S., and Woodruff, R.B. (2005), 'French versus American consumers' attachment of value to a product in a common consumption context: a cross-national comparison', *Journal of the Academy of Marketing Science*, vol. 32, no. 4, pp. 437–60.

39.   Andrews, L., Kiel, G., Drennan, J., Boyle, M.V., and Weerawardena, J. (2007), 'Gendered perceptions of experiential value in using Web-based retail channels', *European Journal of Marketing*, vol. 41, no. 5/6, pp. 640–58.

40.   Finch, J.E. (2006), 'The impact of personal consumption values and beliefs on organic food purchase behavior', *Journal of Food Products Marketing*, vol. 11, no. 4, pp. 63–76.

41.   Sheth, J.N., Newman, B.I., and Gross, B.L. (1991), 'Why we buy what we buy: A theory of consumption values', *Journal of Business Research*, vol. 22, no. 2, pp. 159–70.

42.   Finch, op. cit.

43.   Sheth, Newman, and Gross, (1991), *Consumption Values*, op. cit.; Bansal, H. and Eiselt, H.A. (2004), 'Exploratory research of tourist motivations and planning', *Tourism Management*, vol. 25, pp. 387–96.

44.   Shanka, T. and Pau, I. (2008), 'Tourism destination attributes: what the non-visitors say – higher education students' perceptions', *Asia Pacific Journal of Tourism*, vol. 13, no. 1, pp. 81–94.

45.   Loewenstein, G. (1994), 'The psychology of curiosity: a review and reinterpretation', *Psychological Bulletin*, vol. 116, no. 1, pp. 75–98.

46.   Berlyne, D.E. (ed.) (1974), *Studies in the New Experimental Aesthetics*, Hemisphere, Washington.

47.   Loewenstein, op. cit.

48.   Berlyne, D.E. (1954), 'A theory of human curiosity', *British Journal of Psychology*, vol. 45, pp. 180–91.

49.   Hebb, D.O. (1955), 'Drives and the C.N.S. (Conceptual Nervous System)', *Psychological Review*, vol. 62, pp. 243–54.

50.   Hunt, J.M. (1963). 'Motivation inherent in information processing and action', in Harvey, O.J. (ed.), *Motivation and Social Interaction*, Ronald Press, New York, pp. 35–94.

51.   Raju, P.S. (1980), 'Optimum stimulation level: its relationship to personality, demographics, and exploratory behavior', *Journal of Consumer Research*, vol. 7, pp. 272–82.

52.   Steenkamp, J.B.E.M. and Baumgartner, H. (1992), 'The role of optimum stimulation level in exploratory consumer behavior', *The Journal of Consumer Research*, vol. 19, no. 3, pp. 434–48; Steenkamp, J.B.E.M. and Burgess, S.M. (2002), 'Optimum stimulation level and exploratory

consumer behavior in an emerging consumer market', *International Journal of Research in Marketing*, vol. 19, pp. 131–50.

53. McReynolds, P. (1971), 'The nature and assessment of intrinsic motivation', in McReynolds, P. (ed.), *Advances in Psychological Assessment*, Science and Behavior Books, Palo Alto, pp. 157–77.

54. Bawa, K. (1990), 'Modeling inertia and variety seeking tendencies in brand choice behavior', *Marketing Science*, vol. 9, no. 3, pp. 263–78.

55. Berlyne, D.E. (1954), 'A theory of human curiosity', *British Journal of Psychology*, vol. 45, pp. 180–91.

56. Litman, J.A. and Jimerson, T.L. (2004), 'The measurement of curiosity as a feeling of deprivation', *Journal of Personality Assessment*, vol. 82, pp. 147–57.

57. Litman, J.A. (2005), 'Curiosity and the pleasures of learning: Wanting and liking new information', *Cognition and Emotion*, vol. 19, no. 6, pp. 798–814.

58. Ibid.

59. van Ittersum, K. (2001), *The Role of Region of Origin in Consumer Decision-Making and Choice*, Ph.D. Dissertation, Mansholt Graduate School, Wageningen.

60. Ibid.

61. Stefani, G., Romano, D., and Cavicchi, A. (2006), 'Consumer expectations, liking and willingness to pay for specialty foods: do sensory characteristics tell the whole story?', *Food Quality and Preference*, vol. 17, pp. 53–62.

62. Verlegh, P.W. J. and van Ittersum, K. (2001), 'The origin of spices: The impact of geographic product origin on consumer decision making', in Frewer, L. et al., (eds), *Food, People and Society*, Springer, Berlin, pp. 267–80.

63. Stefani et al., op. cit.

64. van der Lans, I.A., van Ittersum, K., Di Cicco, A., and Loseby, M. (2001), 'The role of the region of origin and EU certificates of origin in consumer evaluation of food products', *European Review of Agricultural Economics*, vol. 28, pp. 451–77.

65. Stefani et al., op. cit.

66. Skuras, D. and Dimara, E. (2004), 'Regional image and the consumption of regionally denominated products', *Urban Studies*, vol. 41 no. 4, pp. 801–815.

67. Sheth, Newman, and Gross (1991), 'Why we buy', op. cit.

68. Gabbai, M., Rocchi, B., and Stefani, G. (2003), 'Pratiche alimentari e prodotti tipici: un'indagine qualitativa sui consumatori', *Rivista di Economia Agraria*, vol. 58, no. 4, pp. 511–52.

69. Pacciani, A. (ed.) (2003), *La Maremma Distretto Rurale*, Il mio amico, Grosseto.

70. Beverland, M. and Lindgreen, A. (2004), 'Relationship use and market dynamism: a model of relationship evolution', *Journal of Marketing Management*, vol. 20, pp. 825–58.

71. Brunori, G., Ceron, F., Rossi, A., and Rovai, M. (2000), 'Plurality of organisational forms in the supply of typical products: Empirical evidence in Italy', in Sylvander, B., Barjolle, D. and Arfini, F. (eds), 'The socio-economics of origin labeled products in agri-food supply chains: Spatial, institutional and co-ordination aspects', *Economie et Sociologie Rurales*, vol. 17, pp. 79–93.

72. van Ittersum, op. cit.

73. Ibid.

74. Giddens, A. (ed.) (1990), *Modernity and Self Identity*, Polity Press, Cambridge.

75. Plailly, J., Tillmann, B., and Royet, J.P. (2007), 'The feeling of familiarity of music and odors: The same neural signature?', *Cerebral Cortex*, vol. 17, November, pp. 2650–2658.

76. De Sousa, R. (2008), 'Epistemic feelings', in Brun, G. et al., (eds), *Epistemology and Emotions*, Ashgate, Aldershot (in press).

77. Litman, J.A., Hutchins, T.L. and Russon, R.K. (2005), 'Epistemic curiosity, feeling-of-knowing, and exploratory behavior', *Cognition and Emotion*, vol. 19, no. 4, pp. 559–82.

78. Ibid.

79. van Ittersum, op. cit

80. van Ittersum, op. cit.; Porter, M.E. (1998), 'Clusters and the new economics of competition', *Harvard Business Review*, November, pp. 77–90.

81. van Ittersum, op. cit.

82. Arfini, F. (2005), 'Segni di qualità dei prodotti agro-alimentari come motore per lo sviluppo rurale', *Agriregionieuropa*, vol. 1, no. 3. Available at http://www.agriregionieuropa.it (accessed September 15, 2008).

83. van Ittersum, op. cit.

84. Ibid.

85. van Ittersum, Candel, and Meulenberg, op. cit.

86. van Ittersum, op. cit.

18 *The Process and Critical Success Factors of Evolving From Product Excellence to Market Excellence: The Case of Mastiha in Chios, Greece*

BY CHRISTOS FOTOPOULOS,* ILIAS P. VLACHOS,† AND GEORGE MAGLARAS‡

Keywords

critical success factors, transformation process, market excellence, Mastiha

Abstract

This case study examines the Chios Mastiha Growers Association (CMGA, or the Association), which exemplifies best practices in transforming from a regional, product-oriented agricultural cooperative to a global, market-oriented food company. Mastiha is a unique gum product that has attracted scientific interest because of its health benefits. However, the Association has accumulated liabilities over the years owing to some common management malpractices and lack of appropriate brand management. A turning point in the history of the Association was the establishment of an affiliated retailing company, called Mediterra S.A., in 2000. Although the Association owns the majority of Mediterra S.A. capital, an independent management team created a new vision and strategy. Several key success factors combined to create a successful case for building a strong

---

*   Professor Christos Fotopoulos, University of Ioannina, Department of Business Administration in Food and Agricultural Enterprises, Seferi 2, Agrinio, 30100, Greece. E-mail: chfotopu@cc.uoi.gr. Telephone: + 30 26410 39523.

†   Dr Ilias P. Vlachos, Agricultural University of Athens, Department of Agricultural Economics and Rural Development, Iera Odos 75, Botanikos, 118 55, Athens, Greece. E-mail: ivlachos@aua.gr. Telephone: + 30 21052 94757.

‡   Mr. George Maglaras, University of Ioannina, Department of Business Administration in Food and Agricultural Enterprises, Seferi 2, Agrinio, 30100, Greece. E-mail: geomag@cc.uoi.gr. Telephone + 30 26410 74151.

brand image and accumulating above-average growth rates. Key success factors include top management support, leadership and strong social cohesiveness among producers and supply chain management. Critical success factors are not discussed separately but in conjunction with the process of transformation from a production orientation to a marketing orientation.

Introduction and Background to the Research

Mastiha is a unique natural product cultivated exclusively on the Greek island of Chios. Mastiha is a resin of *Pistacia lentiscus var*, Chia trees that are found in the southern part of Chios Island. Trees can grow in locations with similar climates, but they fail to produce the same resin, making Mastiha Chiou a unique product and its market a monopoly. Chia trees cover an area of 1,900 hectares in total.

Mastiha is traditionally produced, processed and prepared in Chios. Thus, it was not difficult for the Chios Mastiha Growers Association to certify Mastiha as a 'Protected Designation of Origin' product. The same designation marks products derived from Mastiha, such as Tsikla Chiou (a chewing gum with Mastiha as an ingredient) and Mastihelaio Chiou (an essential oil distilled from the less clean granules of Mastiha).

Current research suggests that Chios mastic (*Pistacia lentiscus var. chia*) possesses beneficial (antimicrobial, antioxidant, hepatoprotective, cardioprotective) properties.[1] The combination of these two factors, market monopoly and strong indication of health benefits, creates real potential for high profit margins. Yet for many decades, that potential was a missed opportunity for the Association, and Mastiha producers accumulated debts rather than profits.[2] In addition, management failures to profit from Mastiha generated several attempts at illegal trading, which were altogether unsuccessful. However, another important factor not commonly found in Greek cooperatives played a role: the strong social cohesiveness among Mastiha producers. The proximity of growers' villages in Chios Island, their social relations, and the common awareness of their unique but unexploited product kept farmers together and created an environment receptive to change.

This research aims to examine the process and critical success factors of the Association's evolution from product excellence to market excellence. Specifically, the objectives are (1) to determine the factors that affected the process of transformation of a production-oriented cooperative to a market-oriented company and (2) to examine the process of evolving from a product- to a market-orientation.

We selected the Chios Mastiha Growers Association as a case study because it is a unique case in Greece that has successfully managed to transform itself from an agricultural cooperative into a market-oriented food company.

The Chios Mastiha Growers Association

The Association is a typical agricultural cooperative; its main activities include the collection, manufacturing, packaging and selling of Mastiha. Its main activities are:

- managing the production of Mastiha;
- supporting producers;

- protecting Mastiha's trading and brand names;
- supporting scientific research regarding the Chia tree and Mastiha's attributes and uses;
- cleaning, packaging and trading natural Mastiha;
- producing and trading Mastiha's gum;
- research into and development of new products

Mastiha producers were active from as early as 1347, but the Chios Mastiha Growers Association was formally established in 1938. For more than six decades, the Association had to deal with the same problems and barriers that most small-scale enterprises do: lack of scale to reduce production costs, financial shortages, difficulty in accessing retail shops, difficulty in competing with large multinationals, elementary education of the personnel, and lack of access to modern technologies. Moreover, the Association's poor management marketing skills allowed intermediates to capture surplus value and exploit profits.[3] In addition, owing to the commodity character of Mastiha, the product was offered at a low price. The end result was poor performance, accumulated liabilities, poor brand image, reduced production prices and cultivation abandonment. As a typical Greek food small-to-medium-sized enterprise (SME), the Association lacked economies of scale and scope. However, by concentrating its interest on consumers' wants and needs, there was an opportunity for the Association to meet latent or evident customer needs and gain a competitive edge over large internationals.[4]

The main prerequisite for exploiting Mastiha's growth potential was the transformation of the Association itself into a more market-oriented union to achieve improved performance outcomes.[5] The first step of the Association's transformation occurred through sporadic personnel hiring in early 2000s. Staff changes accumulated and gave rise to a broader reorganisation plan. The plan's crucial element was the establishment of Mediterra S.A., which aimed at the creation of a strong brand image and helped attract funds from venture capital and institutional investors. Within a decade, the Association managed to create a strong brand image and open 12 retail Mastiha shops (including in Dubai and Saudi Arabia). Mastiha is also available for online purchase (www.mastic.gr). Furthermore, the Association is currently developing 150 new products under the brand names of mastihashop, cultura Mediterra and mastihatherapy. Mediterra S.A. has a steady growth rate of more than 10 per cent, and its sales, 60 per cent from exports, climbed to more than €6 million in 2007. The Greek media has made many positive references to the success of Mastiha, and many Greek food companies already consider it an exemplary case.

Key Success Factors

In order to examine the factors that affected the transformation process of a production-oriented cooperative to a market-oriented company, we used a case study approach.[6] We developed a case protocol to control field work and interviewed top managers of Mastiha Association. We used secondary research to develop the case protocol and pre-tested it by interviewing academics and practitioners. Interviews with the Association's managers took place in Chios Island and Athens. In total, seven interviews were conducted between 20 November 2007, and 20 January 2008.

We examined factors related to business management and compared the Association's behaviour and operation before and after the transformation, as well as how it plans to evolve in the future. This method was useful in uncovering key transformation factors, as well as for shedding light on the process of transformation. Specifically, we examined the following topics:

- business strategy
- culture
- leadership
- organisational policy
- economical policy
- productive process
- supply chain management
- research and development
- marketing strategy.

Table 18.1 presents the critical success factors of Chios Mastiha Growers Association's evolution to market excellence.

Table 18.1 Critical success factors of Chios Mastiha Growers Association evolution to market excellence

Critical Success Factors	Past	Present	Future
Business strategy	Fair price for Mastiha, fair income for the producers	Fair price for Mastiha, fair income for the producers Business growth, new products Increase of Mastiha consumption	Fair price for Mastiha, fair income for the producers Established product
Culture	Conservative perceptions Prejudice toward new ideas and technologies Lack of education in business and marketing	Trust in the Association's staff and their ideas Innovation Decision making in accordance to the economic environment Administrative staff educated in marketing Lack of education in business regarding producers	Continue to innovate Producers' education
Leadership	Lack of leadership, lack of management capabilities	Strong leadership vision Recruitment of administrated staff educated in management and marketing Communication of vision to the producers	Continue leadership vision (doubtful in case of changing of shareholders) Same management decisions raise criticism

Table 18.1 *Continued*

Critical Success Factors	Past	Present	Future
Organisational structure	Decision making by the board of directors of CMGA, influenced by the producers' representatives Simple centralised organisational structure (small number of directors: Gum, Mastiha, Financial)	Establishment of Mediterra S.A. Decision making by the board of directors of CMGA, influenced by Mediterra S.A.'s market knowledge and producer's representatives Mediterra S.A. has its own board of directors: focus on retailing and distribution of Mastiha and its products More flexible, decentralised business chart (add Marketing, Research and Development, and Support directors)	Small changes in the business chart for further flexibility
Financing policy	Focus on cost reductions through no investments in mechanical and human resources Fair price for Mastiha (€19)	Positive financial balance for many years Fair price for Mastiha (88€) Demand increases (new customers in new markets) Difficult sales forecast due to ups and downs of production Revenues decrease due to dollar depreciation Financial aid from subsidies, institutional investors, and strong local financial entities Profits' transfer to the producers	Secure producers' revenues by penetrating markets with stable environment Increase Mastiha's fair price
Production	Mass production without quality focus Traditional production methods No financial capability for plant modernisation Technical consulting to the producers	Production focused in quality Establishment of quality control systems Gum plant modernisation for better packaging and lead times. Little technical support to the producers	Focus in quality Standardisation of producing methods Training of producers Production increase based on market demand, not producers' influence

Table 18.1 *Continued*

Critical Success Factors	Past	Present	Future
Supply chain management	Manual raw material reception Distribution of Mastiha and its products by the CMGA	Computerised raw material and order reception using strict food safety standards CMGA: Mastiha's distribution abroad through agents Mediterra S.A.: Distribution of Mastiha in domestic market and Mastiha's products in domestic market and abroad Outsourcing of gum's distribution in domestic market Modernisation of packaging ERP system establishment	Modernisation of storehouses for quality preservation and traceability
Marketing strategy	Confusion due to lack of marketing education	Focus on marketing Positioning based on the market trend of healthy diets Market and product development Consumer research Research and development Persuade consumers to include Mastiha in their diets Product range expansion with value-added products Participation in rural and agricultural development programmes	Consumers' loyalty increase and loyal base creation Market development within stable environments (Western markets) Scientific research to support health claims Increase Mastiha's uses for market penetration, market development, and product development
Product	Product portfolio: Mastiha and ELMA gum Mastiha: Impractical form	Brand names portfolio: Mastiha, ELMA gum, Mastiha shop, Mastiha shop therapy, Cultura Mediterra Expanded product portfolio due to many uses of Mastiha Products with health claims Launch of Mastiha in a practical form (e.g., yoghurt, gum) Improvement of Mastiha's package	Need for gum package improvement Communication of Mastiha's products in a clear way to consumers Reduction of product portfolio expansion

Table 18.1 *Concluded*

Critical Success Factors	Past	Present	Future
Place	Distribution of Mastiha as a spice	New distribution channels (retailing stores, pharmacy stores)	Distribution in Western markets
Promotion	Infrequent promotion of ELMA gum with television spots	Experimental promotion strategies Outsourcing of promotion mix Promotion through Mediterra's S.A. retailing stores High publicity due to Mediterra's S.A. recent enter in the Greek Stock Exchange Alternative Market Promotion through food fairs, exhibitions, and public relations abroad	Fear of overpromotion of Mastiha and its products Awareness of Mastiha as a part of everyday diet Mastiha's positioning as a premium product
Price	Producers' pressures for price increase	Producers' pressures for price increase Price discounts based on purchasing quantity Price discrimination based on target market Advance payment (lack of liquidity)	Price increase (premium product positioning) Reduce price discounts due to bad practices

Figure 18.1 also depicts the timeline of events as Chios Mastiha Growers Association progressed into market excellence.

STRATEGY

The Association's main objective until the late 1990s was the satisfaction of the Mastiha's producers by selling the total quantity of produced Mastic. A few efforts were also made to get a better price than that of the previous year, though they were not always successful. The goal was to get a price that could provide a fair income to the average Mastic producer. This goal is still an objective for Mediterra S.A., but its strategic orientation is to create a health image for Mastiha that it can use to achieve market expansion and create new products (i.e., launching new products based on scientific research). Future plans include the establishment of Mastiha as leader in the gum product range in domestic and global markets.

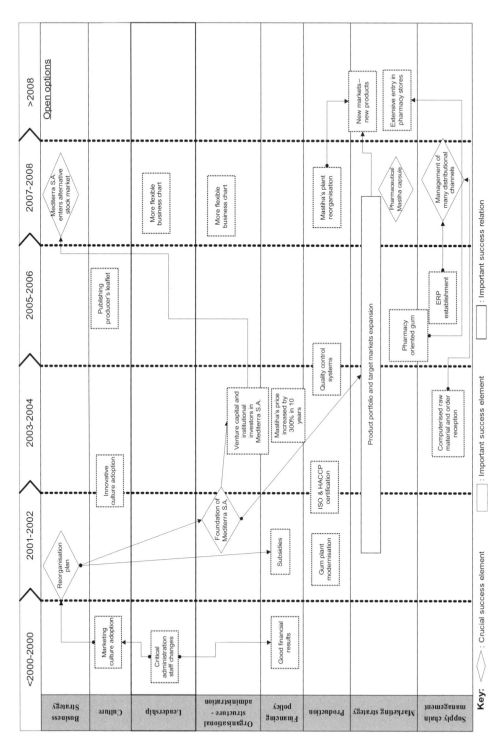

Figure 18.1 Events as Chios Mastiha Growers Association progressed into market excellence

CULTURE

As mentioned, the Association previously operated like many local agricultural associations currently do, characterised by conservative perceptions, prejudice against new ideas and technologies, lack of education in business and marketing and liabilities. A pessimistic atmosphere hindered the development of the Association. Early in 2000, the Association recruited new administration staff that specialised in marketing. At the same time, the financial balance finally was positive. This stability resulted in greater trust in the product and new ideas, both of which helped promote decisions that matched the change in the economic environment. Another profitable year occurred within this positive atmosphere; as a result, an innovative culture was created. This new trend changed views about taking risks and chances on new products.

LEADERSHIP

For many decades, the Association's administration appeared to be in poor shape, lacking strong leadership, vision and management skills. Top managers limited themselves to informative speeches directed at the Mastiha's producers, hesitating to take any kind of action that would bring change. Resistance to change jeopardised every critical decision. Therefore, the transformation toward a more market-oriented union was difficult and required radical changes to long-held practices and beliefs.[7] Today, the differentiation occurs through a distinct, strong, leadership vision that created a standard growth plan for the company, starting in 2001. The growth plan had a clear market orientation and was well communicated and comprehended by the administrative staff. Recruiting capable staff educated in management and marketing was a key factor in understanding and sharing the leadership's strategic vision. The staff made attempts to communicate that vision to producers through regular speeches and informative leaflets. At the same time, management decisions provoked some criticism because Mediterra S.A. had entered the alternative stock market, which caused producers to worry about their future. Despite a strong will to continue the same policies and strategies, there is some doubt about whether leadership alone can overcome producers' hesitations about the pace and direction of Mediterra S.A.'s growth.

ORGANISATIONAL STRUCTURE

In the past, the structure of the Association was quite centralised, with few main directors. One of the changes implemented was the creation of a more flexible and decentralised organisational chart that included more departments: marketing, research and development, and support. At the same the Association decentralised, Mediterra S.A. was established, focusing its efforts on retailing and the distribution of Mastiha and related products.

This structural change altered the former decision-making chain, making Mediterra S.A. a key decision maker. Producers were happy with the new scheme because they received more (income) for less (involvement and subsequent responsibility). This structure is unlikely to change in the future, except for some slight modifications to facilitate an expansion to remote foreign markets.

FINANCIAL POLICY

Until 2000, the Association had no clear financial policy with business growth objectives that could expand its activities and improve its financial figures. Poor financial management skills were easily exposed through a total lack of investments in technology or human resources. Mastiha's prices reflect these policies: Producer prices increased 50 per cent in seven years – from €19 per kilo in 1994 to €30 per kilo in 2000. With the new financial policies, the producer price climbed to €88 per kilo in 2006, an approximate 300 per cent increase within the next seven years. Furthermore, Mediterra S.A. was skilful enough to raise capital from subsidies and institutional investors and to mobilise local ship owners to back up its investment plan by becoming 'business angels.' The last action in Mediterra S.A.'s financial plan was to enter the alternative stock market. What currently concerns its financial management is the US dollar's devaluation, which affects income from exports.

PRODUCTION

The production process of Mastiha has a traditional character, especially in its first stages. The raw mastic's perishability demands initial processing by hand. Further processing and the production of Mastiha-based products (e.g., ELMA gum) now is open to improvement that was not previously possible, owing to financial problems that did not allow for the modernisation of the production units. The Association focused mainly on the mass production of Mastiha and gum without giving special emphasis to quality. One of the few moves toward quality was the Association's provision of technical advice to the producers. Consultation about production methods unfortunately was abandoned, though it could be a good practise to improve the quality of Mastiha.

Presently the Association, through its reorganisation plan, has changed its production philosophy and now follows a quality orientation. A quality control system, which contributes to better cleaning of Mastiha and checks the purity of the raw material the producers deliver to the Association, has been applied. Variance in purity defines the quality of the raw material, and therefore, price differentiation is available to the producers. Based on this policy, the producers are more motivated to provide clean the raw material they sell to the Association. In this way, the quality that the final consumer enjoys improves, and the cost of processing the raw material on the Association's behalf decreases. Furthermore, ELMA's warehouse has been upgraded, resulting in improved lead times as well as better packaging, which makes the product more attractive to end-consumers.

The quality orientation has been supported through better standardisation of the production process (i.e., scientific research), as well as producers' training in technical matters. At the same time, the Association aims to implement new ideas for standardised Mastiha products. In the critical matter of the production quantity level, the decisions are made according to market demand and conditions, less according to pressure from the producers.

SUPPLY CHAIN MANAGEMENT

Regarding past practises, the Association previously collected raw material from every Mastiha-producing village and then transported it to its industrial unit. Today, this process has been fully computerised: producers carry the raw materials to the Association on pre-scheduled days and times. As an immediate result, food safety has increased considerably.

The Association distributes Mastiha abroad through domestic representatives. In the Greek market, Mediterra S.A. is responsible for the distribution of Mastiha and Mastiha-based products (other than ELMA gum). Mediterra S.A. also has improved the packaging and developed a logistics information system that diminishes lead times.

The main future objective is the modernisation of warehouses and plants to improve the final product's quality, as well as trace and track Mastiha products in real time.

MARKETING

Prior to now, the operation of the Association was characterised by few marketing practises. One of the main objectives of the reorganisation was to bring meaning to marketing and make clear that this initiative could help the Association's business. Marketing management thus aimed to create a strong brand image. Mediterra S.A.'s main target group included nutritionally concerned consumers and those willing to try new tastes. The number of Mastiha products (mainly value-added products) also increased, and retail stores were created (Mastiha shop). The Association penetrated the functional foods market (i.e., launching products with health claims) and pharmacy retail stores as well. Emphasis focuses on public relations, mainly through participation in rural, agricultural developmental programs that can support the Association financially. The Association was also able to get most of its promotion campaign subsidised by the Hellenic Foreign Trade Board from 2000 to 2006.

For the immediate future, Mediterra S.A. aims to create a loyal customer base that will ensure repeated purchases, especially in Western countries. However, foreign consumers are not familiar with Mastiha's taste and lack knowledge of its beneficial properties.

Product

Traditionally, the Association has offered only two products: Mastiha and LMA (gum based on Mastiha). Today, a much wider variety exists: Mastiha Chiou, ELMA gum, foods enriched with Mastiha (usually found in Mastiha shops), para-pharmaceutical Mastiha-based products (Mastiha shop therapy brand name), a series of products called Cultura Mediterra (products without Mastiha but with a focus on a Mediterranean diet), nutritional supplements, oral hygiene products, soaps, cosmetics, bakery products, snacks, sauces, pasta, paté, chocolates, sweets, drinks and beverages. This variety of products stems from the various properties and uses of Mastiha, applied in the food and drink industry, as well as the pharmaceutical and chemical industries. Launching these new products has been the reason for new market developments (e.g., Mastiha capsule, Mastiha powder and Mastiha gum offered in pharmacy retail stores), which also differentiate the positioning of Mastiha products, relating them to specific health claims. The newly improved packaging of Mastiha also has attained good market responses.

Place

Until recently, Mastiha was distributed by the Association and its representatives, mainly as a spice for cooking and confectionery use in powder or oil form. The distribution of

Mastiha and its products is now more diverse (e.g. the Association, Mediterra S.A., ELGEKA S.A.), without a particularly substantial increase in the number of export countries. Other competitive companies supplied by the Association use different distribution channels, such as the Internet. In addition, niche Mastiha's products required specialised distribution channels such as pharmacy retail stores, Mastiha shop and Mastiha shop therapy. The Association is optimistic to expand in Western markets such as New York.

Promotion

In the past, the promotion mix was limited to some scarse ELMA television commercials. Today, the Association engages in experimental promotion strategies, and the marketing mix is continuously changing (outdoor, radio, television, magazines), featuring great intensity and outsourcing to another company. The promotional efforts are favoured by Mediterra S.A.'s retail store development, and publicity increased due to the company's entry in the alternative stock market. Regarding promotion in foreign markets, efforts are limited to promotion through exhibitions and public relations. Mastiha targets a niche market and therefore has no need for a massive and costly promotion campaign. Moreover, Mastiha traditionally is well known in Greece, so the mass media informs consumers that it has become an added-value product. Still, there is no marketing strategy to promote the product. The future objective is to insert Mastiha into households' consciousness. There is also fear of over-promotion, which, combined with competitive substitutes, creates some anxiety about consumers' reactions. Thus, the objective is to increase brand loyalty to Mastiha.

Price

Typically, the exclusive interest of the producers was better prices. However, pricing policy is a tricky issue for the Association: exports account for half of the Association's income, and exporting takes place at wholesale prices, which are affected by order quantities. The expansion of exports to many countries creates a complex pricing policy, in which payment in advance is not always guaranteed. This policy creates liquidity pressures. Furthermore, importers conduct their own marketing and therefore press for higher profit margins. The Association needs to create long-term business relations with importers, but it is not in a position to finance them. Therefore, the management faces a dilemma: a conservative pricing policy, which means paying the producer higher prices every year to keep them happy and supportive, but reduced profits, which jeopardise future investments. It alternatively could pursue an aggressive policy, keeping production prices low and saving money to invest in exporting relations, but this latter approach runs the risk of cultivating scepticism among producer bases. Even worse, if this pricing policy is not fruitful, management would be subject to severe criticism and threats of exit. The pricing policy has attempted to find a balance between these two extremes, which suggests it is a moderate policy.

The Process of Transformation

Changes in these business dimensions have resulted in a greater market focus and the rapid growth of the Association. However, each factor has had a specific influence on the Association's objectives and, in many cases, is the basis for some key decisions regarding market excellence. As Figure 18.1 showed, several factors played key roles in the transformation from a product orientation to a market orientation; they related closely during the past decade as the Association implemented its reorganisation.

Staff changes in early 2000 reversed the long-standing negative financial results. At the same time, a more marketing-oriented culture was cultivated. The newly hired staff had the education and capabilities, which the Association and most cooperatives previously lacked, to carry out basic marketing and economic activities. For the first time, after many years of operation, producers realised that marketing could help them charge prices that lead to more income. This realisation increased trust in marketing practices among the Association, which helped encourage the adoption of a stronger marketing culture, because trust in the staff meant trust in the new ideas.[8] Furthermore, the reorganisation plan developed in a very optimistic atmosphere for the Association.

The first key decisions were the gum plant modernisation and the foundation of a subsidiary company called Mediterra S.A., which undertook the promotion and retailing of Mastiha. The new plant met the need for modernised logistics that would satisfy intermediate customers and make the Association more efficient (e.g., decrease lead times). In addition, Mediterra S.A. represented a clear attempt to establish strong brand names for Mastiha products. It was obvious that the raw material was something unique and valuable. The subsidiary's foundation related to the commercial exploitation of Mastiha's uniqueness by applying marketing techniques and creating products that met modern consumer needs. The reorganisation plan and growth vision were financially supported by subsidies. Generally, the decision for a new start, through the organisation plan and foundation of Mediterra S.A., led the Association in the adoption of a more innovative culture, a prerequisite for accomplishing desired goals.[9] The innovative culture resulted in an expanded product portfolio and new target markets that satisfied greater and better consumer needs.

Another way in which the Association tried to achieve market orientation was by paying greater attention to quality improvement. The ISO and HACCP certifications, quality control systems, and Mastiha's plant modernisation trended toward a quality orientation. The quality orientation added value to Mastiha products by increasing food safety, securing the quality of raw material, and launching more user-friendly products (e.g., new packaging for gum).

The new perspectives and potentials that the Association and Mediterra S.A. presented attracted institutional investors and a venture capital's serious investment, which in turn improved the cooperative's economical health and introduced a new chapter in its financial strategy. For the first time, foreign capital supported the Association's growth effort, which was the first step toward a strategy that would attract funding more easily. This investment helped harmonise the economic strategy of the Association with the modern economic environment and its practises, resulting in Mediterra S.A.'s introduction to the alternative stock market. In the meantime, these attempts to move toward a market-oriented Association increased the price of Mastiha, rapidly encouraging decisions to support change.

The ambitious market expansion that began in 2003 required efficient distribution. Therefore, supply chain management was crucial for the Association's operational performance. Previously, the Association's investments in new technologies had resulted in the production of more added-value products. Between 2003 and 2006, the Association put into practice automated procedures for input and outbound logistics, as well as order handling. The Association also established an enterprise resource planning (ERP) system to control and improve the supply chain processes of Mastiha.

Market orientation improves a firm's ability to capitalise on environmental changes, leading to superior performance.[10] A critical success factor, resulting from the exploitation of healthy eating as a consumption trend, was product expansion. In 2005, Mastiha was introduced in pharmacy stores through the launch of a pharmacy-oriented gum. This approach was a strategic move of great importance, because penetration in the specific market created a new target market with high growth and profit margin potentials for Mastiha products. In 2007, a pharmaceutical Mastiha capsule was produced, marking the first step toward a more extensive entry into pharmacy stores. In addition, Mastiha's presence in the specific market improved its market positioning as a premium product with higher profit margins. Mastiha must now compete with functional and pharmaceutical products in pharmacy stores, such as dairy with omega-3 or fruit drops with extracts and herbs. Mastiha cannot yet support a health claim, but strong evidence exists for its health benefits.[11] It differs from functional foods, in the sense that Mastiha is a physical, traditional Greek product, so it can accumulate profits earned from this specific niche as well as the growing Greek market.

Discussion

This case study examines the transition from product orientation to market orientation by the Chios Mastiha Growers Association. The Association suffered long-term financial problems. Although the Association was exclusively trading Mastiha, a unique product with many uses in food, pharmaceutical, and chemistry fields, it suffered problems common to most agricultural associations (e.g., lack of education, prejudice against new ideas, conservatism). As a result, the Association operated without a serious business strategy and under the pressure of producers focusing on cost reduction and trading of Mastiha as a commodity. It is very important to highlight the special importance of the Association for the local community; it provides a fair income for a large number of producers and promotes the entire region of Chios Island. Therefore, the producers are very sensitive in changes, partly due to some previous unsuccessful attempts at unrelated diversification but mainly because of the lack of business orientation.

Various key factors contributed significantly to the transformation of the local cooperative into a market-oriented company with great growth promise. Many of the critical factors reported in this case study are specific to the Chios Mastiha Growers Association and Greece. However, the case also includes some important features of the transformation from product excellence to market excellence that may apply to many agricultural cooperatives in the food industry. Those cooperatives might be facing the same conditions as the Chios Mastiha Growers Association did before applying its organisation plan in the early 2000. Some examples include cooperatives for saffron, dictamnus, raisins, Greek mountain tea, asparagus, dried figs, olive oil and wine. In these examples, we find high potential due to product excellence; the promise of a successful transformation to market excellence is very positive. We describe the necessary features next.

CULTURE OF CHANGE

Market orientation must be understood as a culture,[12] and the common problems of agricultural cooperatives – such as conservative ideas and lack of education – are barriers to this perspective. The Association was affected by these issues for many decades. Some crucial staff changes that had immediate positive financial results helped the Association adopt a more market-oriented and innovation-oriented culture. As a result, an atmosphere of trust grew among the administration staff. New ideas and practices recommended by staff members with basic marketing and financial knowledge were most welcome inside the Association. Their first attempts were successful, and thus, there was less fear of and perceived risk in their implementation. This new climate created a culture of trust that became the basis for the development of a clearer business vision.

BUSINESS GROWTH VISION

The production of a unique valuable raw material, in addition to improved financial results, was the basis for the implementation of an ambitious reorganisation plan. The plan would materialise into a clear vision of business growth for the Association. The vision was communicated and comprehended well by the administrative staff and thus contributed to new technology investments, as well as foreign capital funding that supported the Association's growth intentions. Nevertheless, the crucial matter of communicating the vision to the producers has not been solved. The Association's great emphasis on the producers' income causes the producers to distrust new ideas and practices. For this reason, the vision must be communicated very carefully, in conjunction with a basic marketing and business education that will help producers comprehend it.

OUTSOURCING

Market orientation requires a great deal of effort in consumer research, and the Research and Development division attempts to meet customer needs.[13] Cooperatives do not have the experience or capabilities in retail and promotion because of their product orientation. For this reason, and to achieve a rare, difficult to imitate operation, the outsourcing of specific operations may be reasonable. The foundation of Mediterra S.A. helped the Association focus its efforts in its core business (the production of Mastiha). In contrast, the participation of the Association, Mediterra S.A. and ELGEKA S.A. in the distribution complicates effective supply chain management and should be addressed.

VALUE-ADDED PRODUCTS

The Association, as many agricultural cooperatives do, traded its products mostly as commodities and thus allowed surplus value to be captured by intermediaries. After the implementation of the reorganisation plan, the Association focused on the production of value-added products. Consequently, quality became a critical factor for meeting customer needs, value-added operations (e.g., better packaging) were applied, and consumer research and R&D strategies were developed.

MARKETING MIX

The Association made important changes to its culture, focusing on marketing and innovation. Many efforts were made to launch new brand names, expand target markets, and increase the uses of Mastiha. Nevertheless, the marketing mix is not cohesive, and many firms participate not only in its implementation but in its conduct as well. Distribution is carried out by three firms. The promotion mix of Mediterra S.A. is determined exclusively by the subsidiary company, and the promotion mix of the Association's products is decided by an outsourcing company. There must be mutual planning to communicate Mastiha's positioning in a cohesive way that might attract other brands' sales. In addition, harmonisation with modern market trends must be addressed.

The process of evolving from product excellence to market excellence may take a long time. In many cases, cooperatives completely lack business knowledge, and the application of basic marketing practices can bring immediate results. Such a process requires hard work, commitment and a fair degree of trust among the administrative staff and the producers. Herein lies the difficult part: because producers' income depends exclusively on the cooperatives' performance, they are afraid of radical changes. However, products like Mastiha, which have valuable raw material, must be developed with a marketing-oriented approach to meet consumer needs with value-added products.

References

1. Triantafyllou, A., Chaviaras, N., Sergentanis, Th. N., Protopapa, E., and Tsaknis, J. (2007), 'Chios mastic gum modulates serum biochemical parameters in a human population', *Journal of Ethnopharmacology*, vol. 111, pp. 43–9.
2. Katsikis, I.N. and Kyrgidou, L.P. (2007), 'Sustainable entrepreneurship, global success and local development: The case of Mastiha in Chios, Greece', paper presented at the 47th ERSA Congress, Paris.
3. Ibid.
4. Panigyrakis, G.G. and Theodoridis, P.K. (2007), 'Market orientation and performance: An empirical investigation in the retail industry in Greece', *Journal of Retailing and Consumer Services*, vol. 14, pp. 137–49; Salavou, H. (2002), 'Profitability in market-oriented SMEs: Does product innovation matters?', *European Journal of Innovation Management*, vol. 3, no. 3, pp. 164–71; Voudouris, I., Lioukas, S., Makridakis, S. and Spanos, Y. (2000), 'Greek hidden champions: Lessons from small, little-known firms in Greece', *European Management Journal*, vol.18, no. 6, pp. 663–74; Oustapassidis, K. and Vlachvei, A. (1999), 'Profitability and product differentiation in Greek food industries', *Applied economics*, vol. 31, pp. 1293–8.
5. Harris, C.L. and Ogbonna, E. (1999), 'Developing a market oriented culture: a critical evaluation', *Journal of Management Studies*, vol. 36, no. 2, pp. 177–96; Golan, B. (2006), 'Achieving growth and responsiveness: Process management and market orientation in small firms', *Journal of Small Business Management*, vol. 44, no. 3, pp. 369–85.
6. Yin, R.K. (1993), *Case Study Research: Design and Methods*, Sage Publishing, Newbury Park, CA.
7. Narver, J.C., Slater, S.F., and Tietje, B. (1998), 'Creating a market orientation', *Journal of Market-Focused Management*, vol. 2, no. 1, pp. 241–55.

8. Menguc, B., Auh, S., and Shih, E. (2007), 'Transformational leadership and market orientation: Implications of competitive strategies and business unit performance', *Journal of Business Research*, vol. 60, pp. 314–21.

9. Gebhardt, G.F., Carpenter, G.S., and Sherry, J.,F., Jr. (2006), 'Creating a market orientation: A longitudinal, multifirm, grounded analysis of cultural transformation', *Journal of Marketing*, vol. 70, no. 4, pp. 37–55.

10. Shoham, A., Rose, G.M., and Kropp, F. (2005), 'Market orientation performance: A meta-analysis', *Marketing Intelligence and Planning*, vol. 23, no. 5, pp.435–54; Panigyrakis and Theodoridis, op. cit.

11. Triantafyllou et al., op. cit.

12. Beverland, M.B. and Lindgreen, A. (2007), 'Implementing market orientation in industrial firms: A multiple case study', *Industrial Marketing Management*, vol. 36, pp. 430–42.

13. Green, W.K., Chakrabarty, S., and Whitten, D. (2007), 'Organisational culture of customer care: Market orientation and service quality', *International Journal of Services and Standards*, vol. 3, no. 2, pp. 137–53.

19 *A Study of a High Value Coconut Product: The Midrib Basket Market Chain in Vietnam*

BY MENNO KEIZER* AND NGUYEN THI LE THUY†

Keywords

market chain assessment, product diversification, entry barriers, poverty reduction, livelihoods

Abstract

This chapter investigates the marketing chain of coconut-leaf-based gift baskets, a high-value coconut product made by local communities in Vietnam. The use of the coconut leaf – which is traditionally seen as a waste product – to produce baskets provides livelihoods to many rural people in the Ben Tre province, especially the elderly. Basket producers are experienced craftswomen and -men. Owing to the proximity of the market, the existence of local traders and their competitive edge of producing high-quality baskets, producers in Ben Tre are able to make a decent living. Although local authorities have expressed concern about the reduced productivity of coconut trees due to excessive harvesting of its leaves, presently no hard evidence exists to support that claim. A study of this concern is highly recommended and may lead to improved coconut-leaf harvest practices by the midrib producers without compromising coconut production.

Background

Lack of market information and access to markets is a major cause of poverty in many coconut-growing communities. Because of limited market opportunities, poor coconut

* Mr Menno Keizer, Vredeseilanden/VECO East Africa, P. O. Box 7844 Kampala, Uganda. E-mail: menno.keizer@veco-uganda.org. Telephone: + 256 414 533855.

† Ms Nguyen Thi Le Thuy, Department of Science and Technology, 280, February 3th Street, Ward 3, Ben Tre Province, Vietnam. E-mail: gnlethuy2001@yahoo.com. Telephone: + 84 75 812628.

farmers often have few alternatives than to sell copra (dried, fresh coconut meat from which oil is extracted and used in the food processing industry), which is a low-value commodity.[1] The Poverty Reduction in Coconut Growing Communities (PRCGC) project, funded by the Asian Development Bank (ADB), is a research project that aims to develop, test and evaluate sustainable livelihood options for poor coconut-growing communities and to conserve coconut diversity in eight countries in the Asian, Pacific, and Oceanic region: Bangladesh, Sri Lanka, India, Indonesia, the Philippines, Vietnam, Fiji, and Papua New Guinea. The project is being conducted in 24 coconut-growing communities, or 3 communities in each country. The project intends to test the hypothesis that the coconut tree 'is a tree of life,' for which humankind can derive multifarious uses, and that coconut farmers 'need not be poor.' The project is an offshoot of a previous research project that demonstrated that unless coconut farmers earned more from coconut trees, they would not conserve or plant more trees on their farms. Diversification of coconut products can reverse this situation, as well as address marketing issues in the industry. As part of the PRCGC project implemented by Bioversity International (the world's largest international research organisation, dedicated solely to the conservation and use of agricultural biodiversity[2]), focused market research was carried out in the Hung Phong community and adjacent districts in the Ben Tre province of Vietnam.[3] The research aimed to identify the various marketing strategies deployed by actors in the coconut midrib basket marketing system. The midribs of leaves of the coconut palm are used to make baskets that in Vietnamese society are used to present gifts. This system comprises various actors that perform different services and functions in the production and marketing of the product. In addition, the sustainability of the harvesting system of the coconut leaves is examined herein.

Various case studies of agricultural chain management have been published, most of which describe a detailed marketing system for a particular commodity.[4] This chapter focuses on the way market strategies are developed and negotiated between actors in the coconut chain. It also describes how a group of people in a rural area of Vietnam are adopting a market orientation and improving the efficiency and effectiveness of the market chain by adding value to a waste product, namely, coconut leaves. This venture allows them to improve their livelihoods.

Methodology and Sampling

This research was conducted using rapid reconnaissance techniques, which provide broad and preliminary overviews of the operation and performance of food systems, or components thereof, and are designed to identify constraints and opportunities.[5] Holtzman[6] notes that less formal data collection methods are more effective means to study interrelationships and linkages in farming and marketing systems, as well as to understand system constraints and opportunities. Van Willigen and DeWalt[7] cite Chambers, who notes that the advantage of such surveys is 'that they include searching for and using existing information; identifying and learning from key informants; ... direct observation and asking questions about what is seen; guided interviews and group interviews with informal or selected groups.' Data were collected on four occasions over a two-year period (2005 and 2006). Visits were made to the research area during various times of the year to capture different information and verify marketing trends.

SEMI-STRUCTURED INTERVIEWS

Interviews were unstructured and semi-directed with a largely informal character. During the collection of secondary data, a question list for each actor group – midrib producers, agents, basket producers, and traders – was produced. These question lists provide mental checklists during the interviews rather than a strict format. However, informal interviews were structured in the sense that they intended to cover important topics in a preferred sequence. Topics discussed include a general overview of the business (e.g., short history, resources needed), descriptions of the process of buying raw materials and selling the baskets, current and past prices, payment conditions, working capital required, recent changes in the market system, constraints on the business, visions of how to develop the business further, and the effect of harvesting leaves on coconut yield.

The average interview lasted an hour. Some data collected during the interviews were cross-checked with those from other respondents, and some respondents were interviewed on multiple occasions over the two years. The purpose of this intensive process is to clarify statements presented during prior interviews and to ask more detailed and sometimes more sensitive questions. During the interviews, short notes were taken, then written out the same or the next day. The effort to keep the interviews informal seemed to encourage the respondents to speak open and frankly, which enable the collection of sensitive information. The midrib system in the communities was observed on three occasions at different times of the year: lean months, peak months and the months before the peak period. By spreading the visits, we anticipated that we could observe a mostly accurate reflection of the marketing chain.

SAMPLING PROCEDURE

Actors were identified among the key informants and staff from the Oil Plant Institute of Vietnam (OPI). Snowball sampling was also used. With this method, after the interview was conducted, the respondent could identify traders, producers or midrib agents who could give other insights into the basket trade. This approach made it easier to identify people in the basket trade. Repeat visits also yielded more detailed information by building a certain level of trust that encouraged the interviewees to share trade secrets (e.g., contact details of raw material agents).

Table 19.1 contains an overview of the number of actors interviewed. All these actors were interviewed according to the prepared checklist.

Table 19.1 Number of respondents interviewed per actor group

Actor Group	Number of Respondents
Raw material suppliers	7
Basket producers	23
Midrib agents	9
Traders	9

Discussion of Research Findings

INTRODUCTION

Gift baskets have an important socio-cultural value in Vietnamese society. They are widely used for the presentation of gifts for special events (e.g., Tết Festival, Mothers' Day). Tết Nguyên Đán, or more commonly known by its shortened name Tết, is the most important and popular holiday and festival in Vietnam. The Vietnamese New Year (at the end of January, beginning of February), based on the lunar calendar, lasts for three days. During this festival, gifts are presented to family and friends, and midrib baskets serve to present these gifts. Retailers pack the baskets with food items and wrap them in colourful plastic. The baskets are made from coconut leaf midribs with a bamboo frame. Approximately 20 different designs, shapes and sizes are available in the market, each with its own distinct name, such as *gio cong dua* (basket with handle), *ro cong dua* (rectangular basket without handle), and *lang hoa cong dua* (basket to arrange flowers). Basket-making and -selling provides livelihoods to many families in the Hung Phong community, with approximately 30 per cent of the households involved in producing and/or trading the baskets.

Our research reveals various actors in the basket-marketing chain, including raw material producers, raw material agents, basket producers, traders, retailers and exporters (Figure 19.1). Often, there are no clear distinctions among these various actors. For example, basket producers are often also traders, or traders are simultaneously raw material suppliers.

Most actors in this industry are women, often because women negotiate (bargain) and converse better than men – two skills that are crucial in this business (according to both men and women interviewed). Also, the predominantly female producers mentioned that they prefer to deal with female traders. Traders are usually contacted by their buying clients (retailers, wholesalers, exporters), who place orders. The agreed prices are based on current prices of raw materials and the design of the baskets. Different sizes of the same design often are ordered. The order is then distributed among the various producers working with the trader. After completing the order, the small trader takes the baskets

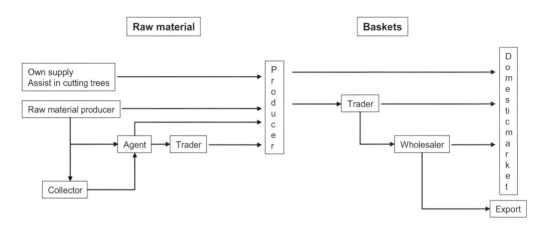

Figure 19.1 Supply chain raw material and baskets

to retailers in Ho Chi Minh City (HCMC), though bigger traders maintain their own warehouses in HCMC.

RAW MATERIAL SUPPLY

Most of the raw material used by Hung Phong basket producers comes from adjacent districts, such as Mo Cay and Giong Trom, due to the limited number of coconut trees within Hung Phong itself. Only a few people in the community sell the raw material. Basket producers usually buy raw material from outside the community and supplement it partly with leaves from their own trees. The latter resource is limited because producers are reluctant to harvest from their own trees, which they believe will decrease their coconut yield. Various harvest practices minimise coconut yield loss, such as cutting four half-leaves of each palm every month, harvesting only two to three leaves per palm per month, or collecting leaves only from old and low-yielding trees (10 leaves produce 1 kg of midrib). Respondents claim that these methods do not have a negative effect on the coconut yield.

As is demonstrated in Figure 19.1, the raw material supply chain consists of various actors:

- Raw material producers. Splitting and cleaning the midribs is an activity mostly performed by the elderly or people who lack the strength to do fieldwork. They usually produce a small volume (1–3 kg/day). The bundles are sold directly to basket producers, collectors, or agents. Depending on the agreement, they either must deliver the raw material to the buyer or have the material collected from their houses.
- Collectors. Collectors usually work in a specific community where they have a small network of around 10 midrib producers. Most collectors are prompted by relatives to organise the midrib supply in the area. Often, they produce the raw material themselves. They trade roughly 35–70 kg of midribs per week.
- Agents. Agents supply the traders who produce baskets (either in-house or by outsourcing) and occasionally deal with basket producers. Depending on the volume being traded, each agent has a network of 40–100 raw material suppliers, each supplying 1–3 kg per day. Larger agents also make use of collectors who provide bulk raw material in a particular community to reduce transaction costs. One agent said she bought around 300 kg per day, with 50 per cent of her income coming from the raw material trade. Because most agents have histories as traders or retailers buying and selling agricultural products (e.g., coconuts, prawns, poultry), they have an existing network of people who deliver to them on a regular basis. Their biggest strength is therefore the contacts they have in the agricultural communities and their easy access to people who wish to sell midribs to them.
- Traders. As mentioned, traders are supplied by agents and involved in supplying midrib to the basket producers. One large trader indicated that she needed around 100 kg of midrib per day. She has five raw material agents who source the midribs from raw material producers and collectors.
- Basket producers. Although some basket producers use their own leaves to produce midribs, this practise is not common in Hung Phong due to the few trees each household has. Another strategy used by some producers is to assist with the cutting

of old trees and receive some leaves as payment. However, these two strategies are only implemented to supplement other raw material sources.

MARK-UPS IN THE RAW MATERIAL CHAIN

The average mark-up by the various actors in the raw material chain is around VND[‡] 500–1000/kg (Figure 19.2). The raw material supplier receives the highest price when dealing directly with the producer. However, this method of selling is not common because of the physical distance and lack of communication or transportation between these actors, as well as because basket producers demand a larger volume of raw material than a single supplier cannot fill. The marketing costs (mainly transport) for a small delivery of raw material to Hung Phong are too high, so raw material collectors and agents are quite important.

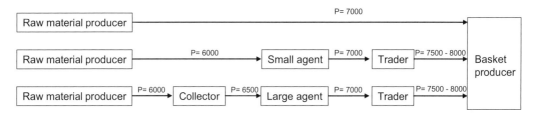

Figure 19.2 Raw material prices in the chain (February 2006)

Based on the mark-ups and volumes traded, we can calculate the gross incomes (Table 19.2). The calculations show that incomes made from producing and trading raw material are not sufficient to make a living (less than US$1 per day). As the interviewees indicated, producing and trading raw materials are side activities to supplement other income sources. The exception is the large agents, who earn high incomes because of the large volume they trade, but even they indicated that they have other sources of income.

Table 19.2 Mark-ups and gross incomes of raw material actors in the chain

	Volume Produced/ Traded per Week (kg)	Selling Price (VND/kg)	Mark-Up (VND/kg)	Gross Income (VND/week)
Raw material producer	4–20	6,000–7,000		24,000–120,000
Collector	35–70		500	17,500– 5,000
Small agent	40–100		1,000	40,000–100,000
Large agent	300–2,000		500	150,000–1,000,000

‡   US$1 = VND 15,700.

As Figure 19.3 shows, the prices for raw material start to decline after the Têt Festival season (mid-February onward), reaching their lowest point (VND 6,000/kg) in May/June. Around August, demand starts to pick up.

Prices steadily increase, peaking at VND 8,500/kg. Actors exploit this trend by buying the raw material during the lean months (April–August) and storing them for use during the peak season. However, only those with sufficient capital or access to capital during these lean months are able to exploit this opportunity fully.

LOW ENTRY BARRIERS FOR BASKET-PRODUCERS

Approximately 30 per cent of Hung Phong households are involved in midrib basket making. Knowledge about making the baskets is readily available in the community, as relatives or neighbours teach newcomers the art. Many women and elderly people are involved. The work is easy and light and can be done between normal domestic tasks. Often members of the entire family, including children, are involved. Each family member specialises in a certain part of the process. Men are usually responsible for the bamboo frames, and women do the actual weaving of the midribs. With this strategy, producers can work more efficiently. Another strategy, employed by single women, is to outsource the frame-making stage, which is very labour intensive.

Producers work both part- and full-time. The part-time producers are only involved in producing baskets during the peak season (October–January). The rest of the year, they engage in farm-related activities. In addition, there are independent and contractual producers. Independent producers receive orders from traders and are responsible for all subsequent steps to produce the baskets. Contractual producers are employed by traders to produce the baskets (the trader supplies the materials) and are paid either per unit of finished products they produce or after completing a certain stage of the product (e.g., bamboo frame).

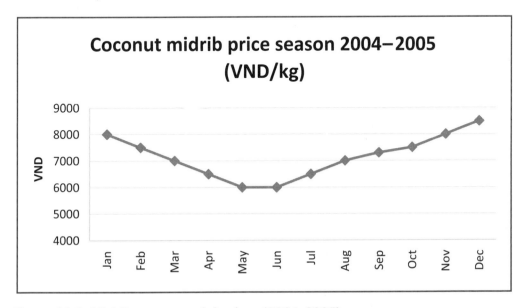

Figure 19.3 Midrib raw material prices (2004–2005)

IMPORTANCE OF NETWORKS AND SOCIAL RELATIONS TO OVERCOME HIGH ENTRY BARRIERS FOR TRADERS

Although new basket producers enter the midrib basket chain relatively easily, entrance for traders is a different story. Becoming a trader requires participation in three networks:

- A new trader must create a network of basket producers to secure supply of the product; to ensure demand, the trader also must establish contacts with potential buyers.
- To guarantee supply and volume, a network of producers who deliver their products on a regular and reliable basis is vital. To ensure this supply, the trader needs producer tie-in, achieved when the traders provide various value-added services such as capital, pick-up from the producers' house, fast payment, and the provision of raw material.
- To provide a constant supply of raw material, the trader needs to create a raw material network, especially big traders that require large volumes of raw material. Usually, these traders have agents who source the raw material for them through collectors.

In Vietnam, the socio-economic relationship of patronage, locally referred to as *moi*, also plays an important role. Having a *moi* relationship means doing business with the same people every time. Such a relationship takes a long time to build; thus, new traders find it difficult to enter the basket trade in Hung Phong.

WINDOW OF OPPORTUNITY

To maximise profit, some basket producers and traders store baskets and raw material to sell during the peak season when prices are highest, usually from December to January. However, we observed that more and more basket producers and traders are now adopting this strategy, which creates some doubt about whether prices will remain as high as they have been. Previously, the window period was 10 days, but in 2004, according to the biggest trader in Hung Phong, the period only lasted 3 days. In 2005, the period was again around 3 days. In this scenario, producers and traders need to plan the selling of their baskets more carefully. Documented cases describe basket producers who were stuck with their stock for too long. To avoid this scenario, some producers set a minimum selling price in their minds, and when the actual market prices reach that price, they begin selling their basket stock. This strategy is more effective for reducing risk than maximising profits. In addition, to reduce the dependency on one product (and therefore minimise risk), some traders are expanding their product range to include handicrafts.

The window of opportunity in January 2006 was completely the opposite of that in previous years, as prices remained high for almost a month. The main reasons for this outcome included (1) economic growth in Vietnam, such that people had more money to spend and consequently spent more on gift baskets, and (2) demand for the baskets from other provinces increased substantially. Traders and producers said that this increase was not a one-time event but rather the start of a new trend.

Traders need to keep a close eye on developments in other production areas as well. Buyers tend to shift their attention to other production areas if those traders can deliver the same products at lower prices. Hung Phong traders should therefore start exploring and building trade relationships with these new basket producers.

PRICES, CREDIT, AND PAYMENT CONDITIONS

Due to the highly fluctuating demand for midrib baskets, prices are also volatile. The highest price a producer can fetch for products is in December and January, during the run up to the Tết Festivities (see Figure 19.3). However, prices of input (raw material) are also high during this period, which considerably affects profits. The design, size and number of baskets in a set are the main factors that affect product price. Figure 19.4 presents the price of a set of two baskets during the peak season.

The price is stable at VND 5,000 during the lean months but starts to climb from November onward due to increases in demand, which peak just before the Tết celebration. After Tết, the price drops rapidly to VND 5,000 again. However, this example comes from one producer; other producers indicate varying price ranges, such as VND 3,500–6,000 during the lean months and up to VND 6,000–9,500 during the peak period.

Raw material producers sometimes have inaccurate price information, and the price offered may be too low. For example, price changes down the chain are not immediately passed on to them and only occur when they start complaining about the price. Producers indicated that prices offered by traders differ slightly, with variations up to VND 500 per set. Seemingly, higher prices are offered by traders who need to fulfil an order quickly. Producers who do not receive credit from traders benefit from this opportunity. Those who operate with source credit from traders may receive the credit without interest but with the obligation to sell products to that specific trader, reducing their bargaining power. At the same time, this credit can be used only to fulfil an existing order and not for buying raw materials to be stored and used at a later time; hence, trader-provided credit is not useful for buying raw material when prices are low.

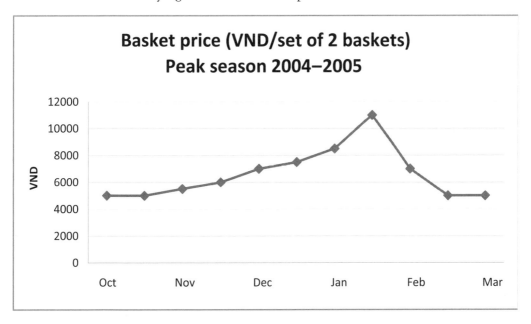

Figure 19.4 Farm-gate price of baskets during the peak season

All traders make use of the same debt-dependent payment system. They often pre-finance the basket producers by providing raw materials; the costs are deducted after each delivery. Traders usually receive 50 per cent of the total value of the order on the first delivery to wholesalers or retailers. The rest of the payment is made during the next delivery to ensure a steady supply.

COMPLETE PICTURE: SWOT ANALYSES

Strength, weakness, opportunity, threat (SWOT) analyses can be conducted for the raw material actors, basket producers, and traders. Although often used as a means to determine a market strategy, these analyses are also useful because they give an overview of the marketing system at each level. Each SWOT analysis is followed by a short discussion regarding the main strengths, weaknesses, and threats. The analysis concludes with opportunities that the specific actor might implement to overcome the main threats.

Table 19.3 SWOT raw material producer

Strengths	Weaknesses
• Alternative source of income for elderly and disabled people • Activity fits within the normal domestic tasks of the household • Immediate cash in hand	• Low return on labour • Weak price negotiation position • Lack of up-to-date price information
Opportunities	**Threats**
• Target group for 'sustainable harvest' study	• Unsustainable harvest practises threaten the coconut trees • Local government discusses possibilities to stop harvesting leaves because it puts pressure on the coconut yield

The production of midrib provides an additional, or sometimes only, source of income. It is therefore an important activity for people's livelihoods. The activity is not popular with people who can work on farms as labourers. The return on labour is low compared with salaries of farm workers: Only after harvesting the crops do people become interested in this kind of work. As noted previously, raw material producers sometimes lack accurate price information and offer a low price. As one raw material agent stated, 'as long as the producer does not complain about the price I do not see any reason of changing the buying price.' Since this actor-group plays an important part in the harvesting of leaves, it is a good target group to involve in future research that investigates the effects of various harvest practices in coconut yield.

This activity is not the sole income source for raw material collectors but instead is more akin to a side activity. The calculated mark-ups show that collectors make $1–3 per week depending on the volume traded. Often collectors have been asked by family members who work as agents to source raw material in their community. Some collectors indicated that, due to their low income, they do not give high priority to this type of business.

Table 19.4 SWOT raw material collector

Strengths	Weaknesses
• Not the sole income, provides additional income	• Poor quality of material (dirty or too wet) • High transaction costs, many small suppliers from various districts
Opportunities	**Threats**
• Deliver to buyers directly, bypass the agents • Increase volume by expanding trade network	• Unsustainable harvest practices threaten the coconut trees • Local government discusses possibilities to stop harvesting leaves because it puts pressure on the coconut yield

Table 19.5 SWOT raw material agent

Strengths	Weaknesses
• Are existing traders, who use their trade network for the midrib business • Easy communication with buyers • Extensive network of suppliers (often relatives) • Strong negotiation position due to high demand for midrib	• High transaction costs, many small suppliers from various districts • Need for working capital, everything is paid on delivery • Poor quality of material (dirty or too wet)
Opportunities	**Threats**
• Forward integration, endeavour in the lucrative basket trade business	• Unsustainable harvest practices threatens the coconut trees • Local government discusses possibilities to stop harvesting leaves because it puts pressure on the coconut yield

Because most agents have a history as traders or retailers in agricultural product markets, their existing network delivers to them regularly. Their biggest strength is the contacts they have in the agricultural communities and their access to people who wish to sell to them. The next logical step is to enter the basket trade. Some of the agents interviewed indicated that they were considering entering this market segment.

Basket producers in the Hung Phong community are experienced craftswomen and -men. The proximity of the market, existence of local traders, and competitive edge they have in producing high-quality baskets enable these basket producers to make a decent living. However, they could lose this competitive edge to other production areas. To remain competitive, they therefore must continue to improve their skills and develop new products. The main threat is the *Brontispa* beetle, which is attacking the palms. If this pest continues to spread and grow, it may affect the coconut industry. Fortunately

Table 19.6 SWOT basket producer

Strengths	Weaknesses
• High quality baskets • Knowledge and skills available in the community • Capable of making new designs on request • Activity fits within the normal domestic tasks of the household • Efficient trading system • Direct contact between producers and traders	• Limited working capital available • Credit tie-up with traders weakens the negotiation position of the producers • High prices of raw material • Volatile prices during peak season • Difficult to predict how long the window of opportunity lasts • Dependency on family labour
Opportunities	**Threats**
• Availability of micro-credit through PRCGC project • Raw material can be stored up to 1–2 years • New (smaller) designs in the market with higher profit margins • Supply straight to retailers • Diversify products	• Pests and diseases affect coconuts • Continuation of raw material price increases • Other communities start to produce high quality baskets • Consumers shift to different gift materials

OPI has begun a research and training program to control *Brontispa* through the release of parasitoids and the training of farmers in rearing parasitoids.

Basket producers use the following business strategies – whether all Hung Phong producers use them or only a few do.

Profit maximisation

• Smaller designs with fewer raw materials but a good selling price. This tactic increases profitability; unfortunately, there is only a limited demand for such baskets. Usually these baskets are produced only on an order basis.

• Buy raw material during the lean months, store it, and sell the baskets during the peak months. Producers said that if raw material were bought during the lean months and the baskets sold during the 'window of opportunity,' profits reached VND 1,500–2,000 per set of two baskets, compared with VND 500 when the raw material is bought in the peak season. To increase profits even more, producers who make all the basket in-house (i.e., no outside labour) can make a 100 per cent profit margin per set of two baskets.

Decrease working capital

• To save on raw material costs, producers can buy uncleaned material and do the cleaning themselves, which requires less capital to buy raw material. However, the cleaning process requires extra labour, and 20 per cent of the raw material is lost.

These additional costs and material losses mean the price of uncleaned material is almost the same as pre-cleaned material. The difference, however, is that the necessary capital is spread over a longer period, which may help reduce cash flow problems.

- Utilise social capital. By assisting other community members in cutting down their coconut trees, they can earn leaves or a small fee.

Specialisation

- Each member of the family specialises in a certain part of the basket making. This strategy enables producers to work very effectively and efficiently and increase product output.
- Outsource labour-intensive stages of the basket. Some households consist of too few members to produce enough baskets to fulfil orders from traders. By outsourcing the bamboo frame-making stage, these households still manage to fulfil the orders and earn income.

Table 19.7 SWOT basket trader

Strengths	Weaknesses
• Strong linkages with buyers (= *moi*) • Make additional income by supplying raw material to producers • Efficient trading system • Ensure quality by physically checking the baskets during the bleaching process • High entry barriers for new entrants (in Hung Phong market) • Integrate backward in the market system	• Dependency on Hung Phong producers for supply
Opportunities	**Threats**
• Expand network to new basket-producing communities • Diversify product range	• Pests and diseases affect coconuts • Consumers shift to different gift materials • Traders in other communities take market share

Traders hold a strong position; they have long-standing relationships with buyers and retailers. The socio-economic relationship of patronage, or *moi*, plays an important role. By providing credit to basket producers, they can secure supply. Another way to tie in basket producers is to provide them with raw material. Traders must keep a close eye on developments in other production areas, because buyers will shift their attention if traders from those other areas deliver at a lower price. Hung Phong traders must therefore focus on new production areas and try to build trade relationships with new basket producers. To reduce the dependency on one product, some traders are expanding their product ranges.

Conclusions

Table 19.8 contains an overview of the various business strategies deployed by the actors in the midrib chain. Both basket producers and traders use a large variety of strategies, as do agents. Raw material producers and collectors make use of social capital, mainly to secure supply. To work more efficiently, each member of a family specialises in a certain part of the basket-making process. By adopting this strategy, producers can work effectively and efficiently, with higher product outputs. By outsourcing certain labour-intensive stages, households with fewer members can still produce enough volume and fulfil orders from traders. To take advantage of high prices during the peak period, actors store either baskets (producers and traders) or raw material (agents) to fetch higher market prices. They also make smaller basket designs with fewer raw materials but higher selling prices. This approach increases profitability but is subject to limited demand. To secure supply, traders may opt to provide credit to basket producers.

Currently, no hard evidence exists to indicate that harvesting leaves has a direct effect on coconut yield, nor have concrete studies been conducted regarding the effect of various harvest methods on coconut yield. With increasing prices for raw material, the effects on the productivity of coconut trees must be monitored. Studies into these effects are highly recommended and may identify improved, sustainable harvest practices that do not sacrifice coconut production.

One of the key suggestions made by producers during the interviews was to bypass middlemen (traders) to maximise profits. However, the important tasks of middlemen are often forgotten: they bring basket producers and buyers together, they provide credit in some instances, and they add the finishing touches to the product. Through the traders'

Table 19.8 Business strategies used by the various actors in the chain

Strategy	Raw material producer	Collector	Agent	Basket producer	Trader
Specialisation				x	
Outsourcing				x	x
Storage of raw material			x	x	x
Storage of baskets				x	x
Reduce transaction costs (assembling)			x		x
Use of social capital	x	x	x	x	x
Use existing trade networks			x		
Product diversification					x
Product development				x	x
Pre-financing					x

network, producers can sell products. Bypassing the trader does not automatically mean more profit for the basket producers, because some costs and risks shouldered by the trader would otherwise be borne by the producers themselves. A buyer network also is vital to the success of a trader and the business.

This market system research provides a holistic, albeit general, view of a specific coconut-based livelihood, aspects of which might be applied to similar livelihoods that utilise the coconut crop. The study identifies opportunities that might be effectively utilised by coconut farmers but that otherwise might have been overlooked. We therefore highly recommend the inclusion of focused market system research as an integral component in livelihood-based poverty-reduction interventions.

Acknowledgements

The authors thank Bioversity International and the International Coconut Genetic Resources Network (COGENT[8]) for providing the financial resources to conduct this research and the Oil Plant Institute of Vietnam for making the necessary logistical arrangements and freeing staff to assist in conducting the research. A special word of thanks goes to Ms. Kieu Duong and Mrs. Be for assisting in the fieldwork. Lastly, we express our gratitude to the various people who constitute the midrib basket market chain for the time they made available to take part in this research.

References

1. Batugal, P. (2003), 'Poverty reduction in coconut growing communities: The framework and project workplan', pp. 39–54, in Batugal, P. and Oliver, J.T. (eds), *Poverty Reduction in Coconut Growing Communities Volume I: The Framework and Project Plan,* IPGRI-APO, Serdang, Selangor, Malaysia; Keizer, M. (2005), 'Increasing livelihood opportunities through market research and strengthening of market channels: Conduct of market research and development of marketing channels', pp. 69–72, in Batugal and Oliver, op. cit.

2. Bioversity International, Available at http://www.bioversityinternational.org.

3. Keizer, M. (2007), *The Coconut Midrib Market Chain in Ben Tre Province, Vietnam, Midrib Baskets: The Art of Giving*. Bioversity International, Rome, Italy.

4. Cadilhon, J.J., Fearne, A.P., Giac Tam, P.T., Moustier, P., and Poole, N.D. (2007), 'Business to business relationships in parallel vegetable supply chains of Ho Chi Minh City (Vietnam): Reaching for better performance', pp. 135–47, in Batt, P. and Cadilhon, J.-J. (eds), *Proceedings of the International Symposium on Fresh Produce Supply Chain Management*, FAO, Thailand. Concepcion, S.B. and Digal, L.N. (2007), 'Alternative vegetable supply chains in the Philippines', pp. 172–83, in Batt and Cadilhon, op. cit. Keizer, M. (2007), 'The fresh sweet potato market chain in Bataan (the Philippines): The importance of interrelationships between actors for chain management', pp 155–65, in Batt and Cadilhon, op. cit.

5. Dijkstra, T., Meulenberg, M., and van Tilburg, A. (2001), 'Applying marketing channel theory to food marketing in developing countries: A vertical disintegration model for horticultural marketing channels in Kenya', *Agribusiness*, vol. 17, No. 2, p. 227–41. Lutz, C. and van Tilburg, A. (1997), 'Framework to assess the performance of food commodity marketing systems in developing countries with an application to the maize market in Benin', in Asenso-Okyere,

W.K., Benneh, G., and Tims, W. (eds), *Sustainable Food Security in West Africa*, Kluwer Academic Publishers, Boston, pp. 264–92.

6. Holtzman, J.S. (1986), 'Rapid reconnaissance guidelines for agricultural marketing and food system research in developing countries', Working paper no. 30, Michigan State University.

7. van Wiligen, J. and DeWalt, B.R. (1985), *Training Manual in Policy Ethnography*, American Anthropological Association, Arlington, VA.

8. Coconut International Genetic Resource Network (COGENT), Available at www.cogentnetwork. org.

20 Old World Wineries and Market Orientation: Empirical Evidence From the Italian Wine Industry

BY CRISTINA SANTINI*, ALESSIO CAVICCHI,† AND VINCENZO ZAMPI‡

Keywords

Old World wineries, market orientation, competitive environment

Abstract

In the growing competition between Old and New World wine-producing countries, the Old World countries are definitely suffering. After a brief description of the competitive environment and the new challenges for Old World wineries, this chapter provides a framework (strategic maps) for understanding wineries' competitive strategies. We focus on Italian wineries to check the validity of the key resources on which their competitive advantage is based. Results of two focus groups with Italian winemakers and professionals outline the typical problems and limitations that occur while formulating a strategy. Recommendations for industry and practitioners conclude this chapter.

Introduction

The global wine market has changed in both its supply and its demand. This change in the competitive environment is essentially due to the globalisation process[1] and the growth of wine production and consumption all over the world. European countries

---

*   Dr Cristina Santini, School of Business Administration, University of Florence, Via delle Pandette 9, 50127 – Firenze, Italy. E-mail: santini.cristina@gmail.com. Telephone: + 39 055 43741.

†   Dr Alessio Cavicchi, Department of Studies on Economic Development, University of Macerata, Piazza Oberdan 3, 62100 – Macerata, Italy. E-mail: a.cavicchi@unimc.it. Telephone: + 39 0733 258 3919.

‡   Professor Vincenzo Zampi, School of Business Administration, University of Florence, Via delle Pandette 9, 50127 – Firenze, Italy. E-mail: vincenzo.zampi@unifi.it. Telephone: + 39 055 4374723.

(e.g., Italy, France, Spain, Germany) that have long traditions in the wine industry ('Old World' wine countries) now face new competitors, such as Australia, Chile, the United States, New Zealand and South Africa ('New World' wine countries).

In the period 2000–05, the growth in global wine consumption (by volume) was driven by a few key markets, including the United Kingdom (+30.1 per cent), United States (+19.8 per cent), the Netherlands (+37 per cent), Australia (+9.4 per cent) and Chile (+23.8 per cent). Old World wine consumption has decreased, especially in France (-5.5 per cent) and Italy (-10.4 per cent). In addition to the increase in consumption, production has almost doubled in some New World countries, such as Australia and Chile.

Old World wineries thus are forced to examine and expand their market-orientation strategies. The concept of market orientation has evolved since its first appearance in the early 1990s.[2] This chapter mainly focuses on the first two of three behavioural components of market orientation outlined by Narver and Slater: customer orientation, competitor orientation, and inter-functional coordination.

The EU policy aimed at increasing wine quality production has fulfilled its goal and led to greater availability of quality wines, as well as increased competition in this market segment. The traditional differentiation strategy, based mainly on the assumption that 'quality is appellation,' has lost much of its effectiveness for Old World wineries. Rather, 'more and more wines, especially those from the New World, have many different cues on the package that influence purchase: the region, subregion and country of origin, the vintage date, the grape variety or blend, the producer or negotiant (blender of the wines), style (e.g., bottled fermented, late harvest), the winemaker, and the specific vineyard.'[3] How can Italian (or Old World) wineries compete in such a complex environment?

This chapter provides a description of the competitive environment, describes the basis on which Old World and New World countries have built their competitive advantage, provides a framework for classifying wineries' strategies, and outlines the main strategic questions that Old World wineries should take into account to build a sustainable competitive advantage.

Competing in the Wine Industry

Competition in the wine industry has assumed a global perspective,[4] and as a consequence, the competitive environment has been reshaped. To describe the competitive environment, we use the model of five competitive forces provided by Porter.[5] Despite some limitations,[6] this model is a useful framework for understanding the structure and dynamics of an industry and its underpinnings for profitability.

A remarkable number of barriers hinder entry into the wine industry: the combination of economies of scale on both supply and demand sides, high capital requirements,[7] and an unequal access to distribution channels.[8] Nevertheless, the threat of new entrants is moderate, mostly coming from big companies (e.g., retailers, multinationals[9]), that can face deal with capital requirements or have easier access to distribution channels. The exception is buyers – referring here to a heterogeneous number of actors such as grocery chains, specialised shops, wholesalers, importers, restaurants and end consumers – whose bargaining power has risen in recent years, mainly due to a change in consumption patterns. The other competitive forces have a low impact on reshaping this industry.

Rivalry among existing firms in the wine industry is high, the growth rate is almost steady, and there are high exit barriers (see Figure 20.1).

Wineries tend to converge toward the same dimension of competition.[10] Briefly, the main factors reshaping the wine industry are:

- Changed consumption patterns: per capita wine consumption is decreasing in Old World countries while increasing in New World countries. Furthermore, the increased bargaining power of retailers, due to consolidation in the retail industry occurring in key wine markets, such as the United States and United Kingdom, led to a high availability of good wines (at reasonable prices) on supermarket shelves. This occurrence influenced consumers to concentrate their purchases in supermarkets.
- Increased export orientation: owing to falling domestic consumption and an increase in home-based competition, wineries are relying more and more on export to better their performance. Being export-oriented is no longer simply a peculiarity of Old World wine countries.[11] The competitive dynamics occurring in the international trade system are also affecting the wine industry, and countries that have reached favourable trade agreements receive a stronger competitive advantage.
- Focus on quality: the overall degree of wine quality has increased in the past decade, and the gap between Old and New World producers has shrunk, thanks to research and development investments and improved skills and competence. The presence of specific educational programmes in the main wine clusters that characterise the New World countries[12] has helped them gain a competitive advantage in production.

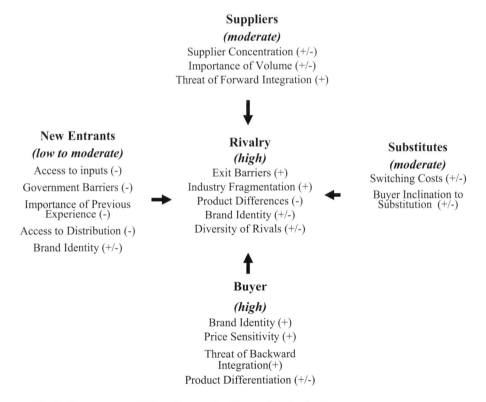

Figure 20.1 Five competitive forces in the wine industry

Old World versus New World: Competitive Positioning

The wine industry has seen growing competition between Old and New World wine producers. In particular, New World wine countries have adopted aggressive marketing strategies and invested in R&D to improve the quality of their production. Old World countries, whose production is regulated locally by appellation systems, local policies, and, on a higher level, European polices, must rethink the basis of their competitive advantage to compete with the new industry entrants. Previous research has highlighted the competitive positioning of New World and Old World countries.[13]

By analysing the relative competitive advantage that countries have in domestic markets (actual and potential), production systems (economies of scales and flexibility), location (geography, climate), and inputs (skills, knowledge, cost of labour), we rank countries' competitive advantages as in Table 20.1.

Table 20.1   Countries' competitive advantages in the wine sector

	Old World				New World				
	France	Germany	Italy	Spain	Argentina	South Africa	Australia	Chile	United States
Domestic market	S	S	S	M	M	W	W	W	S
Domestic market growth potential	W	M	W	W	W	W	W	W	S
Economies of scale	W	W	W	W	M	M	S	S	S
Industry adaptability to change	W	W	M	M	S	S	S	S	S
Potential to attract foreign investment	M	W	M	M	M	M	S	S	S
Geography	M/S	M/S	M/S	M/S	No Ad	No Ad	No Ad	No Ad	M/S
Climate	S	M	M/S	M/S	M/S	S	M/S	S	M/S
Land and raw materials	M/S	M/S	M/S	M/S	S	No Ad	M/S	S	M/S
Labour	No Ad	No Ad	No Ad	No Ad	M/S	M/S	No Ad	M/S	No Ad
Capital	S	S	M/S	M/S	M	M/S	S	No Ad	S
Infrastructure	M/S	M/S	M/S	No Ad	M	M/S	M/S	No Ad	M/S
Knowledge infrastructure	S	S	M/S	M/S	No Ad	M/S	S	M/S	S
Network	S	S	No Ad	No Ad	No Ad	M/S	S	M/S	S
Government	N	N	N	M/S	No Ad	M	M/S	M/S	M/S

Table 20.1 *Concluded*

	Old World				New World				
	France	**Germany**	**Italy**	**Spain**	**Argentina**	**South Africa**	**Australia**	**Chile**	**United States**
Economic variable	M/S	M/S	S	M/S	No Ad	M/S	S	M/S	M/S
Overall Competitive Advantage	**M**	**W**	**W**	**W**	**W/M**	**M**	**S**	**M/S**	**S**

Notes: S = strong; W = weak; M = moderate; M/S = moderate to strong; W/M = weak to moderate; No Ad = no specific advantage; N = neutral.

Italian Wine Industry: A Positioning Map

Italian wine production is highly fragmented due to the differences in climate conditions and a high number of appellations (more than 470 in total), most of them concentrated in northern Italy.[14] Piedmont is the region with the highest number of appellations, followed by Tuscany. Although the number of denominations has grown (in 2006–07, 15 new denominations were introduced), wine acreage has decreased (-8 per cent in 1996–2005). The debate on appellations is a hot issue, as shown by the recent changes in the EU regulation system.

This trend demonstrates the quality orientation that has characterised the Italian wine industry in the past decade. The Italian Institute for Services in Food and Agricultural Markets[15] provides a comprehensive map of Italian regional wine production by combining the importance of quality wine in the regional economy with the share of quality wine for the whole of wine production. According to this map, regions can beare grouped into four sub-categories:

1. Quantity before quality: these regions produce large volumes of wine with a low incidence of appellation wines.
2. Excellence in production: in these regions (e.g., Tuscany, Piedmont), wine plays a key role in the local economy, and producers focus their efforts on realising and promoting quality wines.
3. Toward an equilibrium: this group includes the regions where wine production has a marginal role on the agricultural system, but the situation is headed for a change in the near future (e.g., Umbria).
4. Niche quality: although wine does not play a key role in the rural system, these regions have a quality-oriented production, and the percentage of quality wine of the total wine produced is considerable. In this group, regions such as the Marches, Sardinia, or Lombardy are found.

The high number of appellations and brands operating in the Italian wine industry underlines how difficult it is for wineries to emerge in such a competitive environment and leads to a key question: is appellation a reliable basis for differentiation? Furthermore, can a strategy based on differentiation be effective simultaneously on national and international markets?

Mapping Strategies in the Wine Industry

In this section, we aim to provide a framework for describing wineries' strategies. According to the general strategic framework,[16] companies gain a competitive advantage by choosing a strategy among cost leadership, differentiation and focus. Although some authors have outlined the limits of Porter's general strategic framework,[17] it is still worth implementing in some cases because of its immediacy and convenience.

Product differentiation is based on factors that have been recognised as crucial for consumers; a clear example is the impact that country reputation or *terroir* has on consumers' perceptions. Background research has provided useful indications for product differentiation based on consumers' perceptions.[18] In this chapter, we focus on the following aspects to outline wineries' strategies:

- Width of product portfolio: the product range of a winery depends on the firm's ability to obtain economy of scales in production and marketing. Focusing on one or very few products, though it reduces the multiplication of costs and expenses, does not help reduce market risk.
- Export orientation: the effectiveness of a strong presence in foreign markets depends on wineries' efforts in marketing and sales and its skills and competence in this specific field.
- Price: price has been used to classify products and wineries' positioning.[19] An adequate price policy ensures that wineries have a reasonable mark-up on sales and are able to cover production costs.
- Volume of sales: the quantity of wine sold by a single winery provides useful information about the dimension of the winery itself and its organisation.
- Alignment with direct competitors: every company competes directly with other companies. Geographical proximity in the wine industry must be taken into account when identifying direct competitors. Wineries in close proximity frequently offer similar products, in terms of varietals produced, and in some cases must follow similar producing rules (especially if they belong to the same appellation). The effects of an alignment in wine clusters are outlined by Porter.[20]

These five factors described combine to provide a description of wineries' strategies. A first classification follows the traditional contrast of niche versus mass production, further classified by adopting a segmentation based on price (Figure 20.2).

When combining export orientation with the width of product portfolio, we understand which operating activity the company should invest in to support its strategy (Figure 20.3).

An international orientation requires companies to reinforce sales and marketing, whereas being locally based implies dependence on a strong local network. In contrast, enlarging the width of the product portfolio enables increased consumer loyalty through a strong company brand.

If we consider the degree of alignment with locally based direct competitors, we see that a company can lead the way in introducing a new product (e.g., new varietals) or conceiving of an entire innovative product range. The role of expert knowledge and R&D is crucial in pioneering endeavours. The need to adopt a profile that differs from the one diffused among local direct competitors arises when a company wants to differentiate

Sales (Volume)

	Low	High
High	*High Market Niche*	*Mass Branded*
Low	*Low End Niche*	*Mass Wine*

Price Positioning

Figure 20.2 Price/sales matrix

Export Orientation

	Low	High
High	*Multi Product National Base*	*Market Customization*
Low	*Single Product Specialist on National Base*	*Territorial Brand Going International*

Width Product Portfolio

Figure 20.3 Export orientation/width of product portfolio

among its products or targets a specific geographic market or customer segment. The relationship between customer orientation and new product development through value innovation is not new in the wine business.[21]

The Right Questions for Solving the Strategic Dilemma

How can Italian wineries achieve a competitive advantage? According to the description provided of the competitive environment, fierce competition exists in the wine industry. Italy's overall competitive advantage is weak in comparison with some New World countries, especially in terms of the high cost of inputs. The structure of the industry does not help wineries achieve economies of scale. According to this perspective, the only way

Italian wineries can compete is through differentiation. A generic differentiation based on quality and more specifically on appellations does not seem to guarantee a basis for a successful strategy on a global scale. In such a complex situation, managers play a key role in evaluating market opportunities according to:

- The company's available resources and competitive positioning:
    - What are the key resources the company relies on to pursue its strategic intent?
    - What is its strategic positioning?
    - What segment does the company want to serve?
- Competitors' relative positioning and market strategies:
    - Who are the main competitors?
    - What are their characteristics?
- Market dynamics perceptions:
    - What are the main issues that erode a company's competitive advantage?
    - What are the main opportunities in the industry?

Some short cases follow to reveal the link between empirical evidence and theory.[22]

CASE 1: WHEN AN EXCESSIVE FOCUS ON TERROIR DISTRACTS FROM CUSTOMERS' NEEDS

In March 2006, a focus group was carried out with six wine entrepreneurs involved in the consortia of Chianti Classico (two), Nobile di Montepulciano (two), and Brunello di Montalcino (two) – three of the most important appellations in Tuscany. All the people interviewed were deeply involved in the local wine industry, and they demonstrated a remarkable passion and dedication to their jobs and country life, together with a profound consciousness of the economic risk that characterises this business. They were asked to illustrate the main challenges that their companies must face in the near future.

The main problem for them is remaining competitive, and the only chance they saw to gain a competitive advantage was by focusing on quality. From this perspective, they suggested that each consortia avoid a differentiation of its brand and concentrate its efforts on protecting the overall quality of the wine produced within its boundaries, as other regions (e.g., Piedmont) have done. The entrepreneurs interviewed also underlined the critical nature of understanding customers' needs: a shift in the implementation policy of public funding (from a system based on an immediate benefit for a few companies to financing a long-term programme that would lead to indirect benefits for the whole population of firms) that helps educate the customer and improve involvement in local products. Through consumer education, entrepreneurs aimed at achieving sales targets. The interviews revealed that the key factor of success for an effective marketing campaign combines wine and *terroir*: 'The final customer should know our product. We should sell the wine together with the *terroir*.'

The linkage with the territory is very strong; the product is seen as deeply linked to the area and to the presence of some historical institutions, just as the three consortia have an influence on entrepreneurial choices. Consortia are expected to protect local wine images and ensure adequate quality standards, though some critical remarks emerged from the discussion: 'Big companies in the consortia risk obscuring smaller ones. Under big companies' pressure consortia follow what is in fashion.'

Another issue that emerged was the need to reinforce inter-firm cooperation among consortium members to ensure efficient management of the common brand (the one identified with the appellation): 'The consortium brand should not be parcelled out, for its strength is based on a common tradition shared by all the associated.'

On the basis of the group discussion, we discern the need for *co-opetition*,[23] that is, simultaneous cooperation and competition among consortia members. Differentiation can be pursued by defining the linkage between territory and its tradition (or *terroir*). Entrepreneurs identify a typical trait of their wine as it production following tradition: 'Respecting the identity of the product means making the wine in the same way our fathers did. Every vintage is different and the wine cannot be standardised.'

The entrepreneurs interviewed held deep convictions that Tuscan wine is better than the New World wine and that New World wines excel because they are produced in the New World by Italians. The entrepreneurs interviewed were convinced that California is gaining market share on the global wine market because of its aggressive marketing campaign rather than the quality of its wines. They were also aware of the differences in production systems between the Old and New World systems.

Entrepreneurs emphasised that the only way they could compete was by offering a mix of quality and tradition: 'Consumers should understand that this wine cannot be reproduced. We sell the *terroir*, the tradition and the culture. Tuscany has history and culture and a very long history as a wine producer.' Entrepreneurs also are aware of the need to communicate to consumers that Tuscan wine cannot be reproduced anywhere else; by selling a combination of wine and *terroir*, they also sell Tuscany tradition, culture, and long experience in wine.

In summary, the basic idea of competition, according to these entrepreneurs, is to persuade the customer to adopt a certain product that is unique because of its territory. Such a competitive strategy faces some problems, such as:

- the risk of ineffective consumer educational programs;
- overall rigidity that reduces a company's adaptability to market and demand changes;
- external risks arising from the environment or unexpected crises that may threaten the reputation of the territory.

In particular, the risk of external threats to country reputation recently increased. In April 2008, a scandal about Brunello wine production exploded in the Italian wine industry. Five important companies were accused of failing to respect the production guidelines set by the consortium to guarantee Brunello appellation standards (aging and grapes). Furthermore, the Brunello scandal happened at a very difficult time for the Italian wine and food industry: After problems caused by a high percentage of dioxins in *mozzarella di bufala*, more trouble arose when some wine production companies in the south of Italy were found to be using forbidden chemical agents.

Owing to the international renown of Brunello wine and Montalcino, the news quickly spread all over the world,[24] with serious implications for export performance and wine companies' images. Although the problem affected a very small minority of producers (five wineries, according to news agencies[25]), the whole Brunello brand was affected by the problem. This example shows how reliance on a territory becomes a weakness when unexpected events threaten the reputation of a certain area.

The entrepreneurs interviewed also seem more product-oriented rather than conscious of consumer needs. At the heart of this behaviour is the deep conviction that the product, due to its key traditional features, can be sold to a generic customer.

Building an effective relationship with the final consumer becomes trickier for wineries that suffer difficulties in experimenting (especially those operating in the food service industry):

> 'Restaurants play a key role in our sales: through Italian restaurants we could sell our wines all over the world. Prices are now too high, because of the excessive mark-ups that restaurant actors want to get.'

> 'Another problem is that the people working in the restaurants often don't have adequate background knowledge about how to pour our wine or about food pairing. This penalises our products.'

In conclusion, what is lacking in this case is an identification of a customer segment and a focus on its needs. Education must come as a further step to be more effective.

CASE 2: WHEN AN EXCESSIVE FOCUS ON TERROIR DISTRACTS FROM COMPETITORS' POSITIONING

In June 2007, we conducted a focus group with eight winemakers in Le Marche region, followed by a blind tasting. The aim of this research was to understand how professional winemakers and entrepreneurs perceive their competitors' products (in particular, California wines). The second issue was to assess how prejudice affects the perception of competitors. During the focus group, we asked the interviewees to express their opinion about New World producers and specifically which was the most competitive New World country and why.

What emerged was a clear perception of the competitive advantage that New World producers have in some specific inputs, such as the cost of labour. All the winemakers were convinced that New World wineries offered lower quality and that the success of their wine relied mainly on advertising and promotion. When asked how many wines from the New World they had ever tasted, the answer was none or one; when asked how many Californian brands they knew, all of them answered zero or one. All the winemakers interviewed noted tradition as a key factor for success.

In the second part of the interview, we asked winemakers to take part in a blind tasting. We chose a couple of red wines and a couple of white wines: same vintages, same varietals, same prices, but different countries (one from California and one from Italy). For the white wine, three of the eight winemakers recognised which was the Italian wine, three gave the wrong answer, and two could not answer. For the red wines, four winemakers recognised which was the Italian, three gave the wrong answer, and one was not able to give an answer.

After the blind tasting, we provided winemakers with some market information and tested their knowledge about market dynamics and foreign competitors. In particular, we focused on Cal Italian competition; Cal Italian producers are Californian wineries that use Italian varietals and threaten Italian wines in the US wine market.

In this case, a lack of awareness of the Cal Italian phenomenon, together with a shallow knowledge of international competitors' dynamics and a failure to appreciate foreign products' potential, has led to a superficial analysis of competitors. As a consequence, the strategic belief shared among the eight winemakers – condensed as a working hypothesis of the superior competitive advantage of Italian wine quality – appeared too general and built on weak bases. This specific case therefore illustrates how prejudice can affect an understanding of the market and competitive dynamics and potentially threaten a company's strategic success.

Conclusions

Market orientation affects a firm's organisation and culture[26] and stimulates new dynamics for creating a superior competitive advantage.[27] In competitive marketplaces, growth occurs through the optimisation of product selection and customer management.[28]

Accordingly, what elements make a strategy successful for Italian wineries? Wineries that have shown a remarkable export orientation and low awareness in extending their product range must invest more in building a strong company brand and a trustworthy corporate image, rather than relying too much on the territorial perceived image. Location is very helpful in lowering entry barriers at the start of the business by reducing promotion costs, because the country of origin influences consumer purchasing behaviour,[29] and using, for example, 'Tuscany' as a brand can help prompt early business and first purchases. The brand 'Tuscany' also works as an umbrella brand that represents core values shared among local actors, and it is cohesive because the brand elements are consistent.[30] However, exclusively relying on a territory – or finding shelter under a regional umbrella brand – becomes a weakness when unexpected events, such as food crises or wine scandals, threaten the reputation of an area. The more the core values under the aegis of the umbrella brand are specific and identified, the wider the space is for entrepreneurs to differentiate their products and reduce market risk.

The issue of core value identification as a key factor of success clearly emerges from both cases, as does the problem of communicating product characteristics and verifying whether they fulfil the customer's needs.

Another interesting issue is the level of willingness to widen the product portfolio and change product characteristics. The cases show that a reactive response to market dynamics is slowed by a strategic alignment with local direct competitors. In particular, the importance of obtaining deep knowledge about international competitors and being receptive at both local and international level emerges. Local orientation is recommended to be contiguous to and complimentary with an international orientation.[31] We determine from the two cases that companies need to focus on two main questions: Who is my customer? Who are my competitors?

A deep conviction of superior quality cannot alone lead to a competitive advantage but instead can create misunderstanding in explaining and understanding a company's strategy. What ultimately leads to success is an objective evaluation of a company's resources and potential, a clear individuation of the market segment in which the wineries compete, and a reduction, as much as possible, of prejudice in appraising competitors.

References

1. Beverland, M. (1999), 'Shake-out! Will small wineries survive in the global wine trade?', *New Zealand Strategic Management Journal*, vol.4, no. 2, pp. 31–9; Anderson, K., Norman, D., and Wittwer, G. (2003), 'Globalisation of the world's wine markets', *The World Economy*, vol. 26, no.5, pp. 659–87.

2. Narver, J.C. and Slater, S.F. (1990), 'The effect of a market orientation on business profitability', *Journal of Marketing*, vol. 54, no. 4, pp. 20–35; Kohli, A. and Jaworski, B. (1990), 'Market orientation: The construct, research propositions, and managerial implications', *Journal of Marketing*, vol. 54, pp. 1–18.

3. Lockshin, L. (2003), 'Consumer purchasing behaviour for wine: What we know and where we are going', *WP Bordeaux Ecole de Management*, Marchés et Marketing du Vin, Centre de recherche de Bordeaux Ecole de Management. no. 57–03, Août 2003.

4. Beverland (1999), op. cit.; Anderson et al., op. cit.; Campbell, G. and Guibert, N. (2006), 'Old World strategies against New World competition in a globalising wine industry', *British Food Journal*, vol. 108, no. 4, pp. 233–42.

5. Porter, M.E. (2008), 'The five forces that shape strategy', *Harvard Business Review*, vol. 86, no. 1, pp. 78–93.

6. Thurlby, B. (1998), 'Competitive forces are also subject to change', *Management Decision*, vol. 36, no. 1, pp. 19–24.

7. Folwell, R.J. and Volanti, M. (2003), 'The changing market structure of the USA wine industry', *Journal of Wine Research*, vol. 14, no. 1, pp. 25–30.

8. Thach, E.C. and Olsen, J. (2006), 'Building strategic partnerships in wine marketing: Implications for wine distribution', *Journal of Food Products Marketing*, vol. 12, no. 3, pp. 71–86.

9. Coelho, A.M. and Rastoin, J.L. (2006), 'Financial strategies of multinational firms in the world wine industry: An assessment', *Agribusiness*, vol. 22, no. 3, pp. 417–29.

10. Porter (2008), op. cit.

11. Labys, W.C. and Cohen, B.C. (2006), 'Trends versus cycles in global wine export shares', *Australian Journal of Agricultural and Resource Economics*, vol. 50, no. 4, pp. 527–37.

12. Porter, M.E. (2000), 'Clusters and competition,' in Gordon E. Clark, (ed.), *Oxford Handbook of Economic Geography*, Oxford University Press, Oxford; Giuliani, E. and Bell, M. (2005), 'The micro-determinants of meso-level learning and innovation: Evidence from a Chilean wine cluster', *Research Policy*, vol. 34, pp. 47–68.

13. Cholette, S., Castaldi, R.M. and April, F. (2005), *The Globalization of the Wine Industry: Implications for Old and New World Producers*, working paper, San Francisco State University, San Francisco.

14. ISMEA (2007), *I vini DOC e DOCG.Una mappatura della vitivinicoltura regionale a denominazione di origine*, Ismea.

15. ISMEA (2007), *I vini DOC e DOCG.Una mappatura della vitivinicoltura regionale a denominazione di origine*, Ismea.

16. Porter, M. (1985), *Competitive Advantage*, Free Press, New York.

17. Kay, J. (1993), *Foundation of Corporate Success*, Oxford University Press, Oxford.

18. Lockshin, op. cit.; Beverland, M. and Lindgreen, A. (2002), 'Using country of origin in strategy: The importance of context and strategic action', *Brand Management*, vol. 10, no. 2, pp. 146–67.

19. Rabobank (2003), *Wine is a Business. Shifting Demand and Distribution: Major Drivers Reshaping the Wine Industry*, Rabobank International.

20. Porter (2000), op. cit.

21. Santini, C., Cavicchi, A., and Rocchi, B. (2007), 'Premium bag in box as a strategic choice for small Italian wineries', *International Journal of Wine Business Research*, vol. 19, no. 3, pp. 216–30.

22. Yin, R. (2003), *Case Study Research, Design and Methods*, Sage Publication, Newbury Park, CA.

23. Brandenburger, A. and Nalebuff, B. (1996), *Co-Opetition: A Revolution Mindset That Combines Competition and Cooperation*, Bantam Doubleday Dell Publications, New York; Sheth, J.N., Sisodia R.S., and Sharma, A. (2000), 'The antecedents and consequences of customer-centric marketing', *Journal of Academic Marketing Science*, vol. 28, no. 1, pp. 55–66.

24. 'Brunello seized in Italian wine probe' (2008), *BusinessWeek*, April 3. Available at http://www.businessweek.com/ap/financialnews/D8VQKS901.htm.

25. Associated Press, April 3, 2008

26. Narver and Slater, op. cit.; Werner, U., McDermott, J., and Rotz, G. (2004), 'Retailers at the crossroads: How to develop profitable new growth strategies', *Journal of Business Strategy*, vol. 25, no. 2, pp. 10–17.

27. Hunt, S.D. and Morgan, R.M. (1995), 'The comparative advantage theory of competition', *Journal of Marketing*, vol. 59, no. 2, 1–15.

28. Pleshko, L.P. and Heiens, R.A. (2008), 'The contemporary product-market strategy grid and the link to market orientation and profitability', *Journal of Targeting, Measurement and Analysis for Marketing*, vol. 16, no .2, pp. 108–114; Beverland and Lindgreen (2002), op. cit.

29. Nicholson, J.D. and Kitchen, P.J. (2007), 'The development of regional marketing – have marketers been myopic?' *International Journal of Business Studies*, vol. 15, no. 1, pp. 107–125.

30. Iversen, N.M. and Hem, L.E. (2008), 'Provenance associations as core values of place umbrella brands', *European Journal of Marketing*, vol. 42, no. 5/6, pp. 603–26.

31. Nicholson and Kitchen, op. cit.

Index

If you have found this resource useful you may be interested in other titles from Gower

Creating Food Futures: Trade, Ethics and the Environment
Edited by Cathy Rozel Farnworth, Janice Jiggins, and Emyr Vaughan Thomas
268 pages; Hardback: 978-0-7546-4907-6

The Crisis of Food Brands: Sustaining Safe, Innovative and Competitive Food Supply
Adam Lindgreen, Martin K. Hingley and Joëlle Vanhamme
382 pages; Hardback: 978-0-566-08812-4

Food Fears: From Industrial to Sustainable Food Systems
Alison Blay-Palmer
196 pages; Hardback: 978-0-7546-7248-7

The New Cultures of Food: Marketing Opportunities from Ethnic, Religious and Cultural Diversity
Edited by Adam Lindgreen and Martin K. Hingley
344 pages; Hardback: 978-0-566-08813-1

Regoverning Markets: A Place for Small-Scale Producers in Modern Agrifood Chains?
Edited by Bill Vorley, Andrew Fearne and Derek Ray
248 pages; Hardback: 978-0-566-08730-1

Sustainable Change through Benchmarking in Food and Farming
Edited by Lisa Jack
c. 180 pages; Hardback: 978-0-566-08835-3

Visit **www.gowerpublishing.com** and

- search the entire catalogue of Gower books in print
- order titles online at 10% discount
- take advantage of special offers
- sign up for our monthly e-mail update service
- download free sample chapters from all recent titles
- download or order our catalogue